Asia–Pacific Conference on Science and Management of Coastal Environment

Developments in Hydrobiology 123

Series editor

H. J. Dumont

Contents

Hydrobiologia **352**: vii–viii, 1997.
Yuk-Shan Wong & Nora Fung-Yee Tam (eds),
Asia–Pacific Conference on Science and Management of Coastal Environment.

Preface

Human beings have had a long historical relation with the coast. It provided food and security initially, followed by becoming an important location for industrial and commercial development. Now, the emphasis seems to be shifting towards leisure and conservation, although the former functions remain crucial. However, it was only very recently that people began to view the coast as a common and valuable resource that requires rational utilisation and scientific management in order to sustain its attractiveness. Of course, enlightened management comes only through understanding of the complicated coastal regions, which enables coastal managers to balance pressures from different sectors and to minimize risks. Scientific knowledge will continue to be the most important basis for compromising the conflicts between coastal users and interest groups, such as the developers and the ecologists. Coastal management has also shifted from the traditional restorative or remedial actions towards planned avoidance of these conflicts.

Despite rapid advancement in coastal sciences over the past decades, most of the major coastal issues have remained outstanding in the agenda. Control of shoreline erosion and protecting sea level rise continue to be crucial problems facing coastal scientists. Destructive coastal storms still cause tremendous damage, particularly in low latitudes. Wetland and estuary reclamation have led to loss of most valuable estuary wetlands which are required to sustain biological productivity and biodiversity. For over a century, coastal waters have been utilised as a convenient dumping ground for waste materials, which caused not only serious environmental deterioration but direct threat to human health. Yes, all these are problems; and it is because of these problems that 250 scientists and professionals from all over the world came together at the Asia–Pacific Conference on Science and Management of Coastal Environment to discuss all aspects of coastal issues for making our coastal regions more sustainable.

The Conference was held at the Hong Kong University of Science and Technology on 25–28 June, 1996. There were 17 symposium sessions consisting of about 200 oral and poster presentations in 4 main themes, namely coastal ecology, pollution, marine biotechnology and coastal management. Papers published in this issue were selected and critically reviewed by peers.

We would like to thank the Guest Editorial Board for their hard work in reviewing all the manuscripts and the authors for their contributions to this issue. We should also like to thank the keynote speakers for their contributions to this event, and to all the international advisors who have provided us with most valuable suggestions. The event could not have been realised without the generous support from our sponsors, namely the British Council, Cathay Pacific Airways, Environment and Conservation Fund, Epson Foundation, The Hong Kong Electric Co. Ltd., Intergovernmental Oceanographic Commission in Paris, International Society for Mangrove Ecosystems in Tokyo, the National Natural Science Foundation of China and the United Nations Environment Programme.

THE EDITORS

Asia–Pacific Conference on Science and Management of Coastal Environment

List of Reviewers for Manuscripts to be Published in *Hydrobiologia*

Name	Affiliation	Country
Dr William G. Allaway	The University of Sydney	Australia
Dr Donald Anderson	Woods Hole Oceanographic Institution	USA
Prof. Joe Baker	Commissioner for the Environment	Australia
Mr Graham B. Bremer	University of Portsmouth	United Kingdom
Dr Malcolm Broom	Environmental Protection Department, HK Government	Australia
Dr King-Ming Chan	Chinese University of Hong Kong	Hong Kong
Prof. Fu-Shiang Chia	The Hong Kong University of Science and Technology	Hong Kong
Dr Ka-Hou Chu	Chinese University of Hong Kong	Hong Kong
Prof. Michael Dickman	The University of Hong Kong	Hong Kong
Dr Norman C. Duke		Australia
Prof. Tsu-Chung Hung	Academia Sinica	Taiwan
Dr I. J. Hodgkiss	The University of Hong Kong	Hong Kong
Dr Tak-Seng Leong	Universiti Sains Malaysia	Malaysia
Dr Gonasageran Naidoo	University of Druban-Westville	South Africa
Dr I-Hsun Ni	The Hong Kong University of Science and Technology	Hong Kong
Dr Jin Eong Ong	Universiti Sains Malaysia	Malaysia
Dr Pei-Yuan Qian	The Hong Kong University of Science and Technology	Hong Kong
Dr Frank Sin	University of Canterbury	New Zealand
Prof. Jarl-Ove Strömberg	Kristineberg Marine Research Station	Sweden
Dr Nora F. Y. Tam	City University of Hong Kong	Hong Kong
Prof. Chris C. Tanner	National Institute of Water & Atmospheric Research Ltd.	New Zealand
Dr John Twining	Australian Nuclear Science & Technology Organization	Australia
Dr Lilian Vrijmoed	City University of Hong Kong	Hong Kong
Dr Hein von Westernhagen	Biologische Anstalt Helgoland	Germany
Dr Joseph Wong	The Hong Kong University of Science and Technology	Hong Kong
Prof. Tat-Meng Wong	Open University of Hong Kong	Hong Kong
Dr Yuk-Shan Wong	The Hong Kong University of Science and Technology	Hong Kong

Hydrobiologia **352**: 1–7, 1997.
Y.-S. Wong & N. F.-Y. Tam (eds), Asia–Pacific Conference on Science and Management of Coastal Environment.
©1997 *Kluwer Academic Publishers. Printed in Belgium.*

Acid exposure in euryhaline environments: ion regulation and acid tolerance in larval and adult *Artemia franciscana*

Brian R. McMahon & Jason E. Doyle
Department of Biological Sciences, University of Calgary, Calgary, Alberta, Canada

Key words: Acid-exposure, ion-regulation, mortality, euryhaline environments, Artemia

Abstract

Mortality, and whole-body levels of [Na], [Cl] and [K] were measured in adult and naupliar stages of Brine Shrimp, *Artemia franciscana*, after 24h exposure to a waters of a range of acid (H_2SO_4) concentrations (pH 8.3 to 4.5). Acid effects were tested in both dilute (20%) and full strength artificial seawater (ASW). All stages showed high sensitivity to acid stress whether tested in 20% or 100% ASW. Nauplii were more sensitive than adults in either medium. Mortality in each case was correlated with dramatic changes in whole body [Na] and [Cll but animals in 20% ASW lost [Na] and [Cl] to the environment while animals in 100% ASW gained them from the environment. Clearly animals which ion-regulate in euryhaline media are more sensitive than expected to acid stress which may thus pose a serious threat to sensitive euryhaline habitats.

Introduction

The damaging effects of acidification on freshwater environments are well known, especially for teleost fish, including a variety of salmonid and a handful of other fish species (Wood, 1989). Much less is known about other animals, although mortality from acidification is well known for crayfish and some other invertebrates (McMahon & Stuart, 1989). Almost all the animals tested have been stenohaline freshwater species. Additionally the majority of testing carried out thus far has been on relatively large adult animals, although it seems logical that smaller and larval animals are likely to be more sensitive.

The physiological effects of acid exposure in freshwater fish and crayfish include marked extracellular and intracellular acidosis, but mortality is thought to be associated more strongly with increased permeability of the epithelial tissues, leading to loss of body ions. This passive loss is compounded by damage to the inwardly directed branchial ion pumps which would normally compensate for ion loss. In combination these two factors cause massive ion loss and eventually death.

Rather little work has been conducted to test the effects of acid exposure on marine animals (Knutzen, 1981). The paucity of studies appears to be based on two assumptions, (a) that marine animals are perpetually bathed in a veritable soup of ions and thus unlikely to suffer ion loss and (b) that the greater buffering power of seawater makes significant acidification unlikely. We were concerned by the absence of empirical testing and thus carried out some preliminary experiments on the effects of acid exposure in hypersaline environments.

For this purpose we chose a well studied animal, the brine shrimp *Artemia franciscana*. This crustacean is easy to obtain and culture and is extremely euryhaline – thus allowing testing at a range of salinities. Our rationale for conducting experiments at high and low salinities was to test an initial hypothesis that acid exposure at low salinity should be more lethal since low saline acclimated brine shrimp are dependent on active hyper osmotic-regulation; a process which is strongly affected during acid stress in freshwater animals.

We also included experiments on a range of developmental stages so as to test the hypothesis that smaller and/or larval stages would be more sensitive to the effects of increased acidity.

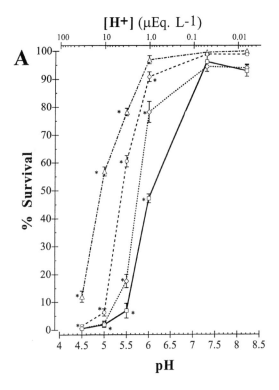

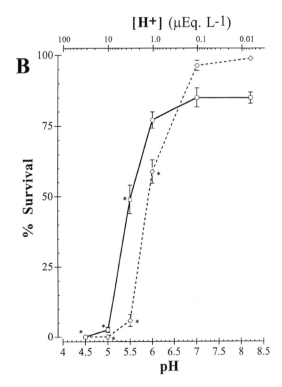

Figure 1. Effects of increased acidity on survival of *Artemia franciscana* in 100% ASW (**A**) and 20% ASW (**B**). Adult △, 4th instar ○, 3rd instar ◇, 1st instar □. Data are mean ± SEM. ∗ denotes significant difference ($p<0.05$) from control animals.

Some of the results presented here have been published previously in Doyle & McMahon (1995) which deals in detail with the effects of acid exposure on hatching survival and ionregulation in seawater. The present paper presents new data on effects of acid exposure in dilute media and compares this with results obtained in 100% seawater.

Materials and methods

Artemia franciscana were cultured from cysts originating from the San Francisco Bay National Wildlife Refuge, pond N-4, purchased from Argent Chemical Laboratories. Animals were cultured in either full strength (100% = 35‰) or diluted (20% = 7‰) artificial sea water (ASW; Instant Ocean) by modification of the method described by De Wachter & Van Den Abbeele (1990). Effects of increased acidity on mortality and hatching success were conducted in this artificial seawater. The effects of variation in pH on whole-body [Na], [Cl] and [K] were conducted in a second artificial seawater of similar ionic composition but without H_3BO_3 and $NaHCO_3$ and with the addition of citric acid. This modified ASW allowed easier adjustment and maintenance of low seawater pH. Whole body [Na] and [K] were measured by Atomic Absorbtion Spectroscopy (Perkin Elmer 4000) following dissolution of whole animals in nitric acid. [Cl] was meassured using coulometic titration (Radiometer CMT 10). Effects of acid exposure were assessed after a 24 h exposure period.

Mortality was assessed as the 24 hour LC_{50} using the Reed-Meunch method described by Woolf (1968). Details of sampling protocols and experimental procedures were described in Doyle & McMahon (1995). Survival times and whole body salt levels were compared using one way analysis of variance (ANO-VA). Significant ANOVA's were further analysed using Tukey's HSD test to detect differences between treatment pH levels.

Results

Survival

All stages of *Artemia franciscana* tested were surprisingly sensitive to acid exposure. First instar nauplii tested in 100% ASW were the most sensitive of all groups (Figure 1, Table 1). Tolerance increased progressively

3

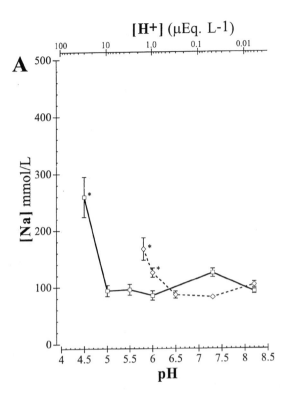

A

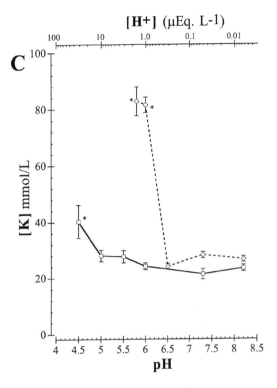

C

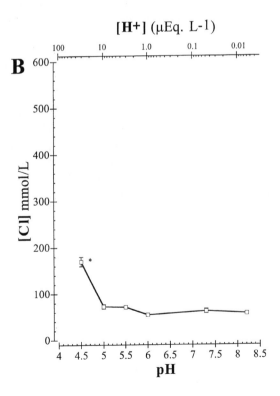

B

Figure 2. Changes in whole body salt concentration for adults and nauplii (1st instar) acclimatised to, and acid exposed 24 h in, 100% ASW (A). [Na] (B). [Cl] (C). [K]. adult □, nauplii ◇. Data are mean ± SEM. ∗ denotes significant difference ($p<0.05$). Dashed lines indicate ambient concentration.

Table 1. 24 h LC$_{50}$ values for *Artemia franciscana* acid exposed in 100% ASW (A) and 20% ASW (B).

LC$_{50}$	pH ± SE	[H$^+$] M×10^{-7} ± SE
A		
1st instar	6.025 ± 0.07	9.42 ± 1.46
3rd instar	5.77 ± 0.04	16.8 ± 1.59
4th instar	5.39 ± 0.03	41.2 ± 2.75
Adult	4.99 ± 0.03	100.8 ± 6.26
B		
1st instar	5.87 ± 0.05	13.4 ± 1.7
Adult	5.62 ± 0.04	23.9 ± 2.5

with each stage tested at either salinity (Table 1), but in 100% ASW even adults were surprisingly sensitive with a 24 h LC$_{50}$ at pH 4.99. Adult animals which were acid exposed in diluted seawater were slightly more sensitive (24 h LC$_{50}$ = pH 5.62; Table 1) but first instar nauplii tested in 20% ASW were essentially equally sensitive when compared with those tested in 100% ASW. As with 100% ASW naupliar stages were less tolerant of acid exposure when compared with adults (Table 1).

Whole-body salt levels

a) Animals tested in 100% ASW
Whole body [Na], [Cl] and [K] were measured after 24 h exposure to pH levels ranging from 4.5–8.3 ($[H^+] = 3.6 \times 10^{-5}$ M–5.02×10^{-9} M, Figure 2). These measurements indicated that all whole body concentrations except [K] (in both adult and naupliar stages) were regulated below those of the external seawater (hypo-ionic regulation). Under conditions of neutral to alkaline pH whole-body levels of [Na], [Cl] and [K] were similar in adults and nauplii (Figure 2).

Acid exposure significantly increased whole-body levels of [Na], [Cl] and [K] (Figure 2). Measured external [Na], [Cl] and [K] remained virtually constant over the pH range tested. Increased whole body [Na], [Cl] and [K] were closely correlated with the LC_{50} value (Table 1) in both naupliar and adult stages (Figure 2).

b) Animals tested in dilute seawater
Whole body [Na], [Cl] and [K] were measured after 24 h exposure to 20% ASW with pH maintained at levels from 5.25 to 8.3 (5.6×10^{-6} M to 8×10^{-9} M [H^+], Figure 3). At neutral pH *Artemia franciscana* tested in 20% ASW maintained [Na], [Cl] and [K] above those of the environment (hyper-ionic regulation). Levels of Na, Cl and K differed only slightly between adult and naupliar stages.

Acid exposure dramatically affected whole-body [Na] and [Cl] but not [K] (Figure 3). The changes were opposite in direction to those observed in full strength seawater since concentrations decreased towards, or fell below, environmental levels. As with the 100% ASW series ambient salt concentrations did not change significantly as a result of acid exposure.

Discussion

The mechanisms by which acid exposure causes mortality are well known only for adult freshwater forms, where lethality is thought to be closely associated with total body ion loss resulting from both increased permeability and from H^+ inhibition of inwardly directed ion pumps (Wood, 1989; McMahon & Stuart, 1989). The limited permeability of the body wall is generally accepted as being one of the major factors involved in the maintenance of internal osmotic and ionic levels. McDonald et al. (1983) and Playle et al. (1989) using rainbow trout, demonstrated that acidification results in an increased permeability of the external epithelium to NaCl and water. This is presumably accomplished through the opening of paracellular channels by displacement of calcium from apical tight junctions. The data from the present study support this reasoning to the extent that permeability apparently increases with acid exposure.

The freshwater ancestry of hypo-hyperosmotic regulators such as *Artemia* (Eriksen & Brown, 1980) suggests that ion regulatory mechanisms remain basically similar to those of freshwater animals but become modified to include utilization of outwardly directed ion pumps which extend the ion-regulatory capability into hypersaline habitats. In adult *Artemia* (as in euryhaline fishes) these pumps are located within specialized 'chloride cells' on the branchial epithelium (Conte, 1984; Holliday et al., 1990). Following this line of reasoning acid exposure could be expected to act in a similar way as for FW animals (i.e. to increase permeability and to decrease the effectivity of ionic compensatory mechanisms).

For adult and larval *A. franciscana* exposed to acid in dilute seawater both whole body [Na] and [Cl] decrease progressively with decrease in pH, supporting the prediction above.

At first sight, the results obtained for whole body [Na] and [Cl] for *A. franciscana* acid exposed in SW are strikingly different. Under conditions of high saline acid expossure [Na], [Cl] and [K] increase especially in response to higher acidity i.e. the opposite response to that observed in dilute seawater. On reflection, however, the two responses are functionally very similar. *A. franciscana* acclimated to seawater are hypo-osmotic regulators maintaining body [Na] and [Cl] at 50% of ambient levels. Thus the ion gradient across the body wall is reversed and thus the same effect of ambient acidity, i.e. increased general permeability and diminished ability to compensate by active ion pumping would lead to an increase in passive ion entry compounded by an inability to remove the additional ions. These would then build up to levels which adversely affect basic cellular metabolism, eventually to the point of mortality. Support for this reasoning in *Artemia* is provided by the work of Croghan (1958a) who found that treatment with $KMnO_4$ destroyed ion regulatory cells on the outer surface of these animals with the result that the animals became isotonic with the medium. Interestingly the upper limit of survival of

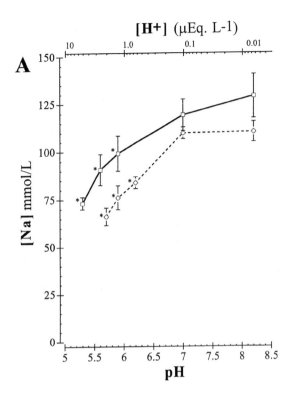

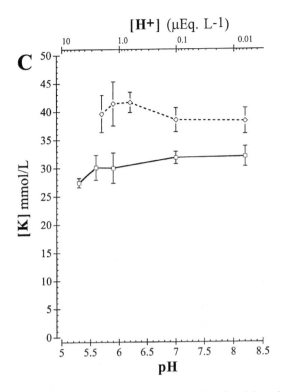

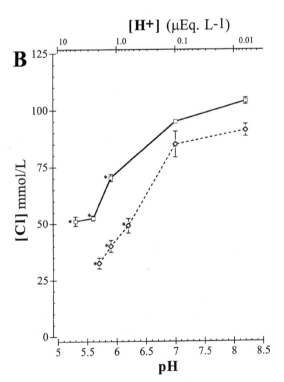

Figure 3. Changes in whole body salt concentration for adults and nauplii (1st instar) acclimatised to, and acid exposed 24 h in, 20% ASW. (**A**) [Na]. (**B**) [K]. (**C**) [Cl]. adult □, nauplii ◇. Data are mean ± SEM. ∗ denotes significant difference ($p < 0.05$). Dashed lines indicate ambient concentration.

such *Artemia* in hypersaline environments was determined by a maximum haemolymph ion concentration equivalent to approximately 75% seawater. The simple scenarios above thus also accurately predicts the observed results in SW acclimated *Artemia* and possibly explains a major portion of the observed mortality in acid exposed *Artemia*.

Nauplii appear more sensitive than adults to acid exposure in either salinity (Table 1). The basis for the increased larval sensitivity may lie in the change in ion-regulatory sites occurring during development.

In the adult, ion regulatory sites include the gut epithelium, the epithelium over the first ten pairs of branchiae (Croghan, 1958a, b) and the maxillary gland (Conte, 1984). In the nauplius, functional ion regulation also occurs over the gut epithelium, but development of branchiae occurs gradually and ion-regulation is initially supplemented by the larval salt gland (neck organ, Conte, 1984) which degenerates once the branchiae are fully functional (Croghan, 1958a). The progressive decrease in sensitivity to acid

6

stress as development progresses (see Figure 1 and Table 1) as well as the differential ability to regulate body ion composition (Figures 2–3) may thus be correlated with these fundamental changes in ionregulatory site occurring between larval and adult *Artemia*. Equally the smaller size and increased surface area of the naupliar stage may also play an important role, especially if acid exposure increases permeability to ions at the surface (see below). Somewhat similar reasoning was put forward by Chulakasem et al. (1989) who studied the effects of acid on development of the teleost fish *Oryzias latipes*.

Doyle & McMahon (1995) also showed strongly negative effects of acid exposure on the success of the hatching process in *A. franciscana*. Serious long-term effects of seawater acidification are suggested by work on other organisms. Several authors have reported reduced calcification of algae at decreased pH in seawater (reviewed in Knutzen, 1981) and decalcification is a serious result of acidification in freshwater crayfish (McMahon & Stuart, 1989). Acidification will also decrease bicarbonate levels and increase PCO_2. Neither of these aspects have received serious study. Lastly significant effects on growth rates have been reported in response to very small increases in acidification (Knutzen, 1981). A final potential area of concern occurs over the effects of increased acidity on solubility of complexed metal ions.

This study has demonstrated lethal effects of acid exposure on the viability of *Artemia franciscana* cultured in either dilute or full strength seawater. Physiological effects of acid exposure on stenohaline marine animals have not yet been assessed but the present results have important ramifications for many animal communities in estuarine or brackish locations close to centres of acid precipitation. Most if not all euryhaline animals are regulators in dilute media; many are hypo/hyperosmotic regulators and thus depend on both low body-wall permeability and active ion pumps to maintain body fluid osmotic and ionic levels different from the environment in either dilute or concentrated media. Since increased environmental acidity both increases permeability and reduces the effectivity of ion regulation, many euryhaline animals are now shown to be highly susceptible to increase in ambient acidity. The levels at which they succumb are not substantially different from those of freshwater species. Extending this reasoning it is clear that, contrary to previous assumptions, many estuarine communities, including many important fisheries, may be seriously threatened by acid precipitation or by drainage from

acid stressed areas. The greater sensitivity of larval stages places these communities at increased risk.

The greater buffering power of seawater does not provide absolute protection against acidification. Knutzen (1981) calculates that proposed seawater scrubbing of acidic gaseous effluents may seriously affect inshore waters over a wide area. It is clear that acidification of coastal and estuarine waters is a area of serious concern which urgently needs further study.

Acknowledgments

The authors would like to thank Mr J. Gudgeon who assisted in the acid stress work in dilute seawater and Dr R. L. Walker for useful discussions and advice. Financial support was provided by NSERC grant A5762 to B. R. McMahon and an NSERC Undergraduate Student Research Award to J. E. Doyle.

References

Chulakasem, W., J. A. Nelson & J. J. Magnuson, 1989. Interaction of effects of pH and ion concentration on mortality during early development of Medaka, *Oryzias latipes*. Can. J. Zool. 67: 2158–2168.

Conte, F. P., 1984. Structure and function of the crustacean larval salt gland. Int. Rev. Cytol. 91: 45–106.

Croghan, P. C., 1958a. The mechanism of osmotic regulation in *Artemia salina* (L): the physiology of the branchiae. J. exp. Biol. 35: 234–242.

Croghan, P. C., 1958b. The mechanism of osmotic regulation in *Artemia salina* (L): the physiology of the gut. J. exp. Biol. 35: 243–249.

De Wachter, B. & J. Van Den Abbeele, 1990. The influence of acclimation on salinity and oxygen on the respiration of the brine shrimp, *Artemia franciscana*. Comp. Biochem. Physiol. 98A: 293–298.

Doyle, J. E. & B. R. McMahon, 1995. Effects of acid exposure in the brine shrimp *Artemia franciscana* during development in seawater. Comp. Biochem. Physiol. 112A: 123–129.

Eriksen, C. H. & R. J. Brown, 1980. Comparative respiratory physiology and ecology of phyllopod crustacea. II. Anostraca. Crustaceana 39: 11–21.

Holliday, C. W., D. B. Roye & R. D. Roer, 1990. Salinity induced changes in branchial Na-K ATPase activity and transepithelial potential difference in the brine shrimp *Artemia salina*. J. exp. Biol. 151: 279–296.

Knutzen, J., 1981. Effects of decreased pH on marine organisms. Mar. Pollut. Bull. 12: 25–29.

McDonald, D. G., R. L. Walker & P. R. H. Wilkes, 1983. The interaction of environmental calcium and low pH on the physiology of the rainbow trout, *Salmo giardneri*. J. exp. Biol. 102: 141–155.

McMahon, B. R. & S. A. Stuart, 1989. The physiological problems of crayfish in acid waters. In Morris, R. W. (eds), Acid toxicity and Aquatic Animals, Cambridge University Press, 171–199.

Playle, R. C., G. G. Gross & C. M. Wood, 1989. Physiological disturbances in rainbow trout (*Salmo giardneri*) during acid and aluminum exposures in soft water of two calcium concentrations. Can. J. Zool. 67: 314–324.

Wood, C. M., 1989 The physiological problems of fish in acid waters. In Morris, R. W. (eds), Acid toxicity and Aquatic Animals, Cambridge University Press: 125–152.

Woolf, C. M., 1968. Principles of Biometry. D. Van Nostrand Co., Toronto.

Hydrobiologia **352**: 9–16, 1997.
Y.-S. Wong & N. F.-Y. Tam (eds), Asia–Pacific Conference on Science and Management of Coastal Environment.
©1997 *Kluwer Academic Publishers. Printed in Belgium.*

Effects of two oil dispersants on phototaxis and swimming behaviour of barnacle larvae

R. S. S. Wu[1], P. K. S. Lam[1]* & B. S. Zhou[2]

[1]*Department of Biology and Chemistry, City University of Hong Kong, Tat Chee Avenue, Kowloon, Hong Kong*
[2]*Institute of Hydrobiology, Chinese Academy of Sciences, Wuhan, China*
(*Author for correspondence; tel: 852 2788 7681; fax: 852 2788 7406; e-mail: bhpksl@cityu.edu.hk)

Key words: Oil dispersant, phototaxis, swimming behaviour, barnacle larvae

Abstract

The effects of two oil dispersants (Vecom B-1425 GL and Norchem OSD-570) mixed with diesel oil on the survival and behaviour of the stage II nauplii of the barnacle *Balanus amphitrite* were investigated. The 24 and 48-hour LC_{50} values for Vecom B-1425 GL:diesel mixture were 514 and 48 mg l^{-1} respectively, while respective values for Norchem OSD-570:diesel mixture were 505 and 71 mg l^{-1}. Under sublethal concentrations, increased levels of the dispersant:diesel mixtures caused a reduction in phototactic responses. *Balanus amphitrite* nauplii failed to exhibit phototactic responses when exposed to Vecom B-1425 GL:diesel mixtures of 400 mg l^{-1} and higher for 24 hours. A longer exposure time of 48 hours further reduced the Lowest Observable Effect Concentrations (LOECs) to 60 mg l^{-1}. The LOECs for Norchem OSD-570:diesel mixtures under exposure periods of 24 and 48 hours were 400 and 80 mg l^{-1} respectively. The curvilinear velocities (VCL) and straight-line velocities (VSL) of the stage II nauplii ranged from 0.7–1.1 and 0.2-0.4 mm s^{-1} respectively. Increased concentrations of dispersant:diesel mixtures caused a significant change in the curvilinear and straight-line velocities. Both oil dispersants, dispersant:diesel mixtures of 20 to 40 mg l^{-1} caused significant increases in VCL, but no significant change in VSL. Dispersant:diesel mixtures of 100 mg l^{-1} and higher resulted in a reduction in VSL for both dispersants.

Introduction

Hong Kong is one of the busiest ports in the world, handling an average of 165,000 ocean-going and river-trade vessels called at the port each year. At least nine major oil spills have occurred in Hong Kong waters since 1968. In addition, minor oil spills are common along coastal waters of Hong Kong. For example, during 1982–1983, 41 minor spills were handled by the Marine Department of the Hong Kong Government, during which some 83,150 litres of oil dispersants were used (Wu, 1988). In 1992, 52 minor oil spills occurred (EPD, 1993). There is clearly a need to understand the ecological effects of chemical dispersants when applied to oil spillages in the local marine environment before a dispersant can be approved for use in oil pollution control. In 1975, a simple screening test was introduced to regulate oil dispersants in Hong Kong under the Dumping at Sea Act 1974 (Overseas Territories) Order 1975. This test is based on acute effects using the rabbitfish, *Siganus oramin* (Bloch and Schneider) and the sea urchin *Anthocidaris crassispina* (A. Agassiz). Such mortality-based tests have been effective in screening different types of oil dispersants for licensing purposes (Thompson & Wu, 1981). However, with an increased awareness of the potential chronic effects of chemical dispersants in the marine environment, it is instructive to study the lethal, as well as sublethal, effects of oil dispersants on important local marine species. Larval stages are often more susceptible to environmental perturbations (Connor, 1972). Behavioural responses, such as orientative and swimming behaviour of marine larvae, have often been used to assess the sublethal impact of environmental toxicants as these responses have important adaptive significance for marine larvae in optimizing feeding efficiency (e.g.

Stearns & Forward, 1984), predator avoidance (e.g. Forward & Hettler, 1992), and larval dispersal and settlement (e.g. Thorson, 1964; Svane & Young, 1989; Dirnberger, 1993; Hurlbut, 1993). Locomotory behaviour, integrating responses at nervous, muscular and energetic levels, clearly have an important influence on the individual's fitness, and ultimately the survival of the species. Barnacles play an important role in the functioning of coastal systems (Wu & Levings, 1978), and the impairment of important behaviour traits of barnacle larvae will have an indirect effect on the coastal system as a whole. This paper reports upon the effects of two oil dispersants (Vecom B-1425 GL and Norchem OSD-570) in combination with diesel oil on the survival, and the orientative and swimming behaviour of the stage II nauplii of the barnacle *Balanus amphitrite* Darwin. The feasibility of developing a test which is ecologically relevant to Hong Kong, based on the lethal and sublethal effects of oil dispersants on the barnacle larvae, is explored.

Materials and methods

Test organisms

The barnacle *Balanus amphitrite* Darwin inhabits the intertidal to sublittoral fringe of circumtropical and subtropical shores (Foster, 1982). Animals used in this study were collected from Ma Lui Siu Ferry Pier in the New Territories of Hong Kong, kept in aerated sea water, and brought back to the laboratory. Brood sacs containing mature nauplii were dissected and removed from the animals, and suspended in sea water to allow the nauplii to hatch. Stage II nauplii actively swimming towards a light source were collected using a pipette, and used for subsequent experimentation.

Acute toxicity ((LC$_{50}$) tests

Stage II nauplii were exposed to two oil dispersants (Vecom B-1425 GL and Norchem OSD-570) in combination with diesel oil for 24 and 48 hours. Both dispersants were mixed with diesel in the ratio of oil dispersant:diesel oil = 1:10 (vol:vol) (the application ratio recommended by the manufacturers). The resulting dispersant:diesel mixtures were subsequently serially diluted with filtered sea water (temperature: 20 ± 1 °C; salinity: 32 ± 1‰; pH 7.8; DO 7 ± 1 mg l^{-1}) over the concentration ranges. Barnacle larvae were exposed to concentrations of 0, 100, 200, 300,

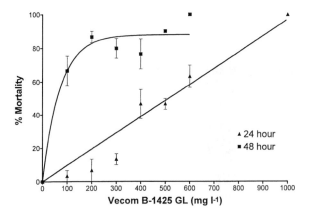

Figure 1. Percentage mortality of *Balanus amphitrite* nauplii exposed to Vecom B-1425 GL:diesel mixtures for 24 and 48 hours. The vertical lines are ± 1 S.E.

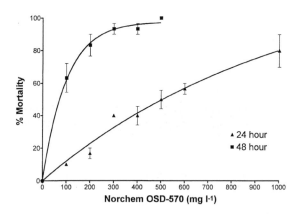

Figure 2. Percentage mortality of *Balanus amphitrite* nauplii exposed to Norchem OSD-570:diesel mixtures for 24 and 48 hours. The vertical lines are ± 1 S.E.

400, 500, 600 and 1000 mg l^{-1} for 24 hours, and 100, 200, 300, 400, 500 mg l^{-1} for 48 hours. All concentrations are nominal and refer to the quantity of dispersant:diesel mixture in sea water. For each of the acute tests, stage II nauplii were exposed to various concentrations of the dispersant:diesel mixtures in 25 ml beakers. The beakers were placed on an orbital shaker set at about 150 oscillations per minute. There were three replicates of ten nauplii each per concentration for each oil dispersant. At the end of the exposure period, the number of swimming, non-motile (with vibrating appendages) and dead animals in each beaker was recorded. LC$_{50}$ values were estimated from equations describing the curves fitted to the mortality data using non-linear regression analyses (Fry, 1993).

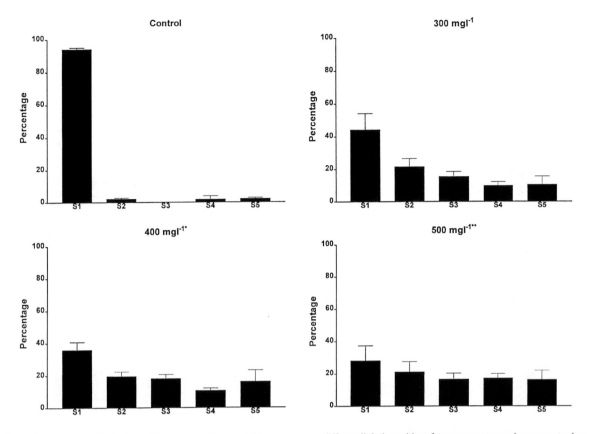

Figure 3. Frequency distribution of *Balanus amphitrite* nauplii in response to different light intensities after exposure to varying concentrations of Vecom B-1425 GL:diesel mixtures for 24 hours. Significance levels: *, $P < 0.05$; **, $P < 0.01$. The bars represent the mean and 1 S.D.

Phototaxis inhibition assay

The phototaxis inhibition assay was carried out in a rectangular ($40 \times 40 \times 150$ mm) plexiglass container with a single transparent window at one end (cf Marsden, 1990). The sides and the other end of the chamber were opaque and non-reflective. The chamber was divided (by an engraved line on the bottom) into five equal sized, non-partitioned sections: S1, S2, S3, S4 and S5. S1 occupied the window end of the chamber, S2 was located at 30 to 60 mm from the window, S3 at 60 to 90 mm, and so on. An artificial light source (a 6.5V, 2.75A tungsten bulb in a MEIJI microscope lamp) was placed at 10 cm in front of the transparent window. Light intensity in seawater in each of the sections (S1 - S5) was determined using a luminance meter (TOPCON BM5). The intensities measured at S1, S2, S3, S4 and S5 were 1.07, 0.47, 0.26, 0.13 and 0.06 (x 10^{-3} Cd m^{-2}) respectively. The barnacle larvae were pre-exposed to 0, 100, 200, 300, 400, 500, 600 mg l^{-1} of the two dispersant:diesel mixtures for 24

hours, and 0, 10, 20, 30, 40, 60, 80, 100 mg l^{-1} for 48 hours as outlined above before their phototactic behaviour was examined. At least sixty actively swimming nauplii were used for each set of observations. Following the appropriate exposure period, the nauplii were transferred to the experimental chamber. The experimental chamber was gently agitated and a 10-minute time lapse was allowed to ensure an initial random distribution of the larvae. The number of test organisms in each section was then recorded. There were five replicates per concentration per exposure period for each dispersant:diesel mixture. All observations were carried out in a darkroom at 20 °C.

Analysis of swimming behaviour

Stage II nauplii were pre-exposed to the two dispersant:diesel mixtures at 0, 20, 40, 60, 80 and 100 mg l^{-1} for 48 hours as outlined above, and then transferred onto a cavity slide for video-taping. A JVC CCD colour camera mounted on a stereo-microscope recorded the

12

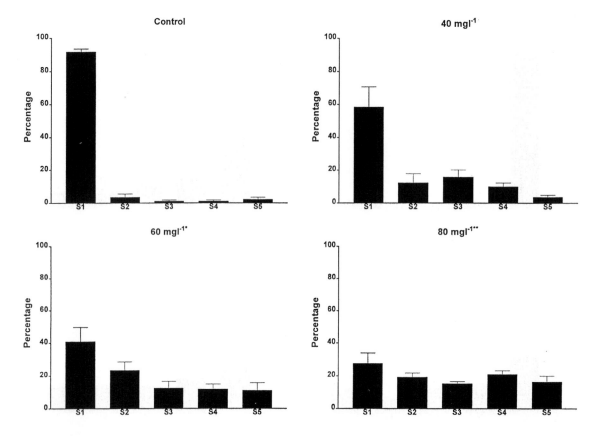

Figure 4. Frequency distribution of *Balanus amphitrite* nauplii in response to different light intensities after exposure to varying concentrations of Vecom B-1425 GL:diesel mixtures for 48 hours. Significance levels: *, $P < 0.05$; **, $P < 0.01$. The bars represent the mean and 1 S.D.

movement patterns of three to five barnacle larvae in a horizontal plane for one minute. Filtered light from a base illuminator was used to minimize effects due to uneven lighting. Recordings were made on six separate batches of nauplii.

Video tapes of naupliar swimming behaviour were played back for analyses by a computer. The computer was equipped with the proprietary hardware and software that constitute the CRISMAS system (Image House, Copenhagen). The programme, which was designed for measuring sperm motility, can be modified to enable the measurement of larval swimming behaviour. The computer acquired and digitized video data at 32 frames per second for two seconds from video images. By following the movement of individual larvae over the 2-second duration, the mean curvilinear velocity (VCL) and mean straight-line velocity (VSL) could be determined. VCL was calculated by summing the distances of the linear path between the position of a nauplius in all successive frames and dividing by the interval during which the nauplius was

tracked, while VSL was calculated from the straight line path between the first and last position of a nauplius (see Budworth et al. 1988 and references therein for details).

Statistical analysis

When test organisms exhibit a positive phototactic response, the distribution of the animals should skew towards the sections with lower light intensities. In this study, we use the distribution pattern of test organisms in sea water (i.e. 0 mg 1^{-1} of dispersant:diesel mixture) as the control and define the No Observable Effect Concentration (NOEC) as the highest concentration of the dispersant:diesel mixture at which no significant deviation from the control distribution pattern is detected. The lowest concentration at which a significant deviation from the control is detected was taken as the Lowest Observable Effect Concentration (LOEC). All comparisons between 'test' and 'refer-

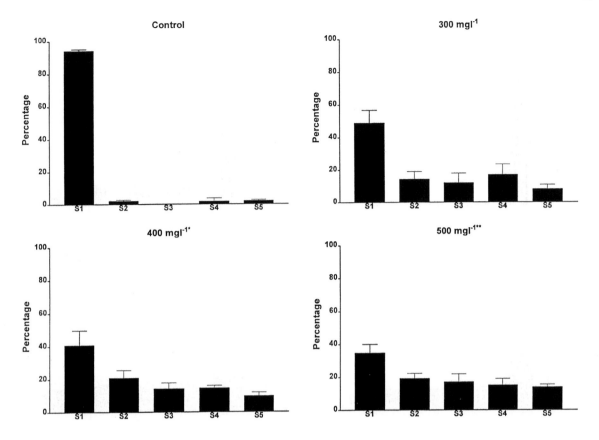

Figure 5. Frequency distribution of *Balanus amphitrite* nauplii in response to different light intensities after exposure to varying concentrations of Norchem OSD-570:diesel mixtures for 24 hours. Significance levels: *, $P < 0.05$; **, $P < 0.01$. The bars represent the mean and 1 S.D.

ence' distributions were carried out using Chi-square Goodness-of-fit tests (Zar, 1984).

Velocities of barnacle larvae exposed to different concentrations of a dispersant:diesel mixture were compared using one-way analyses of variance. Where treatment group means were significantly different, Dunnett's tests were used to compare individual treatment groups with the control to identify NOECs and LOECs (Zar, 1984).

Results

Results of the 24-hour and 48-hour acute tests for Vecom B-1425 GL:diesel and Norchem OSD-570:diesel mixtures are given in Figures 1 and 2. For *Balanus amphitrite* larvae, the LC_{50} values under 24 and 48-hour exposures for Vecom B-1425 GL:diesel mixtures were 514 and 48 mg l^{-1} respectively, while those for Norchem OSD-570 were 505 and 71 mg l^{-1}. At sublethal concentrations, increased levels of the dis-

persant:diesel mixtures caused a reduction in phototactic responses. As examples, the distribution patterns of barnacle larvae in response to different light intensities under 24-hour and 48-hour exposures to Vecom B-1425 GL:diesel and Norchem OSD-570:diesel mixtures are shown in Figures 3–6. *Balanus amphitrite* nauplii failed to exhibit phototactic responses when exposed for 24 hours to Vecom B-1425 GL:diesel mixtures of 400 mg l^{-1} and higher (Figure 3) while longer exposure times of 48 hours reduced the LOECs to 60 mg l^{-1} (Figure 4). The LOECs for Norchem OSD-570:diesel mixture under exposure periods of 24 and 48 hours were 400 and 80 mg l^{-1} respectively (Figures 5 and 6).

The curvilinear velocities (VCL) and straight-line velocities (VSL) of the stage II nauplii ranged from 0.7–1.1 and 0.2–0.4 mm s^{-1} respectively. Exposure to increased concentrations of dispersant:diesel mixtures resulted in a significant change in the curvilinear and straight-line velocities (Vecom B-1425-GL:diesel; $F_{5,30} > 3.00$, $P < 0.05$ and Norchem OSD-570:diesel;

14

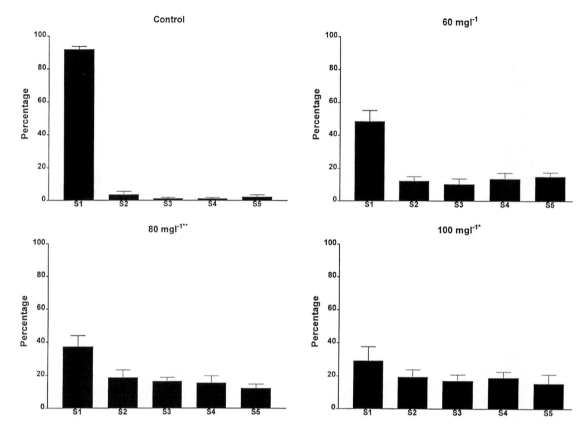

Figure 6. Frequency distribution of *Balanus amphitrite* nauplii in response to different light intensities after exposure to varying concentrations of Norchem OSD-570:diesel mixtures for 48 hours. Significance levels: *, $P < 0.05$; **, $P < 0.01$. The bars represent the mean and 1 S.D.

$F_{5,30} > 2.50$, $P < 0.05$). For both oil dispersants, dispersant:diesel mixtures of 20 to 40 mg l^{-1} caused a significant increase in VCL, but no significant change in VSL (Figures 7 and 8). Dispersant:diesel mixtures of 100 mg l^{-1} and higher resulted in a reduction in VSL for both dispersants (Figures 7 and 8).

Discussion

Average discharge of oil to the world's oceans is between 1.7 and 8.8 million metric tons in a year, most of this occurring in coastal areas (Howarth, 1989 and references therein). In the case of an oil spill, oil dispersants are often used to break up surface oil into small droplets, thus making the oil more vulnerable to degradation by physical and biological processes (Fisher & Foss, 1993). Oil dispersants typically consist of a mixture of a surfactant, a hydrocarbon solvent and sometimes a stabilizing agent. The toxicity of a chemical dispersant is usually a function of the

kind of solvent (Portmann & Connor, 1968), surfactant (Swedmark et al., 1971) or the synergistic effects between the surfactant and the solvent (Nagell et al., 1974). Significantly, the enhanced bioavailability of the oil resulted from the application of chemical dispersants may also lead to an increase in oil toxicity to aquatic organisms. Oil dispersants can cause a range of effects on marine organisms including gill damage (Abel, 1974; Nagell et al., 1974), loss of osmotic stability in gill tissues, and loss of gill cilia (Nuwayhid et al., 1980; Singer et al., 1990). Despite the improvement in the formulation of chemical dispersants from more noxious 'kerosene-based mixtures' to less toxic preparations in recent years (Singer et al., 1990), crucial information on the biological effects of these chemicals on organisms in the receiving environment is still scarce. Despite research efforts into the potential environmental impact of oil-dispersant mixtures during the past 40 years (see reviews by Lewis (1990, 1991, 1992)), general information gaps still exist in certain key areas, for example, the fate and effects of oil and dispersants in the tropical region as well as the Arctic (Sprague et al., 1981).

The present study examines the effect of two locally available oil dispersants (Vecom B-1425 GL and Norchem OSD-570) on the larvae of the barnacle, *Balanus amphitrite*, which occupies the part of the shores vulnerable to oil and dispersant impacts. The phototactic and swimming behaviour of the barnacle larvae has a direct influence on their foraging efficiency and settlement. The positive phototaxis of *B. amphitrite* nauplii allows these animals to move to areas of high primary productivity and thus food abundance. An impairment of such a response would thus have a major deleterious impact on the survival (fitness) of individual larva. Moreover, the swimming velocity can also be of importance as a swimming speed higher than the optimum would lead to an unnecessary waste of energy, while a slow-moving nauplius is likely to have a restricted foraging efficiency and range. Our results showed that the phototactic response of the barnacle larvae are significantly affected by the dispersant:diesel mixtures at concentrations of 400 and 60–80 mg 1^{-1} for 24-hour and 48-hour exposures respectively. Exposure to low concentrations of dispersant:diesel mixtures (20–40 mg 1^{-1}) for 48 hours caused a significant increase in the curvilinear velocities. Lang et al. (1981) also observed a similar initial hyperactivity in *B. improvisus* exposed to metals, and suggested that the effect was a transient response, and that the animals would recover. A further increase in dispersant:diesel mixture concentrations resulted first in an impairment to the phototactic responses, followed by a reduction in the straight-line velocities. By contrast, Lang et al. (1981) reported that the photobehaviour of the *B. improvisus* larvae was less sensitive to metal toxicants than their swimming behaviour.

It should be noted that the range of concentrations reported in this study is much lower than the concentration of 1000 mg 1^{-1} employed in the mortality-based screening tests using *Siganus oramin* for licensing purposes in Hong Kong (Thompson & Wu, 1981). These findings indicate that tests based on organisms such as the barnacle nauplii would be more sensitive than the current tests in providing an assessment of chemical dispersants in the local marine environment. The use of *B. amphitrite* in these ecotoxicological tests in Hong Kong has an additional advantage in that *B. amphitrite* nauplii are available all year round, thus can provide a relatively reliable supply of test organisms.

Although it is clear that the dispersant:diesel mixtures can have a significant impact on the phototac-

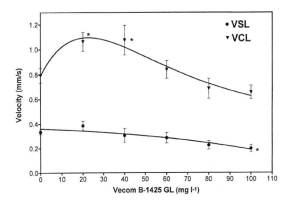

Figure 7. Mean curvilinear and straight-line velocities of *Balance amphitrite* nauplii after exposure to different concentrations of Vecom B-1425 GL:diesel mixtures for 48 hours. The vertical lines are ± 1 S.E.

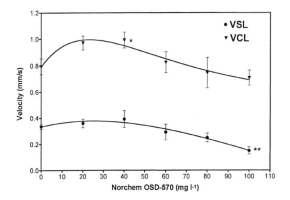

Figure 8. Mean curvilinear and straight-line velocities of *Balance amphitrite* nauplii after exposure to different concentrations of Norchem OSD-570:diesel mixtures for 48 hours. The vertical lines are ± 1 S.E.

tic and swimming behaviour of the barnacle larvae, the mechanism involved is still not entirely clear. The dispersant:diesel mixtures can affect the nauplii in at least two ways: (1) direct toxic effect on the nauplii; and (2) indirect influence through an alteration of the physico-chemical properties, e.g. viscosity and optical characteristics, of the test medium. Previous studies have indicated that the toxicity of oil and chemical dispersants can be influenced by environmental factors such as temperature (e.g. Ordzie & Garofalo, 1981; Fisher & Ross, 1993) and salinity (e.g. Fisher & Ross, 1993). Although the results reported here are obtained at constant temperature and salinity and it is not uncommon to undertake ecotoxicological tests under constant environmental conditions, it would be instructive to assess the toxicities of the compounds at

16

temperature and salinity ranges relevant to local waters so that possible impacts can be predicted in the event of dispersant treatment of oil at sea. Research work is currently being undertaken to examine the effect of dispersant:diesel mixtures on the settling behaviour of barnacle cyprids.

Acknowledgments

The authors acknowledge the support of a research grant from the City University of Hong Kong. We thank also Mr Patrick C. C. Lai for his assistance and an anonymous reviewer for providing helpful comments on the manuscript.

References

Abel, P. D., 1974. Toxicity of synthetic detergents to fish and aquatic invertebrates. J. Fish. Biol. 6: 279–298.

Budworth, P. R., R. P. Mann & P. L. Chapman, 1988. Relationships between computer measurements of motion of frozen-thawed bull spermatozoa and fertility. J. androl. 9: 41–54.

Connor, P. M., 1972. Acute toxicity of heavy metals to some marine larvae. Mar. Pollut. Bull. 5: 171–174.

Dirnberger, J. M., 1993. Dispersal of larvae with a short planktonic phase in the polychaete Spirorbis spirillum (Linnaeus). Bull. mar. Sci. 52: 898–910.

Environmental Protection Department (EPD), 1993. Environment Hong Kong 1993. Hong Kong Government, 183 pp.

Fisher, W. S. & S. S. Foss, 1993. A simple test for toxicity of number 2 fuel oil and oil dispersants to embryos of grass shrimp, Palaemonetes Pugio. Mar. Pollut. Bull. 26: 385–391.

Forward, R. B., Jr. & W. F. Hettler, Jr., 1992. Effects of feeding an predator exposure on photoresponses during diel vertical migration of brine shrimp larvae. Limnol. Oceanogr. 37: 1261–1270.

Foster, B. A., 1982. Hong Kong Barnacles. In Morton, B. S. & C. K. Tseng (eds), Proceedings of the First International Marine Biological Workshop: The Marine Flora and Fauna of Hong Kong and Southern China, Hong Kong, 1980. Hong Kong University Press, Hong Kong: 207–232.

Fry, J. C., 1993. Biological data analysis. IRL Press, Oxford University Press, Oxford, 418 pp.

Howarth, R. W., 1989. Determining the ecological effects of oil pollution in marine ecosystems. In Levin, S. A., M. A. Harwell, J. R. Kelly & K. D. Kimball (eds), Ecotoxicology: Problems and Approaches. Springer-Verlag, New York: 69–97.

Hurlbut, C. J., 1993. The adaptive value of larval behavior of a colonial ascidian. Mar. Biol. 115: 253–262.

Lang, W. H., D. C. Miller, P. J. Ritacco & M. Marcy, 1981. The effects of copper and cadmium on the behaviour and development of barnacle larvae. In Vernberg, F. J., A. Calabrese, F. Thurberg & W. B. Vernberg (eds), Biological Monitoring of Marine Pollutants. Academic Press, New York: 165–203.

Lewis, M. A., 1990. Chronic toxicities of surfactants and detergent builders to algae: a review and risk assessment. Ecotoxicol. environ. Safety 20: 123–140.

Lewis, M. A., 1991. Chronic and sublethal toxicities of surfactants to freshwater and marine animals: a review and risk assessment. Wat. Res. 25: 101–113.

Lewis, M. A., 1992. The effects of mixtures and other environmental modifying factors on the toxicities of surfactants to freshwater and marine life. Wat. Res. 26: 1013–1023.

Marsden, J. R., 1990. Light responses of the planktonic larva of the serpulid polychaete Spirobranchus polycerus. Mar. Ecol. Prog. Ser. 58: 225–233.

Nagell, B., M. Notini & O. Grahn, 1974. Toxicity of four oil dispersants to some animals from Baltic sea. Mar. Biol. 28: 237–243.

Nuwayhid, M. A., S. P. Davies & H. Y. Elder, 1980. Changes in the ultrastructure of the gill epithelium of Patella vulgata after exposure to North Sea crude oil and dispersants. J. mar. Biol. Ass. U.K. 60: 439–448.

Portmann, J. E. & P. M. Connor, 1968. The toxicity of several oil-spill removers to some species of fish and shellfish. Mar. Biol. 1: 322–329.

Ordzie, C. J. & G. C. Garofalo, 1981. Lethal and sublethal effects of short term acute doses of Kuwait crude oil and a dispersant Corexit 9527 on bay scallop, Argopectin irradians (Lamarck) and two predators at different temperatures. Mar. environ. Res. 5: 195–211.

Singer, M. M., D. L. Smalheer, R. S. Tjeerdema & M. Martin, 1990. Toxicity of an oil dispersant to the early life stages of four California marine species. Envir. Toxicol. Chem. 9: 1387–1395.

Sprague, J. B., J. H. Vandermeulen & P. G. Wells, 1981. Oil and dispersants in Canadian seas – recommendations from a research appraisal. Mar. Pollut. Bull. 12: 45–46.

Stearns, D. E. & R. B. Forward Jr., 1984. Photosensitivity of the calanoid copepod Acartia tonsa. Mar. Biol. 82: 85–89.

Svane, I. & C. M. Young, 1989. The ecology and behaviour of ascidian larvae. Oceanogr. mar. biol. Ann. Rev. 27: 45–90.

Swedmark, M., B. Braaten, E. Emanuelsson & A. Granmo, 1971. Biological effects of surface active agents on marine animals. Mar. Biol. 9: 183–201.

Thompson, G. B. & R. S. S. Wu, 1981. Toxicity testing of oil slick dispersants in Hong Kong. Mar. Pollut. Bull. 12: 233–237.

Thorson, G., 1964. Light as an ecological factor in the dispersal and settlement of larvae of marine bottom invertebrates. Ophelia 1: 167–208.

Wu, R. S. S., 1988. Marine pollution in Hong Kong: a review. Asian mar. Biol. 5: 1–23.

Wu, R. S. S. & C. D. Levings, 1978. An energy budget for individual barnacles (Balanus glandula Darwin). Mar. Biol. 45: 225–235.

Zar, J. H., 1984. Biostatistical analysis (2nd edn.) Prentice-Hall International, New Jersey, 718 pp.

Hydrobiologia **352**: 17–23, 1997.
Y.-S. Wong & N. F.-Y. Tam (eds), Asia–Pacific Conference on Science and Management of Coastal Environment.
©1997 *Kluwer Academic Publishers. Printed in Belgium.*

Heavy metal accumulation in tissue/organs of a sea cucumber, *Holothuria leucospilota*

Jun Xing* & Fu-Shiang Chia
Dept. of Biology, The Hong Kong University of Science and Technology, Clear Water Bay, Kowloon, Hong Kong
(*Author for correspondence; tel: (852)23587336; fax: (852)23581559; e-mail: boxjx@usthk.ust.hk)*

Key words: Zinc accumulation, sea cucumber, longitudinal muscle, respiratory tree

Abstract

Holothuria leucospilota Brandt, the large black sea cucumber, is a non-selective deposit feeder, and is commonly found in the bottom of shallow waters in Hong Kong, where the sediments are often polluted with heavy metals. This study was designed to test the possibility of heavy metal accumulation by the sea cucumber at two sites in Hong Kong. Atomic absorption spectrophotometer was used to measure Cu and Zn concentrations in various tissue/organs of the animal as well as in the sediments. The result indicated that *H. leucospilota* accumulated zinc in the longitudinal muscle bands (97.27–98.07 ppm in dry weight) and in the respiratory tree (83.92–89.64 ppm in dry weight). Copper concentrations in these two organs were much lower than that of zinc. After the animals were kept in the aquarium without sediment for 40 days, zinc concentration of the longitudinal muscle and respiratory tree decreased by 48% and 39% respectively whereas copper concentration remained unchanged. The concentrations of zinc and copper in the sediment at the two sites differed significantly but the metal level in the animals from the two sites were similar, suggesting that this sea cucumber was not an ideal bioindicator of heavy metal pollution in the sediment.

Introduction

Many marine animals can accumulate heavy metals such as zinc, mercury, copper, cadmium and lead in their body (Riley & Segar, 1970). Three commercially important groups of marine animals (molluscs, crustaceans and teleosts) have been frequently studied for bioaccumulation of heavy metals. Among the heavy metals studied, zinc and copper show the highest concentration in the animal body; some studies indicate tissue/organ specificity in heavy metal accumulation (see review by Eisler, 1981; Warnau et al., 1995). Gill, digestive gland and kidney of molluscs and hepatopancreas of crustaceans are the heavy metal repository organs while muscle has the lowest metal concentration (Brooks & Rumsby, 1965; Bryan, 1968; 1973); Similarly, liver of fish accumulates high levels of zinc and copper while the muscle of fish represents the least metal-containing tissue (Brooks & Rumsby, 1974; Eisler, 1981).

The heavy metals in the animal body are either originated from food, water or sediment, and their relative importance varies with the metal, the nature and physiological state of the organism (Weeks & Rainbow, 1993). In molluscs, it has been found that heavy metal accumulation is positively correlated with metal concentration in diet and ambient water (Martin, 1979; Thrower & Eustace, 1973; Weeks & Rainbow, 1993; Julshamn & Grahl-Nielsen, 1996). In sea urchin, *Paracentrotus lividus*, accumulation of cadmium in the body is mostly from sea water, and this species can serve as an indicator of cadmium contamination of sea water (Warnau et al., 1995).

Marine sediment acts as a reservoir for heavy metals because most of the trace metals can be trapped by the sediment particles (Luoma, 1990). However, studies of bioindicator animals of sediment quality are rare compared with those of sea water quality indicators. In recent years, deposit-feeding polychaetes and echinoderms have been suggested to be used as indicators of metal level in sediments. However, the relationship between the metal content in body and that in sediment is controversial. For example, Bryan & Hummerstone (1973) reported that copper level in *Nereis diversicolor* directly reflected the copper concentration in sediment but zinc concentration in this animal seemed independent of that in the sediment (Bryan & Gibbs, 1979). Ying et al. (1993) reported that the intertidal snail *Polinices sordidus* did not accumulate zinc from sediments containing less than 10 mg Zn g^{-1}. Eisler (1981) suggested that echinoderms were the 'prime movers of sediments and detritus in the sea and are probably very important in the cycling of trace metals' and 'show promise of becoming suitable indicator'. However, little is known in echinoderms on the relationship between metal content in animal and that in sediment.

The common black sea cucumber, *Holothuria leucospilota* Brandt (Echinodermata, Holothuroidea), ingests sediment non-selectively, and may be a potential indicator animal of heavy metal in sediment. The present study was designed to assess this possibility by examining the heavy metals in various tissue/organs as well as in the ambient sediments, where the animals were collected.

Study sites, materials and methods

Holothuria leucospilota, is a dominant invertebrate in shallow waters of many areas in Hong Kong and can be easily obtained at low tide (1.0 m). Animals (N = 12) in this study were collected in November, 1994 from shallow waters of Wong Shek, which is a public pier and locates at a semi-closed bay, northern-east part of Kowloon, Hong Kong, and Shelter Island, which is more exposed to the sea and is located at the eastern part of Kowloon, Hong Kong (Figure 1). These animals were maintained in an aerated sea water aquarium with sediment from the collecting sites for less than 5 days prior to bioassay processing. Twelve more animals from Wong Shek were maintained as above but without sediment for 40 days before determination of heavy metal concentration. To prevent ingestion of feces, all feces were removed as soon as they were seen in the aquarium. A total of 5 samples (each weighed 250 g in wet weight) of surface sediment (5 cm in depth) from each site were collected at the same time as animal collection by using a plastic spatula. The samples were placed in plastic bags and brought to the laboratory for analysis.

Sea cucumbers were dissected by a longitudinal incision from the anus to the tentacle, after relaxing in a 6.7% $MgCl_2$ solution for 30 min. Longitudinal muscle bands, body wall, respiratory tree, Cuvierian tubule and digestive tract of each individual were carefully removed and washed in filtered sea water. Animal tissue/organs and sediment were dried at 110 °C until constant dry weight was obtained (40–48 h). The dry sediment were passed through a sieve, collecting particles less than 1 mm in diameter for subsequent analysis. For each dry sediment sample, 2 sub-samples (0.5 g each) were prepared (10 sub-samples for each site), each subsample was digested with 5 ml 65% nitric acid in a digestion tube. The digestion tubes were heated at 200 °C on Digestion System 40 1016 Digester (TECATOR) for 30–60 min. After being cooled to room temperature, 1 ml 30% (w/w) hydrogen peroxide was added to each tube, followed by further heating at 200 °C until the color of the liquid was stabilized. Addition of hydrogen peroxide and heating were repeated 2–4 times until the sample solution became clear.

A modified procedure of Agemian et al. (1980) and Thompson & Wagstaff (1980) was followed for tissue/organs digestion. Briefly, 0.05 g–0.5 g dry tissue/organ was placed in a digestion tube containing 5 ml 65% nitric acid and the tube was incubated at 60 °C on the digester for 30 min, followed by an addition of a further 5 ml of 65% nitric acid. The temperature was gradually raised to 130–150 °C, a small funnel was put onto each tube for refluxing, the temperature was again gradually raised and kept at 170 to 180 °C for 10–17 hours (depending on the tissue weight) until the disappearance of brown fume. After being cooled to room temperature, 1 ml 30% (w/w) hydrogen peroxide was added into each tube and the tubes were placed on the digester for a further digestion at 170 to 180 °C until the color of sample solution in the tube was stabilized. Addition of hydrogen peroxide and heating were repeated 2–4 times until the solution in the tube became clear. Each of the digested animal tissue/organs and sediment sample was then diluted to 25 ml using double-distilled water, filtered though a double-layered Whatman No. 41 filter paper and transferred into plastic bottles immediately. Zinc and copper concentrations were determined by using a Z-8100 Polarized Zeeman atomic absorption spectrophotometer (HITACHI).

All statistical analyses were carried out with SAS software, version 6.07. One-way ANOVAs and Tukey's Studentized Range Test were used to compare

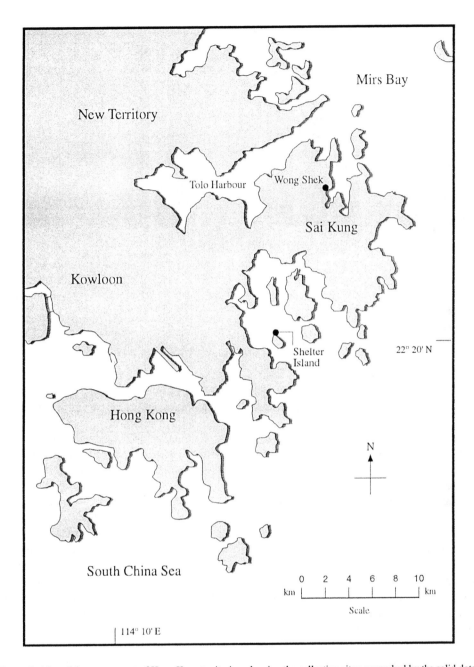

Figure 1. Map of the eastern part of Hong Kong territories, showing the collecting sites, as marked by the solid dots.

the differences of zinc and copper concentrations in various tissue/organs collected from the same site. *t*-test was used to compare the site difference of zinc and copper concentrations in sediment and in various tissue/organs. The comparison of zinc and copper concentrations in freshly collected and lab-kept animals (kept for 40 days without sediment) was also tested by using *t*-test.

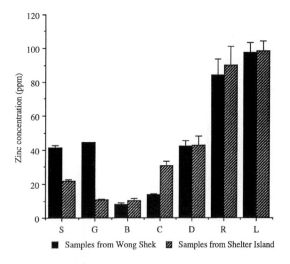

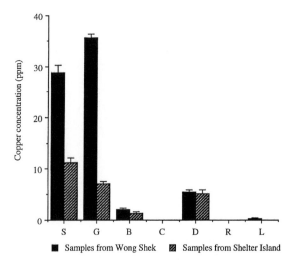

Figure 2. Zinc concentrations (ppm, in dry weight) of sediment and sea cucumbers, *H. leucospilota*, from two sites. Values are mean ± standard error. S: sediment; B: body wall; C: Cuvierian tubule; D: digestive tract; R: respiratory tree; L: longitudinal muscle. Sample size = 10.

Figure 3. Copper concentrations of sediment and sea cucumbers, *H. leucospilota*, from two sites (ppm, in dry weight). Values are mean ± standard error. S: sediment; B: body wall; C: Cuvierian tubule; D: digestive tract; R: respiratory tree; L: longitudinal muscle. Sample size = 10.

Results

Zinc and copper concentrations in longitudinal muscle, respiratory tree, digestive tract, Cuvierian tubule and body wall of *Holothuria leucospilota* are summarized in Figures 2 and 3 respectively. Copper concentrations in various tissue/organs were much lower than that of zinc. In fact, copper level was not detectable in respiratory tree and Cuvierian tubule, and only slightly detectable in some samples in longitudinal muscle (Figure 3).

Results of one-way ANOVAs indicated that there was a significant difference of zinc concentration in various tissue for the samples from Wong Shek and Shelter Island (Table 1).

The Tukey's Studentized Range Test showed that for the samples from Wong Shek, zinc concentration in tissues (ppm in dry weight) decreased as follows: longitudinal muscle (97.27) and respiratory tree (83.92) > digestive tract (42.96) > Cuvierian tubule (13.71) and body wall (8.14). Zinc concentration in longitudinal muscle was 10 times higher than that in body wall. For the samples from Shelter Island, zinc concentration in tissues followed the same trend as those from Wong Shek.

Significant difference among tissues also existed in copper concentration (Table 1). However, the concentration of copper in tissues (ppm in dry weight) followed a different trend from that of zinc, the highest

concentration occurred in digestive tract (5.53), intermediate in body wall (2.11), and the lowest in longitudinal muscle (0.25); copper in Cuvierian tubule and respiratory tree was not detectable.

Comparison of zinc and copper concentrations in tissues of freshly collected and that of lab-kept (40 days) animals were illustrated in Figures 4 and 5 respectively. The result indicated that after 40 days without ingesting sediment, Zn level decreased 48% in longitudinal muscle ($P<0.0001$), 39% in respiratory tree ($P<0.01$) and 57% in Cuvierian tubule ($P<0.01$) respectively, but no significant change was found in digestive tract and body wall ($P=0.41$ and 0.10 respectively) (Figure 4). On the other hand, copper concentrations in digestive tract and body wall did not show significant change ($P=0.12$ and 0.13 respectively), while that in longitudinal muscle, respiratory tree and Cuvierian tubule increased from non-detectable level to 0.51, 2.38 and 0.27 ppm respectively (Figure 5).

Zinc and copper concentrations in sediment from the two sampling sites were significantly different (Figures 2 and 3, $P<0.0001$). Sediment of Wong Shek contained higher level of metals than that of Shelter Island.

Zn concentration in longitudinal muscle, respiratory tree, digestive tract and body wall of animal from Wong Shek was the same as that from Shelter Island ($P=0.93$, 0.70, 0.91 and 0.21 respectively). Zn concentration in Cuvierian tubule of animal from Shel-

Table 1. One-way ANOVAs and Tukey's studentized Range Test (TSRT) of differences of zinc and copper concentrations in various tissue/organs of the sea cucumbers, *Holothuria leucospilota*. Values not differing at 0.05 level are joined by a solid line. B: body wall; C: Cuvierian tubule; D: digestive tract; L: longitudinal muscle; R: respiratory tree. Sample size = 10.

Sampling site	Zn			Cu		
	One-way ANOVA		TSRT	One-way ANOVA		TSRT
	F	P		F	P	
Wong Shek	83.71	<0.0001	LR>D>CB	104.67	<0.0001	D>B>LCR
Shelter Island	41.53	<0.0001	LR>DC>B	73.21	<0.0001	D>B>LCR

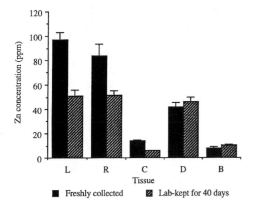

Figure 4. Zinc concentrations (ppm, in dry weight) in various tissues of freshly collected and lab-kept (without ingestion of sediment for 40 days) animals (*H. leucospilota*). Values are mean ± standard error. B: body wall; C: Cuvierian tubule; D: digestive tract; R: respiratory tree; L: longitudinal muscle. Sample size = 10.

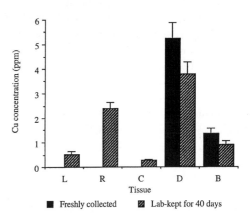

Figure 5. Copper concentrations in various tissues of freshly collected and lab-kept (without ingestion of sediment for 40 days) animals (*H. leucospilota*). Values are mean ± standard error. B: body wall; C: Cuvierian tubule; D: digestive tract; R: respiratory tree; L: longitudinal muscle. Sample size = 10.

ter Island, however, was significantly higher than that from Wong Shek (*P*<0.01) (Figure 2). There was no significant difference in Cu concentration in digestive tract between the two sampling sites (*P* = 0.68). In contrast, Cu concentration of body wall from Wong Shek was significantly higher than that from Shelter Island (*P*<0.05) (Figure 3).

Discussion

Results of this paper showed that zinc concentrations in longitudinal muscle and respiratory tree of the sea cucumber *Holothuria leucospilota* were much higher than those in digestive tract, Cuvierian tubule and body wall. Five longitudinal muscle bands run along the inner surface of body wall and control the contraction and relaxation of the whole animal. Respiratory tree has both circular and longitudinal muscles which account for half of its thickness (Smiley, 1994), con-

tracting and relaxing alternately, allowing water inhalation and exhalation for gas exchange. Digestive tract is made of an epithelium on both surface, a connective tissue compartment (CTC) and a thin muscle layer. Cuvierian tubule, a protective device found in some species of *Holothuria*, has only a few muscle fibers (VandenSpiegel & Jangoux, 1987). The body wall is a mass of connective tissue and epithelium. Thus, the amount of muscle (contractile elements) among the five tissue/organs decreases as follows: longitudinal muscle, respiratory tree, digestive tract, Cuvierian tubule and body wall. This order is consistent with the order of the Zn concentration in these tissue/organs, suggesting a possibility of contractile elements playing an important role in Zn accumulation.

Contrary to findings in other marine animals in which muscle usually has the lowest Zn and other metal concentrations among tissue/organs investigated, muscle bands of *H. leucospilota* had the highest Zn concentration (Figure 2). Furthermore, *H. leucospilota* showed higher zinc concentration than muscle of most marine animals studied. For example, in molluscs, zinc concentration is 25 to 70 ppm in dry weight; in crustacean, it is 10 to 14 ppm in fresh weight, and in teleost, it is 7 to 25 ppm in dry weight (see review by Eisler, 1981), suggesting that the muscle tissue of *H. leucospilota* is unique. Copper concentration in *H. leucospilota,* however, was much lower than that in other marine animals which have been studied (see review by Eisler, 1981).

The fact that the zinc concentrations are the highest in longitudinal muscle and respiratory tree, while the copper concentrations in the same tissue/organs were almost non-detectable indicates an antagonistic interaction between zinc and copper. Our results were consistent with those found in other marine invertebrates, such as polychaete *Nereis diversicolor* (Bryan & Hummerstone, 1973; Saiz-Salinas et al., 1996), oyster *Crassostrea gigas* (Thrower & Eustace, 1973), clam *Serobicularia plana* (Saiz-Salinas et al., 1996), snail *Polinices sordidus* (Ying et al., 1993), sea stars and brittle stars (Riley & Segar, 1970) in which zinc and copper concentrations were always antagonistic.

As shown in Figures 4 and 5, after animals had been kept in laboratory without sediment for 40 days, Zn concentrations of longitudinal muscle, respiratory tree and Cuvierian tubule decreased significantly while that in digestive tract and body wall remained unchanged, indicating that depuration can cause significant zinc release from the metal-accumulated tissue/organs. Meanwhile, copper level remained unchanged (digestive tract or body wall) or slightly increased (longitudinal muscle, respiratory tree and Cuvierian tubules), suggesting that only accumulated metal (zinc) was depurated.

Zinc concentration in the same tissue/organ from two sampling sites was virtually identical with the sole exception of Cuvierian tubule, yet Zn concentration in the sediment of the two sites were significantly different (Figure 2), indicating that *H. leucospilota* was not sensitive enough to reflect the metal level in the sediments, suggesting that this species is not a good indicator of heavy metal contamination in sediment. It was found that the metal content of animal feces was almost identical to that of the sediment (data not shown), indicating that little or no metal was absorbed through sediment ingestion. This is probably because the metals in the sediment is not in a bioavailable form. It is possible that this animal may uptake metals from ambient sea water through respiratory tree, not unlike the uptake of trace metals by aquatic invertebrates through gills and other permeable surfaces (Weeks & Rainbow, 1993). On the other hand, metal concentration in a marine organism do not necessarily depend on bioavailability alone, it may also depend on the balance between uptake and excretion (Ying et al., 1993). It should be noted that zinc is an essential metal which plays a variety of roles in biochemical functions such as serving as enzyme cofactor (Rainbow, 1990), endogenous control of zinc is not well studied among invertebrates.

Bohn (1979) reported that Zn concentration in the soft tissues of the green sea urchin, *Stronngylocentrotus droebachiensis*, was positively related to that in the ambient sediment. Our result indicated that Zn concentration in animals of the two populations was about the same although Zn concentration in the sediment of the two sites differed significantly.

Acknowledgments

We thank Miss Lisa Soo for the technical help. Drs Arthur Lau, Pei-yuan Qian and Daniel Rittschof reviewed the manuscript for us. We appreciate the comments from the two anonymous reviewers.

References

Agemian, H., D. P. Sturtevant & K. D. Austen, 1980. Simultaneous acid extraction of six trace metals from fish tissue by hot-block digestion and determination by atomic-absorption spectrometry. Analyst 105: 125–130.

Bohn, A., 1979. Trace metals in fucoid algae and purple sea urchins near a high Arctic lead/zinc ore deposit. Mar. Pollut. Bull. 10: 325–327.

Brooks, R. R. & M. G. Rumsby, 1965. The biogeochemistry of trace element uptake by some New Zealand bivalves. Limnol. Oceanagr. 10: 521–527.

Brooks, R. R. & D. Rumsby, 1974. Heavy metals in some New Zealand commercial sea fishes. N.Z.J. mar. Res. 8: 155–166.

Bryan, G. W., 1968. Concentrations of zinc and copper in the tissues of decapod crustaceans. J. mar. biol. Ass. U.K. 48: 303–321.

Bryan, G. W., 1973. The occurrence and seasonal variation of trace metals in the scallops *Pecten maximus (L.)* and *Chlamys opercularis (L.)*. J. mar. biol. Ass. U.K. 53: 145–146.

Bryan, G. W. & L. G. Hummerstone, 1973. Adaption of the polychaete *Nereis diversicolor* to estuarine sediments containing high concentrations of zinc and cadmium. J. mar. biol. Ass. U.K. 53: 839–857.

Bryan, G. W. & P. E. Gibbs, 1979. Zinc – a major inorganic component of nereid polychaete jaws. J. mar. biol. Ass. U.K. 59: 969–973.

Eisler, R., 1981. Trace metal concentrations in marine organisms. Pergamon Press, New York, 687 pp.

Julshamn, K. & O. Grahl-Nielsen. 1996. Distribution of trace elements from industrial discharges in the Hardangerfjord, Norway:

a multivariate data analysis of saithe, flounder and blue mussel as sentinel organisms. Mar. Pollut. Bull. 32: 546–571.

Luoma, S. N., 1990. Processes affecting metal concentrations in estuarine and coastal marine sediments. In Furrness, R. W. & P. S. Rainbow (eds) Heavy Metals in the Marine Environment, CRC Press, Inc. Boca Raton: 51–66.

Martin, J. L. M., 1979. Schema of lethal action of copper on mussels. Bull. envir. Contamin. Toxicol. 21: 808–814.

Rainbow, P. S., 1990. Heavy metal levels in marine invertebrates. In Furrness, R. W. & P. S. Rainbow (eds) Heavy metals in the Marine Environment, CRC Press, Inc. Boca Raton: 67–81.

Saiz-Salinas, J. I., J. M. Ruiz & G. Frances-Zubiliaga, 1996. Heavy metal levels in intertidal sediments and biota from the Bidasoa Estuary. Mar. Pollut. Bull. 32: 69–71.

Riley, J. P. & D. A. Segar, 1970. The distribution of the major and some minor elements in marine animals. I. Echinoderms and coelenterates. J. mar. biol. Ass. U.K. 50: 721–730.

Smiley, S., 1994. Holothurioidea. In Harrison, F. W. & F. S. Chia (eds), Microscopic anatomy of invertebrates, Vol. 14, Echinoder-mata, Wiley-Liss Publication, New York: 401–471.

Thompson, K. C. & K. Wagstaff, 1980. Simplified method for the determination of cadmium, chromium, copper, nicked, lead and zinc in sewage sludge using atomic-absorption spectrophotome-try. Analyst 105: 883–896.

Thrower, S. J. & I. J. Eustace, 1973. Heavy metal accumulation in oysters grown in Tasmanian waters. Food Technol. Aust. Nov.: 546–553.

VandenSpiegel, D. & M. Jangoux, 1987. Cuvierian tubules of the holothuroid *Holothuria forskali*: a morphofunctional study. Mar. Biol. 96: 263–275.

Weeks, J. M. & P. S. Rainbow, 1993. The relative importance of food and sea water as sources of copper and zinc to talitrid amphipods (Crustacea; Amphipoda; Talitridae). J. appl. Ecol. 30: 722–735.

Warnau, M., G. Ledent, A. Temera, M. Jangoux & P. Dubois, 1995. Experimental cadmium contamination of the echinoid *Paracentrotus lividus*: influence of exposure mode and distribution of the metal in the organism. Mar. Ecol. Prog. Ser. 116: 117–124.

Ying, W. M., M. Ahsanullah & G. E. Batley, 1993. Accumulation and regulation of heavy metals by the intertidal snail *Polinices sordidus*. Mar. Biol. 116: 417–422.

Hydrobiologia **352**: 25–37, 1997.
Y.-S. Wong & N. F.-Y. Tam (eds), Asia-Pacific Conference on Science and Management of Coastal Environment.
©1997 *Kluwer Academic Publishers. Printed in Belgium.*

Mapping and characterization of mangrove plant communities in Hong Kong

Nora F. Y. Tam[1]*, Yuk-Shan Wong[2], C. Y. Lu[3] & R. Berry[2]
[1]*Department of Biology and Chemistry, City University of Hong Kong, Hong Kong*
[2]*Research Centre/Biology Department, The Hong Kong University of Science & Technology, Hong Kong.*
[3]*Environmental Science Research Centre, Xiamen University, PRC.*
(*Author for correspondence; tel: 852 2788 7793; fax: 852 2788 7406; e-mail: bhntam@cityu.edu.hk)*

Key words: mangrove, plant community, Hong Kong, mapping, characteristics

Abstract

Ecological surveys were carried out to investigate the distribution and characterization of remaining mangrove stands in Hong Kong. The field studies indicate that 43 mangrove stands, excluding Mai Po Nature Reserve, still remained along the coastline of Hong Kong despite tremendous reclamation and development which occurred in the past 40 years. Most mangrove stands were found in Deep Bay (western part) and Sai Kung District (eastern coasts). The total areas occupied by these mangrove stands were 178 ha, varying from a very small stand (with 1–2 mangrove shrubs) to fairly extensive mangroves in Deep Bay (>10 ha). It appeared that mangrove stands located in Deep Bay area were larger than those in the eastern coasts. Twenty plant species were identified from these stands, with 13 being exclusive or associate mangrove species. The major constituent species were *Kandelia candel*, *Aegiceras corniculatum*, *Excoecaria agallocha* and *Avicennia marina*. Rare species such as *Heritiera littoralis* were only found in a few mangrove stands. Out of the 43 remaining mangrove stands, 23 were more worthwhile for conservation and their plant community structures were further investigated by transect and quadrat analyses. The importance values (sum of relative abundance, frequency and dominance) show that *K. candel* was the most dominant species. Species richness and Simpson's indices together with tree height, tree density and canopy area fluctuated significantly between mangrove stands. These values were used to prioritize the conservation potential of the remaining mangrove stands in Hong Kong.

Introduction

The mangrove ecosystem is one of the most important types of natural wetlands found in the intertidal zone of tropical and subtropical regions. It fringes 70% of the coastline in these regions and has been recognized for its ecological and socioeconomic significance. Mangroves are highly productive and provide suitable habitats and food sources for invertebrates, fish and other coastal wildlife. However, the increasing exploitation of mangrove forests and other human activities have led to a rapid decline of mangrove habitats. In Hong Kong, mangrove swamps have been destroyed due to reclamation and infrastructural development over the past decades. Historically, the largest areas of Hong

Kong mangrove stands were in Deep Bay and inside Tolo Harbour, but both places have been greatly altered over the years. Part of the Deep Bay mangrove stand, now called the Mai Po Nature Reserve, was first used as *gei wais* (traditional shrimp farming ponds), then as fish ponds (Irving & Morton, 1988), until recently, only approximately 15% of the original mangrove stands remained at the seaward fringe (Yipp et al., 1995). Similarly, more than 42% of the original mangroves along the coastline of the Tolo Harbour were destroyed due to the construction of new highways, reclamation into large 'satellite' town areas and race course (Holmes, 1988). Moreover, about 50 ha of mangrove habitats were destroyed by construction of the new airport, the associated port and airport develop-

26

Table 1. Size (in hectares) of the mangrove stands remaining in Hong Kong (The stands were arranged in a descending order according to the size).

| \multicolumn{3}{c}{Recommended for detailed study} | | | \multicolumn{3}{c}{Not worthwhile for further study} | | |
No.	Stand	Area (ha)	No.	Stand	Area (ha)
13	Lut Chau	56.36	42	Yeun Long Ind. Estate	7.22
35	Tsim Bei Tsui	35.48	12	Luk Keng	3.84
36	Ting Kok	8.77	21	Sam A Tsuen / Wan	3.00
15	Nam Chung	8.72	39	Tung Chung	2.21
27	Sheung Pak Nai	6.34	7	Kau Sai Chau	2.10
19	Sai Keng	3.84	20	Sai Kung Hoi	2.07
11	Lai Chi Wo	2.70	25	Sham Chung	1.90
33	Tai Tan	2.68	3	Discovery Bay	1.01
24	Sha Tau Kok	2.51	4	Ha Pak Nai	0.71
9	Kei Ling Ha Lo Wai	2.45	28	Shui Hau	0.70
5	Ho Chung	2.37	30	Tai O	0.61
23	San Tau	2.14	34	Tan Ka Wan	0.50
37	To Kwa Peng	2.09	17	Pak Tam Chung	0.47
29	Tai Ho Wan	1.86	14	Nai Chung	0.40
22	Sam Mun Tsai	1.82	26	Shum Wat	0.38
41	Wong Yi Chau	1.78	16	Pak Sha Wan	0.33
18	Pui O Wan	1.71	10	Lai Chi Chong	0.31
38	Tolo Pond	1.41	40	Wong Chuk Wan	0.26
43	Yi O	1.21	2	Chi Ma Wan	0.22
32	Tai Tan	1.19	31	Tai Tam	NA
8	Kei Ling Ha Hoi	0.83			
1	Chek Keng	0.69			
6	Hoi Ha Wan	0.53			
	Sub-total	149.49			28.22
	Mean area per stand	6.50			1.41
	Number of stands	23			20

Total number of stands equalled to 43, and total area was 177.71 ha;
NA: Those stands which contained less than 1 or 2 mangrove shrubs were neglected.

ment in the northern shore of Lantau Island (Greiner Maunsell, 1991). Due to the limited land space and the rapid expansion in industry and population, the remaining mangrove stands in Hong Kong are continuously threatened by urban development. The importance and rapid decline in mangrove stands have led to public concern for their future stability and survival, and to calls for the conservation of the remaining mangrove stands.

For effective management of the mangrove ecosystem, basic information such as species composition, ecology, extent, land use, human intervention, etc. is required. However, such information is rarely available. The present study therefore aims (1) to identify the mangrove stands still remaining in Hong Kong; and (2) to provide baseline information on the structural characteristics of the plant communities in these

remaining mangrove stands in Hong Kong. Such information can then be used as a basis for assessing the conservation potential of each stand.

Materials and methods

Mapping the extent and distribution of Hong Kong mangrove stands

Hong Kong has an area of 1067 km^2 with a hot and humid summer and a cool and mild winter. The annual mean temperature is 23 °C (ranges from 0.2–36 °C) and the annual mean rainfall is around 2214 mm (Howlett, 1996). The mangrove stands are mostly inundated by incoming tides twice a day and the largest tidal range at high spring tide is about 2.8 m. The remaining mangrove stands in Hong Kong were identified by aerial photographs, 1:10 000 maps and field surveys. A total of 43 mangrove stands, except Mai Po Nature Reserve, were identified. Mai Po was not included in this study as a lot of ecological data has already been collected in this area which has been designated as 'Site of Special Scientific Interests' since 1976 and has recently been included in the 'List of Wetlands of International Importance' under the Ramsar Convention. The extent of each mangrove stand (including the seedlings in the foreshore region) was determined by 1:5000 scale map and aerial photographs. Based on the preliminary surveys, 20 sites were found to be relatively small and did not show high conservation values. Detailed ecological study was then focused on the other 23 sites.

Floristic composition and community structure of Hong Kong mangrove stands

The mangrove stands were visited during the low ebb tide period when exposure of the mudflat was maximum. Plant species of each site were identified and categorized into exclusive and associate mangrove species based on the list produced by International Society of Mangrove Ecosystem (Clough, 1993). Other plant species such as seagrass were also identified. The structural characteristics of the plant community were investigated by transect and quadrat (fixed area plots) techniques. Two to five transects were laid perpendicularly to the shoreline covering the extent of mangrove vegetation from land to sea. Three to five quadrats (either 3 m × 3 m or 5 m × 5 m in size) were sampled per transect, depending on the complexity and extent of the mangrove stand. Within each quadrat, number of trees and young plants of each species, tree height, stem basal diameter (calculated from the circumference at the base of each stem), canopy width and length were measured. The representative species were defined as those with more than 50% occurrence and rare species were those with less than 25% occurrence. Species diversity was indicated by Simpson's and richness indices. The relative importance values (the sum of relative abundance, frequency and dominance) of the major constituent species in each stand were calculated according the method described by Cintron & Schaefer-Novelli (1984).

Results and discussion

Distribution and extent of mangrove stands in Hong Kong

A total of 43 remaining mangrove stands was found along the coastline of Hong Kong despite the tremendous reclamation work and infra-structural development carried out in the past 40 years. Most sites were found in the western part (the Deep Bay area) and the eastern part (the Sai Kung District) of the New Territories (Figure 1). The sizes of these mangrove stands varied significantly (Table 1), and all these stands covered a total area of 178 ha (excluding Mai Po which is the biggest stand covering 172 ha by itself, Young & Melville, 1993). The 23 mangrove stands selected for further ecological study occupied around 150 ha and accounted for 86% of the total area. The sizes of most stands were between 1 and 3 ha with only two sites (in Deep Bay area) having an area of more than 10 ha.

Distribution of mangroves is mainly governed by topography, tidal height, substratum and salinity. Mangrove stands in the Deep Bay areas were rather muddy and greatly affected by freshwater flushing from the Pearl River, the People's Republic of China, especially during summer time (Morton & Morton, 1983). Heavy siltation occurred in this area, resulting in an extensive gradual intertidal slope. The salinity also fluctuated from 1 ppt. (parts per thousand, almost all freshwater) to >30 ppt, with an average of around 10–15 ppt. On the contrary, the stands distributed in the eastern coastline of Hong Kong were more oceanic and had more steady salinity. The substratum of these stands was stony and pebbly with a relatively shallow soil layer. Mangroves located in the western part of Hong Kong also received high anthropogenic pollution sources such as nutrients from cities along the Pearl River Delta than those in the eastern coast of Hong Kong. This might explain why mangrove stands located in the Deep Bay areas were significantly larger than the others in Hong Kong (Figure 2).

Floristic composition of Hong Kong mangroves

A total of 20 plant species with 8 exclusive (40%) and 5 associate (25%) mangrove species were found in Hong Kong (Table 2). The reason for including the non-mangrove species in this study was either due to their high conservation value and rarity (such as *Halophila ovata* and *Zostera japonica*, Morton & Morton, 1983) or due to their common occurrence in

No.	Site	True mangrove								Associate mangrove							Other species					No. Sp.
		KC	AC	AM	BG	EA	LR	HL	AI	HT	TP	CI	CM	AH	DT	SC	PT	SA	LS	HO	ZJ	
1	Chek Keng	1	1	1	1	1	1		1	1		1					1	1				12
2	Chi Ma Wan	1	1		1				1			1					1					6
3	Discovery Bay	1	1	1	1	1			1	1		1				1	1	1				10
4	Ha Pak Nai	1	1	1	1				1	1		1				1	1	1		1		7
5	Ho Chung	1	1	1	1	1	1		1	1		1	1	1	1		1			1		11
6	Hoi Ha Wan	1	1	1	1	1			1	1		1					1					8
7	Kau Sai Chau	1	1	1	1	1	1		1	1		1	1		1		1					10
8	Kei Ling Ha Hoi	1	1	1	1	1	1		1	1	1	1	1				1	1	1			14
9	Kei Ling Ha Lo Wai	1	1	1	1	1	1		1	1		1				1	1	1				11
10	Lai Chi Chong	1	1			1				1								1				5
11	Lai Chi Wo	1	1	1	1	1			1	1	1	1	1	1			1	1	1		1	15
12	Luk Keng	1	1	1	1	1			1	1		1	1	1	1		1					12
13	Lut Chau	1	1	1	1	1			1	1		1										8
14	Nai Chung	1	1	1	1	1			1	1	1	1	1		1	1						10
15	Nam Chung	1	1	1	1	1	1		1	1	1	1	1	1			1	1	1			14
16	Pak Sha Wan	1	1	1	1	1				1		1										7
17	Pak Tam Chung	1	1	1	1	1			1	1		1	1	1			1					9
18	Pui O Wan	1	1	1	1	1			1	1		1										7
19	Sai Keng	1	1	1	1	1			1	1	1	1	1				1	1				13
20	Sai Kung Hoi	1	1	1	1	1			1	1		1										8
21	Sam A Tsuen / Wan	1	1	1	1	1				1		1	1	1								8
22	Sam Mun Tsai	1	1	1	1	1			1	1		1	1	1			1	1	1			12
23	San Tau	1	1	1	1	1	1	1	1	1	1	1	1		1	1	1	1	1	1	1	18
24	Sha Tau Kok	1	1	1	1	1			1	1	1	1	1	1	1		1					12
25	Sham Chung	1	1	1	1	1				1		1					1					7
26	Sham Wat	1	1	1	1	1			1	1		1					1					9
27	Sheung Pak Nai	1	1	1	1				1	1		1			1		1			1		9
28	Shui Hau	1	1	1	1	1			1	1		1	1	1			1					11
29	Tai Ho Wan	1	1	1	1	1	1		1	1		1	1	1			1					12
30	Tai O	1	1	1	1	1			1	1		1		1								7
31	Tai Tam	1	1							1												3
32	Tai Tan	1	1	1	1	1			1	1	1	1	1	1	1	1	1	1				13
33	Tai Wan	1	1	1	1	1			1	1		1		1	1	1	1		1			10
34	Tan Ka Wan	1	1			1				1		1					1					4
35	Tsim Bei Tsui	1	1	1	1	1			1	1		1	1		1		1					10
36	Ting Kok	1	1	1	1	1			1	1		1	1	1	1		1	1				12
37	To Kwa Peng	1	1	1	1	1			1	1		1	1				1					7
38	Tolo Pond	1	1	1	1	1			1	1		1			1		1					7
39	Tung Chung	1	1	1	1	1	1		1	1		1	1	1	1		1	1				14
40	Wong Chuk Wan	1	1	1	1	1			1	1		1	1	1	1		1		1			10
41	Wong Yi Chau	1	1	1	1	1	1		1	1		1	1	1	1		1	1				11
42	Yeun Long Ind. Est	1	1	1		1				1		1										6
43	Yi O	1	1			1			1	1		1	1	1	1	1	1	1				14
		43	42	28	27	39	12	5	32	37	9	39	19	14	13	6	34	15	8	4	2	

Table 2. Plant species list of the 43 mangrove stands in Hong Kong (KC: *Kandelia candel*, AC: *Aegiceras corniculatum*, AM: *Avicennia marina*, BG: *Bruguiera gymnorrhiza*, EA: *Excoecaria agallocha*, LR: *Lumnitzera racemosa*, HL: *Heritiera littoralis*, AI: *Acanthus ilicifolius*, HT: *Hibiscus tiliaceus*, TP: *Thespesia populnea*, CI: *Clerodendrum inerme*, CM: *Cerbera manghas*, AH: *Acrostichum aureum*, DT: *Derris trifoliata*, SC: *Scaevola* spp., PT: *Pandanus tectorius*, SA: *Suaedea australis*, LS: *Limonium sinense*, HO: *Halophila ovata*, ZJ: *Zostera japonica*).

29

The locations of Hong Kong's remaining mangrove stands.

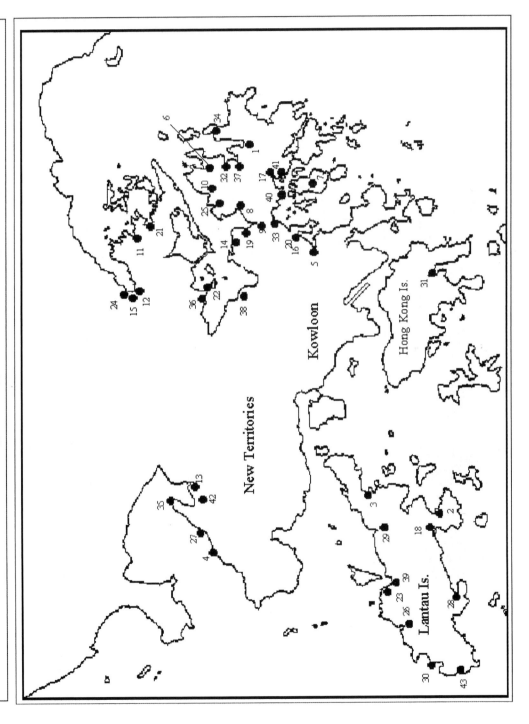

No.	Site
1	Chek Keng
2	Chi M a Wan
3	Discovery Bay
4	Ha Pak Nai
5	Ho Chung
6	Hoi Ha Wan
7	Kau Sai Chau
8	Kei Ling Ha Hoi
9	Kei Ling Ha Lo Wai
10	Lai Chi Cheong
11	Lai Chi Wo
12	Luk Keng
13	Lut Chau
14	Nai Chung
15	Nam Chung
16	Pak Sha Wan
17	Pak Tam Chung
18	Pui O Wan
19	Sai Keng
20	Sai Kung Hoi
21	Sam A Tseun / Wan
22	Sam Mun Tsai
23	San Tau
24	Sha Tau Kok
25	Sham Chung
26	Sham Wat
27	Sheung Pak Nai
28	Shui Hau
29	Tai Ho Wan
30	Tai O
31	Tai Tam
32	Tai Tan
33	Tai Wan
34	Tan Ka Wan
35	Tsim Bei Tsui
36	Ting Kok
37	To Kwa Peng
38	Tolo Pond
39	Tung Chung
40	Wong Chuk Wan
41	Wong Yi Chau
42	Yeun Long Ind. Est.
43	Yi O

Figure 1. Location of the mangrove stands remaining in Hong Kong (refer to Table 2 for the number and name of the mangrove stands).

30

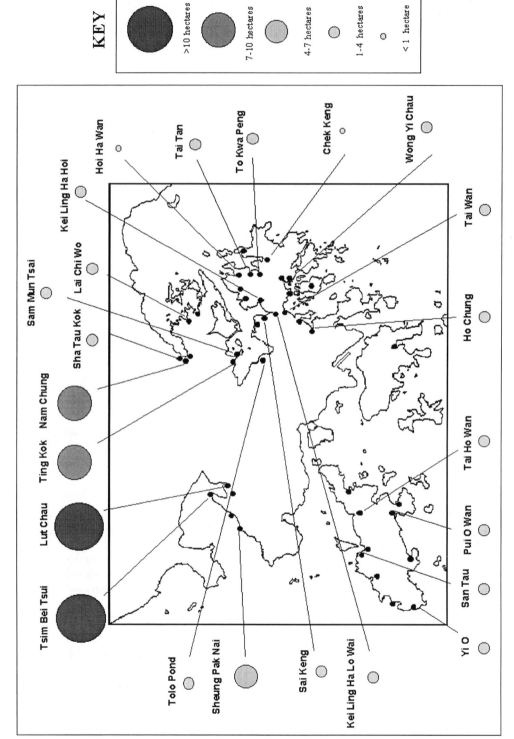

Figure 2. Area of the 23 remaining mangrove stands recommended for conservation in Hong Kong.

Table 3. Relative abundance of the major constituent plant species in the 23 mangrove stands recommended for conservation. (Mangrove stands were arranged in alphabetical order).

No.	Stand	A. corniculatum	A. marina	K. candel	Others
1	Chek Keng	42.4	15.3	39.0	3.4
5	Ho Chung	9.1	0	90.9	0
6	Hoi Ha Wan	96.6	0	3.4	0
8	Kei Ling Ha Hoi	4.4	0	45.6	50.1
9	Kei Ling Ha Lo Wai	16.3	0	83.7	0
11	Lai Chi Wo	69.8	0	10.3	19.6
13	Lut Chau		NA		
15	Nam Chung	0	0	100	0
18	Pui O Wan		NA		
19	Sai Keng	18.9	0.9	64.6	15.6
22	Sam Mun Tsai	5.6	52.8	31.9	9.7
23	San Tau		NA		
24	Sha Tau Kok	0	0	98.7	1.4
27	Sheung Pak Nai		NA		
29	Tai Ho Wan	58.9	1.3	35.1	4.6
32	Tai Tan	29.2	14.6	50.0	6.2
33	Tai Wan	0	0	98.4	1.6
35	Tsim Bei Tsui		NA		
36	Ting Kok	50.0	9.7	33.1	7.2
37	To Kwa Peng	39.3	0.6	57.3	2.8
38	Tolo Pond	15.8	1.9	72.8	9.5
41	Wong Yi Chau	10.5	0	73.7	15.8
43	Yi O	3.8	92.5	3.8	0
	Average	27.4	11.2	55.7	5.7

Relative abundance = (No. of individuals of a species in a stand) / (total No. of all species in a stand) × 100;
NA: No quantitative quadrat analyses were carried out for mangrove stands that were difficult to access.

Hong Kong's mangrove community (such as *Derris trifoliata*, Thrower & Cheng, 1975). The dominant species found in most sites were *Kandelia candel* and *Aegiceras corniculatum*, followed by *Excoecaria agallocha* (Table 2). *Avicennia marina*, a species often considered as an important and dominant species in Hong Kong (Yipp et al., 1995), was found in 28 out of 43 mangrove stands. *Bruguiera gymnorrhiza*, a species considered rare in the past, was also quite widespread and occurred in 27 stands. The rare mangrove species were *Heritiera littoralis* and *Thespesia populnea*. *Lumnitzera racemosa* was also restricted to only 12 out of the 43 stands (Table 2).

The total number of mangrove plant species found in Hong Kong (located at 21°30′N) was similar to that found in Taiwan at 22–25°N (Walter & Breckle, 1986) but was higher than those found in Fujian Province (the PRC) at 23°46′N (8 species, Lin & Fan, 1992). When compared with those found in equatorial regions such as Malaysia (27 species, Chapman, 1984) and Hainan Island, the PRC (35 species, Lin & Fan, 1992), the species diversity recorded in Hong Kong was significantly lower. The distribution of mangrove species is governed mainly by climate, in particular, the temperature and rainfall. It has been reported that species diversity is always highest in equatorial regions especially the humid tropics with an annual rainfall >2000 mm (Semeniuk, 1983), and declines with increasing latitude and aridity. Thus only 1 species was found in southern Japan at 31°31′N (Walter & Breckle, 1986). The number of plant species found in different mangrove stands in Hong Kong varied, from very low species diversity (e.g. only 3 species found in Tai Tam) to stands containing 18 species such as

Table 4. Relative dominance of the major constituent plant species of the 23 mangrove stands recommended for conservation. (Mangrove stands were arranged in alphabetical order).

No.	Stand	A. corniculatum	A. marina	K. candel	Others
1	Chek Keng	17.9	56.5	14.9	10.8
5	Ho Chung	1.3	0	98.7	0
6	Hoi Ha Wan	93.5	0	6.5	0
8	Kei Ling Ha Hoi		NA		
9	Kei Ling Ha Lo Wai	7.9	0	92.1	0
11	Lai Chi Wo	12.0	0	1.0	87.0
13	Lut Chau		NA		
15	Nam Chung	0	0	100	0
18	Pui O Wan		NA		
19	Sai Keng	15.5	2.7	53.5	28.3
22	Sam Mun Tsai	12.4	21.2	15.2	51.1
23	San Tau		NA		
24	Sha Tau Kok	0	0	96.3	3.7
27	Sheung Pak Nai		NA		
29	Tai Ho Wan	66.0	0.8	28.8	4.4
32	Tai Tan	33.7	24.4	32.9	9.0
33	Tai Wan	0	0	98.7	1.3
35	Tsim Bei Tsui		NA		
36	Ting Kok	31.6	16.1	38.3	14.0
37	To Kwa Peng	30.9	0	26.5	42.6
38	Tolo Pond	2.5	4.5	49.2	43.7
41	Wong Yi Chau	22.3	0	39.1	38.6
43	Yi O	0.3	92.7	6.9	0
	Average	20.5	12.9	46.9	19.7

Relative dominance = (total basal area of a species in a stand) / (total basal area of all species in a stand)×100;
NA: No quantitative quadrat analyses were carried out for mangrove stands that were difficult to access.

San Tau in Lantau Island. The mangrove stands on the western coastlines did not contain more species than those in eastern side despite their difference in extent. The majority of stands, especially those recommended for conservation, contained 10 to 12 species.

The number of plant species occurring in a mangrove stand reflects the habitat conditions and degree of disturbance. Cintron & Schaefer-Novelli (1984) have shown that the magnitude and periodicities of tides, nutrients, hydro-period and stresses such as cyclones, drought, salt accumulation and frost may largely determine the flora and fauna composition as well as the community structure. Amarasinghe & Balasubramaniam (1992) found that the most constant conditions of salinity and nutrients as well as the exposure to unexpectedly higher kinetic energy regimes are relatively unfavourable conditions and may restrict the

number of species occurring in fringing habitats. On the other hand, estuarine (riverine) habitats with oscillating nutrients and salinity conditions may be more favourable and thus able to support more species. The variation in species number and diversity of different mangrove stands in Hong Kong might reflect their differences in habitat. The number of plant species that a mangrove stand can support also depends on the soil characteristics and the degree of disturbance. In Hong Kong, human disturbance and site development are often the major factors limiting the development of a mangrove stand. For instance, two plant species were found in the Tai Tam stand (the only site found in Hong Kong Island) at the beginning of this study (September 1993) but *K. candel* had totally disappeared during our recent visit (May 1996). Another two mangrove species, *E. agallocha* and *A. marina*, were recorded

Table 5. Relative frequency of the major constituent plant species of the 23 mangrove stands recommended for conservation. (Mangrove stands were arranged in alphabetical order).

No.	Stand	*A. corniculatum*	*A. marina*	*K. candel*	Others
1	Chek Keng	19.2	26.9	42.3	11.5
5	Ho Chung	18.2	0	81.8	0
6	Hoi Ha Wan	75.0	0	25.0	0
8	Kei Ling Ha Hoi		NA		
9	Kei Ling Ha Lo Wai	43.5	0	56.5	0
11	Lai Chi Wo	31.8	0	22.7	45.5
13	Lut Chau		NA		
15	Nam Chung	0	0	100	0
18	Pui O Wan		NA		
19	Sai Keng	20.0	10.0	45.0	25.0
22	Sam Mun Tsai	8.7	34.8	30.4	26.1
23	San Tau		NA		
24	Sha Tau Kok	0	0	90.9	9.1
27	Sheung Pak Nai		NA		
29	Tai Ho Wan	38.9	5.6	44.4	11.1
32	Tai Tan	33.3	28.6	28.6	9.5
33	Tai Wan	0	0	91.0	9.0
35	Tsim Bei Tsui		NA		
36	Ting Kok	33.3	24.1	33.3	9.3
37	To Kwa Peng	37.5	6.3	43.7	12.5
38	Tolo Pond	29.2	4.2	45.8	20.8
41	Wong Yi Chau	28.6	0	42.8	28.6
43	Yi O	18.2	63.6	18.2	0
	Average	25.6	12.0	49.6	12.8

Relative frequency = (No. of quadrats containing a species in a stand) / (total No. of quadrats sampled in a stand)$\times$100;

NA: No quantitative quadrat analyses were carried out for mangrove stands that were difficult to access.

in Tai Tam in the early 1980s but were lost due to human destruction (Morton & Morton, 1983). In fact, most of the mangrove stands found in Hong Kong now appeared to be secondary growth as the primary forest has probably been destroyed by cutting, dredging and filling.

Structural characteristics of plant communities in the 23 important mangrove stands

The relative abundance, dominance and frequency as well as the importance values of three major constituent species, *K. candel*, *A. corniculatum* and *A. marina* were calculated and are shown in Tables 3 to 6. Although *A. marina* did not occur in all sites, this species was also considered as the major constituent species when calculating the importance values, main-

ly due to its abundance and significance as reported by previous researchers (Thrower & Cheng, 1975; Yipp et al., 1995). The importance values varied considerably between mangrove stands. In general, *K. candel* had highest importance value, followed by *A. corniculatum* and *A. marina* (Table 6). In certain sites such as Nam Chung, *K. candel* was the only dominant species with the relative dominance value of 100 (Table 4). The Simpson's index of Nam Chung was 0 (Table 7) as there was only one single species (*K. candel*) found in the quadrats despite the fact that it had 10 species and a richness index of 0.5. These results indicate the dominance and abundance of *K. candel* in Hong Kong. Actually, out of the 23 stands recommended for conservation, 14 were dominated by *K. candel* and 2 were co-dominated by *K. candel* and *A. corniculatum* (Table 7). The dominance of *K. candel* in Hong Kong

Table 6. Importance values of the major constituent plant species in the 23 mangrove stands recommended for conservation (Mangrove stands were arranged in alphabetical order).

No.	Stand	*A. corniculatum*	*A. marina*	*K. candel*	Others
1	Chek Keng	79.5	98.7	96.2	25.9
5	Ho Chung	28.6	0	271.4	0
6	Hoi Ha Wan	265.1	0	34.9	0
8	Kei Ling Ha Hoi		NA		
9	Kei Ling Ha Lo Wai	67.7	0	232.3	0
11	Lai Chi Wo	113.6	0	34.0	152.3
13	Lut Chau		NA		
15	Nam Chung	0	0	300	0
18	Pui O Wan	NA			
19	Sai Keng	53.7	13.2	162.4	70.3
22	Sam Mun Tsai	26.7	108.8	77.6	86.9
23	San Tau		NA		
24	Sha Tau Kok	0	0	285.9	14.1
27	Sheung Pak Nai		NA		
29	Tai Ho Wan	163.8	7.7	108.3	20.1
32	Tai Tan	96.2	67.6	111.5	24.8
33	Tai Wan	0	0	288.1	11.9
35	Tsim Bei Tsui		NA		
36	Ting Kok	114.9	49.8	104.7	30.5
37	To Kwa Peng	107.7	6.9	127.5	57.9
38	Tolo Pond	47.5	10.6	167.8	74.1
41	Wong Yi Chau	61.4	0	155.6	83.0
43	Yi O	22.3	248.8	28.9	0
	Average	73.5	36.0	152.2	38.3

might be related to its tolerance to low temperature in winter and the disturbed substrate. Ability of *K. candel* to tolerate a wide range of environmental conditions such as salinity, air temperature, stony substrates, and desiccation also favoured its wide distribution in the mangrove stand, not only in the intertidal mudflat but also in infrequently flooded upper inter-tidal areas.

In most mangrove stands still remaining in Hong Kong, the zonation of plants from landward to seaward regions was not obvious. General zonation of species within a mangrove stand actually reflects the gradient in flooding, drainage, salinity and soil characteristics (Kenneally, 1982). Whether a species can occupy a specific position or not depends upon its range of tolerance to salinity, desiccation, tidal flushing as well as relative competitive ability to other species (Thrower & Cheng, 1975). Most mangrove stands in Hong Kong had a very narrow band of vegetation (between 20–50 m from landward to seaward regions), the gradients in tidal flooding and salinity would not be very significant thus the zonation pattern was not obvious

(Tam et al., 1993). Nevertheless, in some sites with a wider spread of mangrove plants from landward to seaward regions such as Lai Chi Wo, zonation was observed. In Lai Chi Wo, *Heritiera littoralis* was found in the very landward region with little tidal inundation and low salinity, *Excoecaria agallocha* dominated the landward zone, *Aegiceras corniculatum* and *Kandelia candel* colonized the intertidal mudflat, and *Avicennia marina* and *Zostera japonica* (the seagrass) was localized at the seaward edge.

The average height of the dominant trees in most mangrove stands varied from 0.7 to 4 m (Table 7), with an average height of 1.8 m. This height was significantly lower than in Mai Po Nature Reserve and in other mangrove stands in Southeast Asia such as Malaysia, Philippines and the South China coasts (Hodgkiss, 1986; Lee, 1989). The well-preserved mangroves in the PRC can reach 15–20 m in height (Lin & Fan, 1992). Because of the limited height, plants in Hong Kong mangrove stands were often simple with only a single layer and little undergrowth. There was no sig-

Table 7. Characteristics of plant community in the 23 mangrove stands recommended for conservation in Hong Kong.

No.	Site	Spp No.	Richness Index	Simpson's Index	Tree Density (No. m^{-2})	Canopy per tree (m^2)	Tree Height (m)
1	Chek Keng	12	0.60	0.64	1.1	0.9	1.0
5	Ho Chung	10	0.50	0.22	0.6	2.3	2.9
6	Hoi Ha Wan	8	0.40	0.07	2.0	0.2	1.3
8	Kei Ling Ha Ho	14	0.70	0.50	1.5	0.7	1.5
9	Kei Ling Ha Lo Wai	11	0.55	0.18	2.4	0.6	1.9
11	Lai Chi Wo	15	0.75	0.64	0.6	1.3	1.6
13	Lut Chau	8	0.40	0.25	0.5	2.0	2.8
15	Nam Chung	10	0.50	0.00	2.6	0.7	3.1
18	Pui O Wan	7	0.35	0.08	1.2	0.2	1.2
19	Sai Keng	13	0.65	0.53	1.8	0.6	1.2
22	Sam Mun Tsai	12	0.60	0.61	1.5	0.4	0.7
23	San Tau	18	0.90	0.59	1.0	0.7	1.5
24	Sha Tau Kok	12	0.60	0.03	0.8	2.4	4.3
27	Sheung Pak Nai	9	0.45	0.15	0.5	1.4	1.8
29	Tai Ho Wan	12	0.60	0.51	4.0	0.6	1.7
32	Tai Tan	13	0.65	0.64	0.7	1.0	1.2
33	Tai Wan	10	0.50	0.03	0.5	1.6	1.7
35	Tsim Bei Tsui	10	0.50	0.20	1.2	1.5	1.5
36	Ting Kok	12	0.60	0.62	2.8	0.5	1.1
37	To Kwa Peng	12	0.60	0.52	2.9	0.8	0.8
38	Tolo Pond	7	0.35	0.44	1.6	4.1	3.9
41	Wong Yi Chau	11	0.55	0.42	0.4	3.9	1.9
43	Yi O	14	0.70	0.11	0.6	1.5	0.9
	Average	11	0.57	0.35	1.4	1.3	1.8

Richness Index = No. of species at a site / Total no. of species, total no. of species = 20;
Simpson's Index = $1 - \sum (Pi)^2$, where Pi is the proportion of individuals of species *i* in the community.

nificant trend in changes of tree height from landward to seaward regions in Hong Kong mangroves although it has been reported that mangrove tree height normally decreases with increasing salinity and decreasing nutrient availability, forming a characteristic tree-height gradient (Lugo, 1989). The tree density was around 1.4 individual m^{-2}, ranging from 0.4–4 m^{-2}. The average canopy area per tree were 1.3 m^2, with minimum and maximum area of 0.23 (in Pui O Wan, Lantau Island) and 4.12 m^2 (in Tolo Pond, New Territories), respectively.

Based on these plant community data, the overall conservation values of each stand were calculated. Table 8 shows that the most important stand was San Tau (in Lantau Island), followed by Kei Ling Ha Hoi (in Sai Kung District) and the least significant mangrove was Pui O Wan in Lantau Island. Such priority ranking can be considered as a preliminary guideline to formulate a conservation strategy for mangrove stands as plants are a very important component in the overall conservation evaluation programme. However, other attributes such as fauna distribution, current protection measures, disturbance and threats, scientific and recreational values, etc. should also be considered in determining the ultimate conservation priorities of mangrove stands in Hong Kong.

Conclusion

The present ecological study shows that the 43 mangrove stands, not including Mai Po Nature Reserve, still remained in Hong Kong and fringed about 178 hectares along the coastlines. Twenty-three out of these 43 stands, accounted for more than 80% of the total area covered by mangroves, showed potential conser-

36

Table 8. Conservation value of the 23 mangrove stands based on ranking the various attributes of plant community structures.

No.	Site	Spp Rich	Spp Rep	Spp Rarity	Tree Ht	Tree Area	Tree Den	Sum Value	Cons Rank
23	San Tau	1	6.5	1	13	15	14	50.5	**1**
15	Nam Chung	16.5	6.5	7	3	15	4	52	**2**
8	Kei Ling Ha Ho	3.5	6.5	7	13	15	9.5	54.5	**3**
11	Lai Chi Wo	2	14.5	2.5	11	10	18	58	**4**
32	Tai Tan	5.5	6.5	2.5	17.5	11	16	59	**5**
37	To Kwa Peng	9.5	6.5	7	22	13	2	60	**6.5**
24	Sha Tau Kok	9.5	14.5	17	1	3	15	60	**6.5**
29	Tai Ho Wan	9.5	6.5	17	9.5	18	1	61.5	**8.5**
19	Sai Keng	5.5	6.5	7	17.5	18	7	61.5	**8.5**
1	Chek Keng	9.5	6.5	7	20	12	13	68	**10**
41	Wong Yi Chau	13.5	6.5	17	6.5	2	23	68.5	**11**
38	Tolo Pond	22.5	19	17	2	1	8	69.5	**12**
35	Tsim Bei Tsui	16.5	6.5	17	13	7.5	11.5	72	**13**
9	Kei Ling Ha Lo Wai	13.5	22.5	7	6.5	18	5	72.5	**14**
43	Yi O	3.5	6.5	17	21	7.5	18	73.5	**15**
5	Ho Chung	16.5	14.5	17	4	4	18	74	**16**
36	Ting Kok	9.5	6.5	17	19	20	3	75	**17**
27	Sheung Pak Nai	19	19	7	8	9	21	83	**18**
33	Tai Wan	16.5	19	17	9.5	6	21	89	**19**
13	Lut Chau	20.5	22.5	17	5	5	21	91	**20**
22	Sam Mun Tsai	9.5	14.5	17	23	21	9.5	94.5	**21**
6	Hoi Ha Wan	20.5	19	17	15	22.5	6	100	**22**
18	Pui O Wan	22.5	19	17	17.5	22.5	11.5	110	**23**

Spp Rich: rank based on Richness index; Spp Rep: rank based on number of representative species in a site; Spp Rarity: rank based on number of rare species; Tree Ht: average height of tree; Tree Area: average canopy area per tree; Tree Den: density of tree; Sum value: summation of all attributes; Cons Rank: overall conservation rank.

vation values and were studied in detail. These stands differed significantly in terms of plant species and community structure. In general, most of the mangrove stands remaining in Hong Kong were small, with narrow strips of mangrove plants. The dominant plant species were *Kandelia candel* and *Aegiceras corniculatum*. In most stands, the plants were dwarf and scattered, probably due to frequent human disturbance, lack of proper mud substrate and the cold winter climate. These features were unique and were different from other mangrove stands in tropical and subtropical regions, indicating that the mangroves in Hong Kong should be protected. The conservation ranks, based on the plant data, provide useful preliminary guidelines for formulating the overall strategy of conservation and protection of Hong Kong mangroves.

Acknowledgments

The authors would like to express their appreciation to all who assisted with field studies, in particular, Mr S. H. Li from Zhongshan University, PRC and Mr Yim Ming Wai, City University of Hong Kong. We would like to thank Mr Frank S. P. Lau, Mr C. C. Lay and Mr Y. K. Chan of the Agriculture and Fisheries Department of Hong Kong Government (AFD) for their continuous support and stimulating discussion during the course of the present study. A research grant awarded by the AFD is also acknowledged.

References

Amarasinghe, M. D. & S. Balasubramaniam, 1992. Structural properties of two types of mangrove stands on the northwestern coast of Sri Lanka. Hydrobiologia 247: 17–27.

Howlett, B. (ed.), 1996. Hong Kong 1996. Government Information Service Department, H. Myers Government Printer, Hong Kong Government.

Chapman, M. L., 1984. Mangrove biogeography. In Por, F. D. & I. Dor (eds), Hydrobiology of the Mangal: The Ecosystem of the Mangrove Forests. Dr W. Junk Publishers, The Hague: 15–24.

Cintron, C. & Y. Schaefer-Novelli, 1984. Methods for studying mangrove structure. In Snedaker, S. C. & J. G. Snedaker (eds), The Mangrove Ecosystem: Research Methods. UNESCO: 91–115.

Clough, B. F., 1993. The Economic and Environmental Values of Mangrove Forests and their Present State of Conservation in the South-east Asia/Pacific Region. International Society for Mangrove Ecosystem, Japan, 178 pp.

Greiner Maunsell, 1991. New Airport Master Plan: Environmental Impact Assessment – Final Report. Provisional Airport Authority, Hong Kong.

Hodgkiss, I. J., 1986. Aspects of mangrove ecology in Hong Kong. Memoirs of the Hong Kong Natural History Society 17: 107–116.

Holmes, P. R., 1988. Tolo Harbour – the case for integrated water quality management in a coastal environment. J. Inst. Wat. envir. Mgmt 2: 171–179.

Irving, R. & B. Morton, 1988. A Geography of the Mai Po Marshes. Hong Kong University Press, Hong Kong.

Kenneally, K. F., 1982. Mangroves of Western Australia. In Clough, B. F. (ed.), Mangrove Ecosystems in Australia: Structure, Function and Management, Australian Institute of Marine Sciences, Australia: 95–110.

Lee, S. Y., 1989. Litter production and turnover of the mangrove *Kandelia candel* (L.) Druce in a Hong Kong tidal shrimp pond. Est. Coast. Shelf Sci. 19: 75–87.

Lin, P. & H. Q. Fan, 1992. Development of the natural reservations of mangroves along China coast. In Wang, Y. & C. T. Schafer (eds), Island Environment and Coastal Development. Nanjing University Press, PRC: 305–319.

Lugo, A. E., 1989. Fringe Wetlands. In Lugo, A. E., M. Brinson & S. Brown (eds), Forested Wetlands: Ecosystems of the World 15, Elsevier, Amsterdam, The Netherlands: 143–169.

Morton, B. & J. Morton, 1983. The Seashore Ecology of Hong Kong. Hong Kong University Press, Hong Kong.

Semeniuk, V., 1983. Mangrove distribution in Northwestern Australia in relationship to regional and local freshwater seepage. Vegetatio 53: 11–31.

Tam, N. F. Y., L. Vrijmoed & Y. S. Wong, 1993. The chemical characteristics of soil and its association with standing litter biomass in a sub-tropical mangrove community in Hong Kong. In Morton, B. (ed.), The Marine Biology of the South China Sea. Proceedings of the First International Conference on the Marine Biology of Hong Kong and the South China Sea, Hong Kong University Press, Hong Kong: 521–541.

Thrower, L. B. & D. W. K. Cheng, 1975. Mangroves. In Thrower, L. B. (ed.), The Vegetation of Hong Kong: its Structure and Change, Proceedings of a Week-end Symposium of the Royal Asiatic Society, Hong Kong Branch: 64–77.

Walter, H. & S. W. Breckle, 1986. Ecological Systems of the Geobiosphere. Vol. 2. Tropical and Subtropical Zonobiomes. Springer-Verlag, Berlin.

Yipp, M. W., C. H. Hau & G. Walthew, 1995. Conservation evaluation of nine Hong Kong mangals. Hydrobiologia 295: 323–333.

Young, L. & D. S. Melville, 1993. Conservation of the Deep Bay Environment. In Morton, B. (ed.), The Marine Biology of the South China Sea, Proc. First Int. Conf. Mar. Biol. of Hong Kong and the South China Sea, Hong Kong University Press, Hong Kong: 211–231.

Hydrobiologia **352**: 39–47, 1997.
Y.-S. Wong & N. F.-Y. Tam (eds), Asia–Pacific Conference on Science and Management of Coastal Environment.
©1997 *Kluwer Academic Publishers. Printed in Belgium.*

Gas exchange responses of a mangrove species, *Avicennia marina*, to waterlogged and drained conditions

G. Naidoo[1], H. Rogalla[2] & D. J. von Willert[2]
[1]*Department of Botany, University of Durban-Westville, Private Bag X54001, Durban 4000, South Africa*
[2]*Institut für Ökologie der Pflanzen, Westfalische Wilhelms Universität, Hindenburgplatz 55, 48143 Munster, Germany*

Key words: Avicennia marina, gas exchange, mangroves, photosynthesis, waterlogging

Abstract

This study was undertaken in summer on fully expanded leaves of *Avicennia marina* trees in the Beachwood Mangroves Nature Reserve, Durban, South Africa. Data sets were obtained over 5–7 days of relatively dry conditions and over two periods of 5 days during which the swamp was continuously inundated with dilute seawater (<150 mol m^{-3} NaCl). Gas exchange responses were strongly influenced by photosynthetic photon flux density (PPFD), leaf temperature and leaf to air vapour pressure deficit (Δw). Carbon dioxide exchange was saturated at a PPFD of about 800 μmol m^{-2} s^{-1}. Maximal CO_2 exchange rates ranged from 8.5 to 9.9 μmol m^{-2} s^{-1} with no differences between drained and waterlogged conditions. Under drained conditions, leaf conductance, transpiration and internal CO_2 concentrations were generally lower, and water use efficiencies higher, than during waterlogging. Continuous waterlogging for 5 days had no adverse effect on CO_2 exchange. Xylem water potentials ranged from -1.32 to -3.53 MPa during drained and from -1.02 to -2.65 MPa during waterlogged conditions. These results are discussed in relation to anatomical and metabolic adaptations of *A. marina* to waterlogging stress.

Introduction

Mangroves, which thrive and grow luxuriantly in tidal saline wetlands along tropical and subtropical coasts, are especially adapted to salinity and waterlogging stresses. In South Africa, one of the most dominant mangroves is *Avicennia marina* (Forsk.) Vierh., a highly salt tolerant pioneer species. Adaptations to salinity in *A. marina* include the presence of salt glands (Boon & Allaway, 1986; Waisel et al., 1986) and a highly efficient salt exclusion mechanism in the roots (Waisel et al., 1986; Ball, 1988a, b). Adaptations to waterlogging in mangroves include the production of aerial roots that vary widely in form (Tomlinson, 1986) and include pneumatophores in *A. marina*, knee roots in *Bruguiera gymnorrhiza*, stilt roots in *Rhizophora mucronata* and plank roots in *Xylocarpus grantum*. Roots and stems of mangroves also exhibit extensive aerenchyma development to cope with waterlogging (Curran, 1985).

Gas exchange characteristics of mangroves have been investigated extensively in the laboratory (Ball & Farquhar, 1984; Pezeshki et al., 1990; Lin & Sternberg, 1993; Naidoo & von Willert 1994, 1995) and in the field (Moore et al., 1972, 1973; Attiwill & Clough, 1980; Andrews & Muller, 1985; Clough & Sim, 1989; Cheeseman et al., 1991). Very little information, however, is available on gas exchange responses of mangroves to waterlogging stress (Naidoo, 1984), especially under field conditions.

Changes in environmental conditions around roots, induced by waterlogging, influence biochemical and physiological processes which may be evident in photosynthetic metabolism. Understanding waterlogging effects on photosynthesis is important for the development of more comprehensive and predictive modelling of mangrove ecosystem dynamics (Ball, 1986). The Beachwood Mangroves Nature Reserve was subjected to frequent, prolonged, flooding episodes during early 1995 as a result of changes in the hydrology of the

Mgeni River. This provided a unique opportunity to investigate gas exchange characteristics of mangroves to waterlogging stress. We tested the hypothesis that gas exchange responses in *A. marina* are not affected by waterlogging.

Materials and methods

Study site

The Beachwood Mangroves Nature Reserve is situated at the Mgeni estuary (29°53′S, 31°00′E) and has been described previously (Naidoo, 1980). The mangroves are dominated by *A. marina* and *B. gymnorrhiza* while occasional individuals of *R. mucronata* also occur. During 1995, there was frequent deposition of a sand bar across the river mouth resulting in flooding of the swamp with dilute sea water (<150 mol m^{-3} NaCl). The sand bar was mechanically breached to drain the swamp. The duration of flooding prior to measurements varied from 5 to 10 d. After drainage, the swamp remained dry for about 5 to 7 d before the next flooding episode. The area is regularly inundated with sea water at high spring tides.

This study was undertaken during the hot, wet summer (February–March). Two separate gas exchange data sets were obtained. The first set was collected over a 5–7 d period when the swamp was relatively dry with no rain or tidal influence. The second set was obtained after the first, over two periods of 5 d when the swamp was continuously inundated to a depth of 0.5 to 1 m. The data set during periods of waterlogging was greater than that during drained conditions. Gas exchange was measured continuously on two leaves per day. The data set for each treatment represents measurements on 8 to 12 leaves.

Measurements

Diurnal gas exchange measurements were made using an open circuit microcuvette system (Walz, Effeltrich, Germany) similar to that described by Schulze & Küppers (1979). The difference in water vapour and carbon dioxide concentrations between the inlet and outlet of the leaf chamber was measured with a Binos infra-red gas analyser (IRGA) (Leybold-Heraeus, Hanau, Germany). Humidity was measured with high-precision dew-point mirrors (Heinz Walz, Effeltrich, Germany) having stability and accuracy within 0.1 °C. Leaf temperature was monitored with

a 0.05 mm copper-constantan thermocouple in contact with the abaxial leaf surface.

An individual mature leaf was introduced horizontally into the chamber and sealed with Terostat plastic putty (Teroson, Gmbh., Heidelberg, Germany). Photosynthetic photon flux density (PPFD) was measured inside and outside the chamber by quantum sensors (Licor Li 190S). The temperature of the chamber air was monitored by a radiation-shielded NTC thermistor. The leaf chamber air temperature was controlled to follow ambient, as measured by a ventilated PT 100 resistance sensor. Heating and cooling of the chamber was via Peltier elements.

Access to the top of the *A. marina* canopy was achieved by the construction of a 5 m multiplatform tower. Data were collected at three different sites. Gas exchange measurements were also made on mature leaves of young trees that were 1 to 2 m in height. The gas exchange system was controlled automatically by a data logging unit which permitted manipulation of the magnetic switches. IRGA data were downloaded at the end of each day and the data set transferred to a personal computer. All gas exchange parameters were calculated using the equations of von Caemmerer and Farquhar (1981). Gas exchange rates were calculated on a leaf area basis using both leaf surfaces. Xylem water potentials were measured at hourly intervals from predawn to dusk with a pressure chamber (Scholander et al., 1965), using a subset of at least three leaves.

Data were subjected to *t*-tests to detect differences between drained and waterlogged treatments.

Results

During drained and waterlogged periods, environmental conditions were similar, the days being sunny and hot with frequent intermittent clouds. Mean values of environmental conditions during data collection are shown in Table 1. During measurements, maximum PPFD was about 1700 μmol m^{-2} s^{-1}, leaf temperature ranged from 21–34 °C and the leaf to air vapour pressure deficit (Δw) was less than 20 kPa MPa^{-1} (Figure 1).

Generally, PPFD increased rapidly from dawn to a maximal value of 1700 μmol m^{-2} s^{-1} between 12 h to 14 h and subsequently declined (Figure 1a, b). Leaf temperature (Figure 1c, d) and Δw profiles (Figure 1e, f) were similar to that of PPFD. Diurnal trends in CO$_2$ exchange, conductance and transpiration were

41

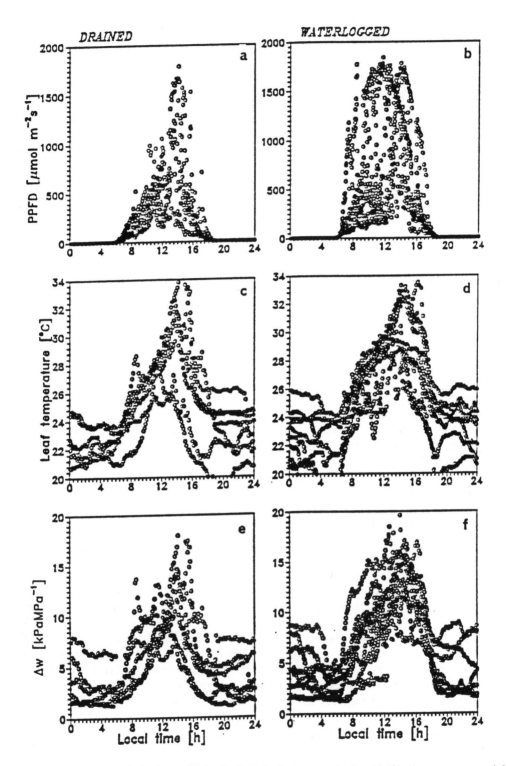

Figure 1. Diurnal course of photosynthetic photon flux density (a, b), leaf temperature (c, d) and leaf to air vapour pressure deficit (e, f) in *A. marina*. The data set, from young and old trees, represents 5–7 d of measurements during drained and two periods of 5 d of waterlogged conditions.

Table 1. Mean light intensity, day temperature, vapour pressure deficit (VPD), external CO_2 concentration, leaf temperature and leaf to air vapour pressure deficit (Δw) during drained and waterlogged conditions (mean $\pm$ SD are shown. Data were selected from 6 representative days). There were no significant differences between drained and waterlogged conditions.

Variable	Drained	Waterlogged
Light intensity (mol Photons m^{-2})	18.75 ± 4.97	22.64 ± 7.72
Temperature (°C)	25.81 ± 1.78	26.70 ± 1.56
VPD (kPa MPa^{-1})	0.76 ± 0.17	1.04 ± 0.26
External CO_2 (μl l^{-1})	353 ± 4.54	352 ± 5.22
Leaf temperature (°C)	26.4 ± 1.87	26.82 ± 1.62
Δw (kPa MPa^{-1})	6.99 ± 2.12	8.96 ± 2.62

Table 2. Mean daily CO_2 uptake, water loss, conductance, internal CO_2 concentration (c_i) and water use efficiency (WUE) in *A. marina* during drained and waterlogged conditions (mean $\pm$ SD are shown. Data were selected from 6 representative days. Asterisk indicates significant difference.

Variable	Drained	Waterlogged
Net CO_2 uptake (mmol m^{-2} d^{-1})	150 ± 35	195 ± 22
Water loss (mmol m^{-2} d^{-1})	23 ± 5	32 ± 6
Conductance (mmol m^{-2} s^{-1})	78 ± 8*	$98 +10$*
c_i (μl l^{-1})	260 ± 22	258 ± 28
WUE (μmol mmol^{-1})	8.0 ± 2.4	7.8 ± 2.1

strongly influenced by PPFD and the accompanying leaf temperature and Δw (Figure 2). There were no differences in gas exchange between young and adult trees. Maximal CO_2 exchange rates were 8.5 μmol m^{-2} s^{-1} for drained and 9.9 μmol m^{-2} s^{-1} for waterlogged conditions (Figure 2a, b). Although differences in CO_2 exchange between drained and waterlogged conditions were not significant, maximal values appeared to be achieved earlier and maintained over a longer period during waterlogging. The mean net daytime CO_2 uptake (Table 2) during drained conditions (150 mmol m^{-2} d^{-1}) was lower than that for a comparable period of waterlogging (195 mmol m^{-2} d^{-1}). The duration of waterlogging stress from 1 to 5 days appeared to have no effect on diurnal CO_2 exchange.

Trends in diurnal conductance (Figure 2c, d) were similar to those for CO_2 exchange (Figure 2a, b). The mean daily conductance during waterlogging (98 mmol m^{-2} s^{-1}) was significantly higher than that during drained (78 mmol m^{-2} s^{-1}) conditions (Table 2). The mean daily water loss during the photoperiod was also higher during waterlogging (Table 2). Maximal values of leaf conductance (Figure 2c, d) and transpiration (Figure 2e, f) were higher during waterlogging.

The response of CO_2 exchange to PPFD was characterized by an initial, steep, linear region, broad convexity at higher irradiance and saturation at about 800 μmol m^{-2} s^{-1} (Figure 3a, b). Maximal CO_2

exchange was achieved at a leaf temperature of about 31 °C and at a Δw of 10 kPa MPa^{-1} in both treatments (results not presented).

Under drained conditions, c_i decreased gradually from ambient to about 190 to 270 μl l^{-1} in the late afternoon (Figure 4a, b). Under waterlogging, c_i values were generally higher. Maximal CO_2 exchange was achieved at a c_i of about 250 μl l^{-1} during drained and 270 μl l^{-1} under waterlogged conditions. However, there were no differences in mean daily c_i between treatments (Table 2).

The relationship between CO_2 exchange and conductance (Figure 4c, d) and CO_2 exchange and transpiration (Figure 4e, f) at saturating PPFD were similar for drained and waterlogged conditions. Maximal CO_2 exchange occurred at a conductance of 150 mmol m^{-2} s^{-1} during drained and at 190 mmol m^{-2} s^{-1} during waterlogged conditions, while transpiration rates were 1.6 and 2.0 mmol m^{-2} s^{-1} respectively.

Under drained conditions, WUE increased rapidly from dawn till midday followed by great variability in the latter part of the day. Under waterlogging, WUE appeared to be lower and less variable (Figure 3c, d). Mean daily WUE between treatments, however, was not significant (Table 2). Xylem water potentials ranged from -1.32 to -3.53 MPa during drained conditions and from -1.02 to -2.65 MPa during waterlogging (Figure 5).

Discussion

Field measurement of gas exchange is subject to many difficulties and when undertaken at different times,

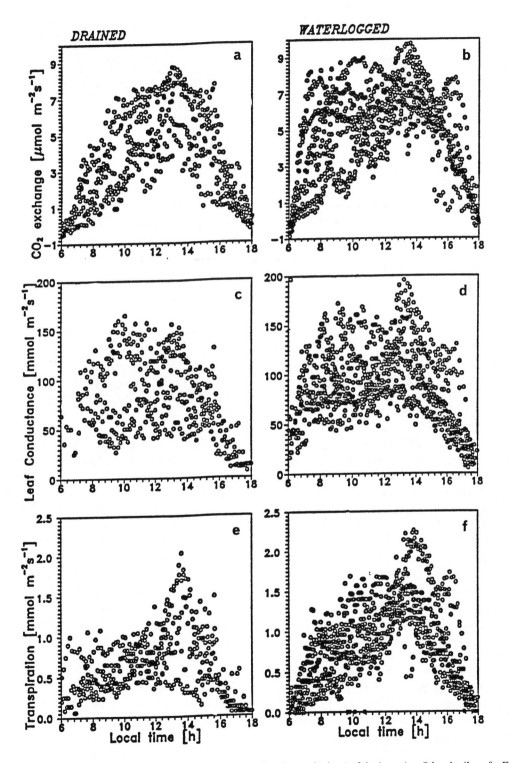

Figure 2. Diurnal course of CO_2 exchange (a, b) leaf conductance (c, d) and transpiration (e, f) in *A. marina*. Other details as for Figure 1.

44

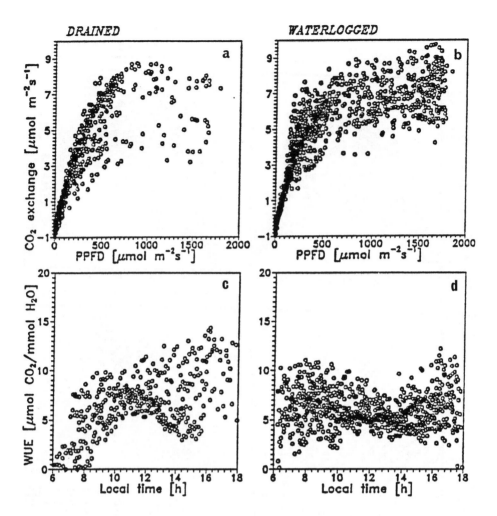

Figure 3. The relationship between CO_2 exchange and PPFD (a, b), and diurnal course of water use efficiency (c,d) in *A. marina*. Other details as for Figure 1.

introduces additional variability. Data sets were collected from three sites over several days from young and adult trees and plotted over all days to emphasize general trends. Despite the limitations of this study, the data indicate some interesting information on gas exchange characteristics in *A. marina* and its adaptation to short-term waterlogging stress.

The strong effect of PPFD on leaf temperature and the accompanying Δw was indicated by similar diurnal profiles (Figure 1). Trends in CO_2 exchange, conductance and transpiration were similar to those of PPFD, indicating the strong influence of irradiance on gas exchange characteristics in *A. marina* (Figure 2). The strong dependence of mangrove photosynthesis on PPFD was also reported for *R. stylosa* (Andrews & Muller, 1985) and *B. parviflora* (Cheeseman et al.,

1991). The response of CO_2 exchange to PPFD was similar for drained and waterlogged conditions (Figure 3a, b). Light compensation point (30 μmol m^{-2} s^{-1}) and light saturation (800 μmol m^{-2} s^{-1}) values determined from light response curves (Figure 3a, b) compare favourably with those obtained for other mangroves (Moore et al., 1972, 1973; Attiwill & Clough, 1980; Andrews et al., 1984; Andrews & Muller, 1985).

An unpaired *t*-test indicated no significant difference in mean net CO_2 uptake between drained and waterlogged conditions (Table 2), although maximal CO_2 exchange rates were 16% higher during waterlogging (Figure 2a, b). Diurnal CO_2 exchange profiles after 1 and 4 days of waterlogging (results not presented) showed that CO_2 exchange was not comprised by the duration of waterlogging. Maximal CO_2 exchange

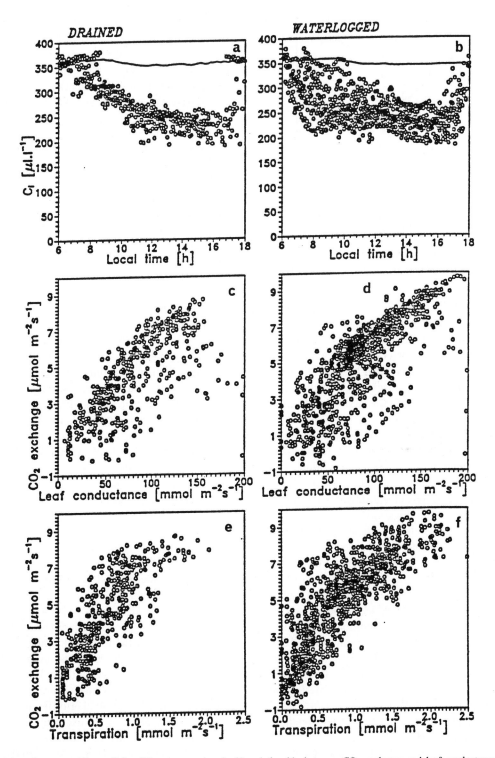

Figure 4. Diurnal course of intercellular CO_2 concentration (a, b), relationship between CO_2 exchange and leaf conductance (c, d) and relationship between CO_2 exchange and transpiration (e, f) in *A. marina*. In a, b the continuous line represents ambient CO_2 concentration. In c, d, e, f data were restricted to saturating PPFD. Other details as for Figure 1.

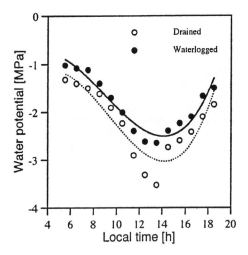

Figure 5. Diurnal course of xylem water potential in *A. marina* during drained (open circles) and waterlogged (closed circles) conditions (*n* = 3).

rates for *A. marina* (8.5 to 9.9 μmol m^{-2} s^{-1}) are higher than those (5.3 μmol m^{-2} s^{-1}) reported by Attiwill & Clough (1980), but considerably lower than that (19 μmol m^{-2} s^{-1}) recorded by Clough & Sim (1989). Our values compare favourably with the results of Andrews & Muller (1985) for *R. stylosa* (11 μmol m^{-2} s^{-1}).

Analysis of photosynthetic conductance responses indicated that CO_2 exchange was strongly correlated with leaf conductance. Under drained conditions, CO_2 exchange was conductance limited, as maximal photosynthesis was achieved at a considerably lower conductance than under waterlogging (Figure 4c, d). Mean daily conductance was also significantly lower under drained conditions (Table 2). The decrease in CO_2 exchange at high rates of transpiration under drained conditions is a reflection of the relationship of both variables to conductance (Figure 4e, f). Limitation in the supply of water probably reduced conductance and consequently c_i under drained conditions. The consistently lower xylem water potentials during drained conditions suggests that water flux through the plant was reduced (Figure 5). Under drained conditions, water loss was more conservative than during waterlogging and contributed to high WUE (Figure 3c, d), which is a characteristic feature of mangrove photosynthesis (Andrews & Muller, 1985; Ball, 1986).

The results of this study clearly indicated that continuous waterlogging with dilute sea water for periods up to 5 days did not adversely affect photosynthetic performance of *A. marina*, despite the complete sub-

mergence of all pneumatophores. In waterlogged soils, O_2 is rapidly depleted as a result of respiration by roots and soil microorganisms (Turner & Patrick, 1968). Since the diffusion of O_2 in water is one thousand times slower than in air (Gambrell & Patrick, 1978) anaerobic and chemically reduced conditions develop around waterlogged roots. Survival of wetland species under waterlogging is dependent on the efficiency of O_2 transport down partial pressure gradients from the atmosphere to sites of respiration in roots (Armstrong, 1979).

There is considerable evidence in the literature that in many wetland species with well developed aerenchyma, diffusion and pressurized convective throughflow of gases occur (Dacey, 1980, 1987; Beckett et al., 1988). These convective flows not only supply sufficient O_2 to maintain aerobic root respiration, but also release O_2 into the soil to oxidise chemically reduced soil phytotoxins (Mendelssohn & Postek, 1982).

In *A. marina*, root ventilation is highly efficient as aerenchyma accounts for about 70% of the root volume (Curran, 1985). Low O_2 concentrations within the aerenchyma tissue could produce large diffusion gradients thereby promoting rapid oxygen movement into roots to support aerobic respiration (Curran et al., 1986). Measurements of soil redox potentials in mangrove swamps showed that leakage of O_2 from roots of *A. germinans* and *R. mangle* caused the rhizosphere to be more oxidized than would be expected in reduced sediments (Nickerson & Thibodeau, 1985; Thibodeau & Nickerson, 1986; McKee et al., 1988). In addition to anatomical adaptations, mangroves are also adapted metabolically to waterlogging. The capacity for anaerobic respiration, as indicated by alcohol dehydrogenase activity, increased in *A. germinans* seedlings subjected to hypoxia for 96 h (McKee & Mendelssohn, 1987). Stomatal closure and reduced water uptake and transpiration were reported for *B. gymnorrhiza* (Naidoo, 1984) and for several other woody angiosperm seedlings (Kozlowski & Pallardy, 1979) subjected to waterlogging. These short term experiments with seedlings probably result in greater perturbations of metabolism. Long term adaptations of more mature trees which have well developed aerenchyma may be modified or different.

This investigation on gas exchange characteristics has clearly indicated that *A. marina* is well adapted to tolerate continuous waterlogging for periods up to 5 d. This adaptation is probably achieved by a well developed aerenchyma system that permits efficient root ventilation. Further research is needed to determine whether convective throughflow of gases is significant in the oxygenation of waterlogged *A. marina* roots.

Acknowledgments

This work was sponsored by the Foundation for Research Development, South Africa and the Alexander von Humboldt Foundation. The authors thank Birgit Rossa, Trevor Govender and Mannie Naicker for technical assistance and the Natal Parks Board for permission to undertake research in the Reserve.

References

Andrews, T. J. & G. J. Muller, 1985. Photosynthetic gas exchange of the mangrove, *Rhizophora stylosa* Griff., in its natural environment. Oecologia 65: 449–455.

Andrews, T. J., B. F. Clough & G. J. Muller, 1984. Photosynthetic gas exchange properties and carbon isotope ratios of some mangroves in North Queensland. In Teas, H. J. (eds), Physiology & Management of Mangroves, Dr W. Junk Publishers, The Hague: 15–23.

Armstrong, W., 1979. Aeration in higher plants. In Wooolhouse, H. W. (ed.), Advances in Botanical Research. Academic Press, Lond. 7: 226–332.

Attiwill, P. M. & B. F. Clough, 1980. Carbon dioxide and water vapour exchange in the white mangrove. Photosynthetica 14: 40–47.

Ball, M. C., 1986. Photosynthesis in mangroves. Wetlands, Australia 6: 12–22.

Ball, M. C., 1988a. Ecophysiology of mangroves. Trees 2: 129–142.

Ball, M. C., 1988b. Salinity tolerance in the mangroves *Aegiceras corniculatum* and *Avicennia marina*. 1. Water use in relation to growth, carbon partioning, and salt balance. Aust. J. Plant Physiol. 15: 447–464.

Ball, M. C. & G. D Farquhar, 1984. Photosynthetic and stomatal responses of two mangrove species, *Aegiceras corniculatum* and *Avicennia marina*, to long term salinity and humidity conditions. Plant Physiol. 74: 1–6.

Beckett, P. M., W. Armstrong, S. H. F. W. Justin & J. Armstrong, 1988. On the relative importance of convective and diffusive gasflows in plant aeration. New Phytol. 110: 463–469.

Boon, P. I. & W. G. Allaway, 1986. Rates and ionic specificity of salt secretion from excised leaves of the mangrove *Avicennia marina* (Forsk.) Vierh. Aquat. Bot. 26: 143–153.

Cheeseman, J. M., B. F. Clough, D. R. Carter, C. E. Lovelock, Ong Jin Eong & R. G. Sim, 1991. The analysis of photosynthetic performance in leaves under field conditions. A case study using *Bruguiera* mangroves. Photosynth. Res. 29: 11–22.

Clough, B. F. & R. G. Sim, 1989. Changes in gas exchange characteristics and water-use efficiency of mangroves in response to salinity and vapour pressure deficit. Oecologia 79: 38–44.

Curran, M., 1985. Gas movements in the roots of *Avicennia marina* (Forsk.) Vierh. Aust. J. Plant Physiol. 12: 97–108.

Curran, M., M. Cole & W. G. Allaway, 1986. Root aeration and respiration in young mangrove plants (*Avicennia marina* (Forsk.) Vierh.). J. exp. Bot. 37: 1225–1233.

Dacey, J. W. H., 1980. Internal winds in water lilies. An adaptation for life in anaerobic sediments. Science 21: 1017–1019.

Dacey, J. W. H., 1987. Knudsen-transitional flow and gas pressurization in leaves of *Nelumbo*. Plant Physiol. 8: 199–203.

Gambrell, R. P. & W. H. Patrick, 1978. Chemical and microbiological properties of anaerobic soils and sediments. In Hook, D. D. & R. M. M. Crawford (eds), Plant life in anaerobic environments, Ann Arbor Science, Ann Arbor, Michigan: 375–425.

Kozlowski, T. T. & S. G. Pallardy, 1979. Stomatal responses of *Fraxinus pennsylvanica* seedlings during and after flooding. Physiol. Pl. 46: 155–158.

Lin, G. & Leonel da Silveira Lobo Sternberg, 1993. Effect of salinity fluctuation on photosynthetic gas exchange and plant growth of the red mangrove (*Rhizophora mangle* L.) J. exp Bot. 258: 9–16.

McKee, K. L. & I. A. Mendelssohn, 1987. Root, metabolism in the black mangrove (*Avicennia germinans* (L.) L): Responses to hypoxia. Envir. Exp. Bot. 27: 147–158.

McKee, K. L., I. A. Mendelssohn & M. W. Hester, 1988. Reexamination of pore water sulphide concentrations and redox potentials near the aerial roots of *Rhizophora mangle* and *Avicennia germinans*. Am. J. Bot. 75: 1352–1359.

Mendelssohn, I. A. & M. T. Postek, 1982. Elemental analysis of deposits on the roots of *Spartina alterniflora* Loisel. Am. J. Bot. 69: 904–912.

Moore, R. T., P. C. Miller, D. Albright & L. L. Tieszen, 1972. Comparative gas exchange characteristics of three mangrove species during the winter. Photosynthetica 6: 387–393.

Moore, R. T., P. C. Miller, J. Ehleringer & W. Lawrence, 1973. Seasonal trends in gas exchange characteristics of three mangrove species. Photosynthetica 7: 387–394.

Naidoo, G., 1980. Mangrove soils of the Beachwood area, Durban. J.S. Afr. Bot. 46: 293–304.

Naidoo, G., 1984. Effects of flooding on leaf water potential and stomatal resistance in *Bruguiera gymnorrhiza* (L.) Lam. New Phytol. 93: 369–376.

Naidoo, G. & D. J. von Willert, 1994. Stomatal oscillations in the mangrove *Avicennia germinans*. Funct. Ecol. 8: 1–7.

Naidoo, G. & D. J. von Willert, 1995. Diurnal gas exchange characteristics and water use efficiency of three salt secreting mangroves at low and high salinities. Hydrobiologia 295: 13–22.

Nickerson, N. H. & F. R. Thibodeau. 1985. Association between pore water sulfide concentrations and the distribution of mangroves. Biogeochem. 1: 183–192.

Pezeshki, S. R., R. D. DeLaune & W. H. Patrick, 1990. Differential response of selected mangroves to soil flooding and salinity: gas exchange and biomass partitioning. Can. J. Forest Res. 20: 869–874.

Scholander, P. F., H. T. Hammel, E. D. Bradstreet & E. A. Hemingsen, 1965. Sap pressure in vascular plants. Science 148: 339–346.

Schulze, E-D. & M. Küppers, 1979. Short term and long term effects of plant water deficits on stomatal response to humidity in *Corylus avellana* L. Planta 146: 319–326.

Thibodeau, F. R. & N. H. Nickerson, 1986. Differential oxidation of mangrove substrate by *Avicennia germinans* and *Rhizophora mangle*. Am. J. Bot. 73: 512–516.

Tomlinson, P. B., 1986. The Botany of Mangroves, Cambridge University Press, Cambridge: 25–32.

Turner, F. T. & W. H. Patrick, 1968. Chemical changes in waterlogged soils as a result of oxygen depletion. Transactions of the 9th Int. Congr. Soil Sci., Sydney 4: 53–65.

von Caemmerer, S. & G. D. Farquhar, 1981. Some relationships between the biochemistry of photosynthesis and the gas exchange of leaves. Planta 153: 376–387.

Waisel, Y., A. Eshel & M. Agami, 1986. Salt balance of leaves of the mangrove, *Avicennia marina*. Physiol. Pl. 67: 67–72.

Hydrobiologia **352**: 49–59, 1997.
Y.-S. Wong & N. F.-Y. Tam (eds), Asia–Pacific Conference on Science and Management of Coastal Environment.
©1997 *Kluwer Academic Publishers. Printed in Belgium.*

Mangrove wetlands as wastewater treatment facility: a field trial

Y. S. Wong[1], N. F. Y. Tam[2*] & C. Y. Lan[3]

[1]*Research Centre/Biology Department, The Hong Kong University of Science & Technology, Hong Kong*
[2]*Department of Biology and Chemistry, City University of Hong Kong, Kowloon, Hong Kong*
[3]*School of Biological Sciences, Zhongshan University, Guangzhou, China*
(*Author for correspondence; tel: (852) 2788 7793; fax: (852) 2788 7406; e-mail: bhntam@cityu.edu.hk)

Key words: mangrove, treatment, sewage, plant, sediment, wetland

Abstract

Field work has been conducted in a 300-hectare natural mangrove intertidal wetlands in Shenzhen, a newly developed city in southern China, to study the feasibility of using mangrove wetlands as a sewage treatment facility. The present paper reports the results obtained in the recent year, between December 1994 and December 1995. Two parallel elongated sites (Sites A & B, each 180 m × 10 m) extending from land to sea were chosen for study. Since September 1991, Site A has received settled municipal sewage three times a week during the low ebb tide period when sediments at landward regions were dry. The hydraulic loading was 20 m^3 per discharge and wastewater was soaked into the sediments within 50 m of the discharge points before the next incoming tide. Site B served as a control. Over the past months in 1994 and 1995, surface sediments and plant leaves were collected at identified locations in two sites at every six month intervals. The impact of sewage on mangrove plant growth was assessed by monitoring plant height, diameter and number of trees using the fixed plot technique.

The plant density, stem diameter and tree height of two dominant mangrove species, *Kandelia candel* and *Aegiceras corniculatum*, found in Site A were comparable with those of Site B. No significant difference was detected between two sites in terms of plant growth and death rates. These results indicate that sewage discharge over a period of about two years did not exhibit any apparent effect on plant growth. The nutrient and organic matter concentrations of surface sediments in Site A were also not significantly different from those found in Site B, except at the very landward regions (2 to 40 m away from landlands). The nutrient concentrations of sediments collected in sampling locations near the discharge points of Site A were however significantly higher than that of the control. In both sites, the organic C, total N and P, NH_4^+-N and NO_3^--N concentrations in the surface sediments exhibited a descending trend from landwards to seaward regions, with notably higher values found in the landward locations. Seasonal variation in NH_4^+-N content was obvious, and more ammonium nitrogen was recorded in July than in December. Leaf samples of the two dominant plant species collected from Site A had similar total N and organic C concentrations as those from Site B. These findings suggest that mangrove intertidal wetlands are of great potential for natural wastewater treatment, and are unlikely to produce any harmful effect on the higher plant communities.

Introduction

Constructed and natural wetlands have been employed an alternative low cost biological treatment systems for municipal sewage and have been extensively studied over the past decades (Kadlec, 1987; Richardson & Davis, 1987). It has been reported by the US Environmental Protection Agency that wetlands appear to perform, to at least some degree, all of the biochemical transformations of wastewater constituents that take place in conventional wastewater treatment systems, in septic tanks and their drain fields, and in other forms of land treatment (US EPA, 1987). Wetlands remove pollutants from aquatic systems, through a complex variety of biological, physical and chemical processes. The nutrient uptake by higher plants and retention of heavy metals in roots are considered to be important processes as the higher plants are the most obvious biological components of wetland ecosystems (Orson et al., 1992; Rai et al., 1995). In addition to plant uptake, bacterial transformations and physico-chemical processes including adsorption, precipitation and sedimentation in soils and in the rhizosphere (the root zone area) are very major mechanisms for pollutant removal (Dunbabin & Browmer, 1992; Gale et al., 1993).

The mangrove ecosystems, as important natural wetlands often occupy the intertidal mudflats of tropical and subtropical regions, have been utilized as a sewage treatment system (Henley, 1978; Clough et al., 1983). The special adaptation of mangrove plants to stressed environments, including unstable substratum, alternating aerobic and anaerobic conditions, fluctuations in salinity and periodic drying (desiccation) and wetting (waterlogging), suggests that these plants should be able to tolerate the pollutants contained in sewage. Henley (1978) studied the effects of sewage discharge on the mangrove communities in the Darwin area in Australia and reported that mangrove trees had considerable capacity to accept waste loads from sewage effluent without suffering any damage to their growth. Similar studies performed by Clough et al. (1983), Dwiedi & Padmakumar (1983), and Por (1984) also found that sewage discharges had no adverse effects on mangrove community structures. The low inherent nutrient concentration recorded in Australian mangrove sediments and the high primary production of this ecosystem indicate that the system is often nutrient deficient and has a large capacity to retain nutrients (Boto & Wellington, 1983; Clough et al., 1983; Boto, 1992). Addition of sewage, with concomitant influx of nutrients and sediments, would be beneficial to the ecosystem by stimulating biomass production and subsequent soil formation (Breaux et al., 1995).

In the past, mangrove intertidal wetlands were used as tertiary treatment facilities to remove nitrogen and phosphorus after aeration, trickling filters or activated sludge processes (Corredor & Morell, 1994, Breaux et al., 1995). Our research group attempted to employ mangrove ecosystems as secondary treatment processes. Wastewater was collected and discharged to an experimental strip in a mangrove wetland in Futian National Nature Reserve, Shenzhen Special Economic Zone, the PRC from September 1991 till October 1992. The preliminary one-year results were promising and showed that discharge of weak sewage for a short period of time (one year) did not cause any significant change in the biotic communities of the man-

grove ecosystem, and the nutrients and heavy metals content of the sediment were also not altered (Wong et al., 1995). In order to understand the longer-term effect of sewage discharge on mangrove ecosystems and their capacity in purifying wastewater, the study was continued from December 1994 onwards. The present paper reports the investigation of the effects of sewage discharge on growth and production of mangrove plant communities, the nutrient concentrations in plant leaves and the nutrient status in sediments of the Futian mangrove intertidal wetlands.

Materials and methods

The field trial was carried out in Futian National Nature Reserve, Shenzhen Special Economic Zone, the People's Republic of China. A detailed description on this study site has been reported earlier (Wong et al., 1995). Within the mangrove ecosystems, two elongated sites 150 m apart, Sites A (the experimental) and B (the control), were chosen. Each site was 10 m wide and 180 m long, extending from land to sea. Sewage wastewater collected from local premises had been discharged to the landward regions of Site A for one year from September 1991 to October 1992, then from December 1994 onwards to June 1996. The sewage was discharged three times a week during the low ebb tide period when the mangrove floor was dried and exposed. The hydraulic loading of each discharge was $20 \, m^3$ and wastewater was observed to infiltrate into the sediments within 50 m of the discharge points. The total volumes of wastewater discharged into the experimental site in September 1991 to October 1992 and December 1994 to June 1996 were 2600 and 5072 m^3, respectively.

Triplicate samples of surface sediments (0–10 cm) were collected at regular intervals from landward to seaward regions in both Sites A and B in December 1994 (prior to the second discharge of wastewater), July and December 1995 (6 and 12 months after the second discharge of wastewater, respectively). Leaf samples of the two most dominant plant species, *Kandelia candel* and *Aegiceras corniculatum,* were also collected from both sites in locations near to the land (5, 10 and 20 m from landward). The sediment and leaf samples were analysed for organic carbon and nitrogen concentrations according to the methods described by this laboratory (Wong et al., 1995).

The structure of the plant community was surveyed in December 1994 and January 1996 using permanent plot techniques. Fixed quadrats (5 m × 5 m) were set at 10 (Quadrat 1), 40 (Quadrat 2), 60 (Quadrat 3) and 100 m (Quadrat 4) from landward to sea in both sites. In each quadrat, each tree was labelled with a tag, with the species identified and the numbers of live and dead individuals of each tree species counted. The height, breast height diameter and crown size of each tree were also measured during each survey. The growth and death rates of trees in each quadrat in both sites from December 1994 to January 1996 were calculated.

Results and discussion

Pollutant loading from municipal sewage

The sewage collected from the premises of the residential area around Futian Nature Reserve has changed significantly throughout the last five years. The 'strength' (both organic matter and nutrients) of sewage as measured in 1994–96 was significantly higher than that measured in 1991–92 (Table 1) and there was also an increase in hydraulic flow. This might explain why there were no significant changes in the nutrient and organic matter content of both plants and sediments in the experimental site (Site A) during the 1991–92 study (Wong et al., 1995). This also reflects the rapid development and urbanization in the Shenzhen city especially the Futian District. The average concentrations of BOD (biological oxygen demand), TKN (total Kjeldahl nitrogen), NH_4^+-N and total P recorded in Futian sewage collected from 1994–96 were similar to those of Hong Kong sewage, with an even higher BOD:COD ratio, suggesting that the sewage wastewater from Futian was more biodegradable. Despite the increases in organic matter and nutrient content in sewage collected in 1994–96, the heavy metals concentrations in this sewage were still below the detection limits of the flame atomic absorption spectrophotometry. This may be due to the fact that Futian District is a residential area with no or little industrial effluent discharged.

Impacts of sewage discharge on the mangrove plant community

Plant growth

The number of *Aegiceras* trees found in December 1994 was higher than that found in January 1996 in all quadrats of both sites (Figure 1), indicating that the trees died naturally to provide gaps for young seedlings

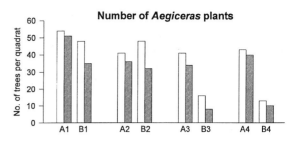
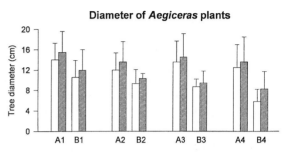
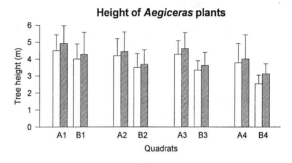

Figure 1. Effect of wastewater discharge on the growth of *Aegiceras corniculatum.* Mean and standard deviations of all trees in the same quadrat were shown. A1, A2, A3 and A4 were the quadrats from landward to seaward regions in Site A; B1 to B4 were the corresponding quadrats in Site B. Empty bars represented data collected in December 1994, before the discharge; and shaded bars represented data in the beginning of 1996, about one year after wastewater discharge.

to develop. The death percentage of *Aegiceras* in Site A was lower than that in Site B (the control), particularly in Quadrat 1 (very close to the discharge point). Only 6% *Aegiceras* died in Quadrat 1 of Site A while a 27% death rate was recorded in Site B (4.5 times of that found in Site A) at the same distance from the land. This finding suggests that addition of wastewater might reduce the death rate of the trees in the mangrove ecosystem especially at the location near the source of sewage discharge. The natural die back throughout 1994–96 was however less obvious in *Kandelia candel*, another dominant species, although its density was generally lower than that of *A. corniculatum* (Table 2).

52

Table 1. Properties of settled municipal sewage prior to the discharge into the experimental site (Site A) of Futian National Nature Reserve.

	Futian (91–92)		Futian (94–96)		Total	Daily	Hong
	Sewage Conc[a]	Loading Quantity[b]	Sewage Conc	Loading Quantity	Loading Quantity[c]	Loading Quantity[d]	Kong Sewage
pH	7.52 (0.28)	NC	7.55 (0.07)	NC	NC	NC	7.09 (0.19)
Cond	0.10 (0.02)	NC	0.95 (0.08)	NC	NC	NC	ND
SS	ND	NC	57.1 (8.9)	289.6	>300	>0.09	71.7 (6.1)
COD	122.5 (11.9)	318.5	268.3 (42.5)	1360.8	1679.3	4.85	351.3 (43.7)
BOD	55.9 (21.4)	145.3	194.2 (41.2)	984.9	1130.3	3.27	158.1 (19.1)
Total N	24.6 (9.0)	63.9	38.5 (5.3)	195.3	259.2	0.75	38.4 (0.5)
NH_4^+-N	ND	NC	28.1 (0.3)	142.5	>143	>0.41	26.9 (2.4)
NO_x^--N	ND	NC	0.29 (0.01)	1.47	>1.5	NC	1.31 (0.01)
Total P	1.23 (0.77)	3.2	4.84 (0.78)	24.6	27.8	0.08	5.82 (0.69)

Mean and standard deviation values (in brackets) of 18 samples were shown;
Cond: electrical conductivity; SS: suspended solids;
ND: not determined; NC: not calculated;
[a] Sewage Conc: pollutant concentration in sewage wastewater, all measured in mg l^{-1} except conductivity which was measured in terms of mS cm^{-1};
[b] Loading quantity: amount of pollutants discharged into mangrove wetlands, calculated as pollutant conc times total hydraulic loading, and presented in Kg dried weight; and total hydraulic loading between Sept 91 to Oct 92 and Dec 94 to June 96 were 2,600 and 5,072 m^3, respectively;
[c] Total loading quantity was the sum of the two discharge periods and was presented in Kg dried weight;
[d] Daily loading quantity was expressed in terms of g dried weight per m^2 per day, assuming sewage was retained within 50 m of the discharge points.

Table 2. Effect of sewage wastewater discharge on growth of two dominant mangrove plants in Futian National Nature Reserve from 1994 to 1996.

Species	Quadrat 1		Quadrat 2		Quadrat 3		Quadrat 4	
	Site A	Site B	Site A	Site B	Site A	Site B	Site A	Site B
Aegiceras corniculatum								
Death%	5.6	27.1	12.2	33.3	17.1	50	6.9	23.1
Increase in Diameter (%)	10.3	13.1	13.4	11.1	6.9	8.8	8.8	12.1
Increase in Height (%)	9.3	6.8	5.7	5.4	7.9	8.4	5.8	22.92
Kandelia candel								
Death%	NP	6.3	0	0	NP	0	0	23.9
Increase in Diameter (%)	NP	12.2	10.6	7.6	NP	16.9	9.8	42.0
Increase in Height (%)	NP	7.2	13.6	7.8	NP	15.2	10.9	28.9

NP: No such species identified in the site; Site A: Experimental site with sewage wastewater discharged; Site B: control site.

The average breast height diameter of *A. corniculatum* in Site A in December 1994 appeared to be higher than that in Site B (Figure 1). However, due to the great variability between individual trees in the same quadrat, there was no significant difference between the two sites according to the analysis of variance (ANOVA) statistical test. The two sites were thus considered as comparable in 1994. Increases in tree diameters were observed in all sampling quadrats one-year later (Figure 1), and no significant difference was found

between Sites A and B (Table 2). A similar pattern was observed when growth of *A. corniculatum* was measured in terms of average tree height and annual growth increments. The response of *K. candel* to sewage discharge was similar to that of *A. corniculatum* (Figure 2, Table 2). Based on the field observation and survey data, there was no change in tree species diversity, dominant species and importance values of the mangrove plant communities of Site A when compared with Site B. These results indicate that sewage dis-

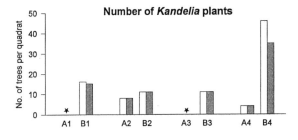

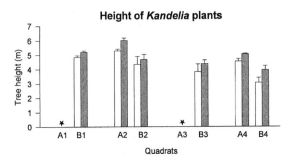

Figure 2. Effect of wastewater discharge on the growth of *Kandelia candel* before and after the experiment (Same legend as Figure 1). ★ represents that no such plant species were identified in that site.

charge did not reduce the growth of mangrove trees and did not alter the plant community structure in this mangrove ecosystem. Mangrove plants have been known to be very tolerant to stressed environments (Por, 1984) and to toxic heavy metals (Lin & Chen, 1990; Zheng et al., 1992). Clough et al. (1983) pointed out that sewage discharge produced a beneficial effect on plant growth and productivity of mangrove ecosystems due to the nutrient supply from wastewater. Boto (1992) summarized that input of inorganic nutrients to mangrove forests are unlikely to be detrimental to the trees and in most circumstances are likely to be beneficial, leading to increased plant growth.

Plant nutrient content
The leaf nitrogen and carbon content of the two dominant species, *K. candel* and *A. corniculatum*, measured in December 1994 were the same as those of Decem-

ber 1995 (Table 3). The values at Sites A and B were comparable, suggesting that the additional nutrients and organic matter from sewage wastewater discharge were not reflected in the nutritive status of the mangrove leaves. These results, however, differed from those obtained by previous workers. Henley (1978) reported that the nutrient level in the foliage of mangrove plants receiving wastewater was significantly higher than that in the control sites, and that the foliage nutrient levels were in equilibrium with those in the soil surrounding the roots, thus reflecting higher levels of available nutrients in the environment. Boto and Wellington (1983) also observed a significant increase in foliar N (from initial 1.20% to 1.43%) and P (from initial 0.088% to 0.095%) for new leaves of *Rhizophora* collected from the fertilized site (after ammonium and phosphate enrichment for one year) in Australia. Similarly, Clough et al. (1983) reported a foliar N concentration of 2.04% for a mangrove forest which received long-term treated sewage effluent, compared with a value of 1.15% at nearby undisturbed control sites. The increase in plant growth and the nutrient status in plants will lead to the immobilization of a substantial amount of the added nutrients, up to at least 300 Kg N and 30 Kg P per hectare annually (Boto, 1992). In the present study, the relatively constant nutrient concentrations in leaf tissues between the treated and the control sites suggested that the increases in nutrient status of plant tissue due to sewage addition were subtle. Nevertheless, the total input of N from wastewater between December 1994 and January 1996 was only about 200 Kg, a value which was much lower than the estimated retention capacity of nitrogen by a mangrove forest (Boto, 1992). This might also explain why the differences between tissue N levels of the treated and control mangrove plants were not detected.

Effect of sewage discharge on nutrient status of mangrove sediments

Sediment nitrogen concentration
The ammonium nitrogen concentrations in mangrove sediments were similar between Sites A and B in December 1994 (before the second discharge of wastewater). The nutrient content of the sewage wastewater added in 1991–92 was a relatively small amount; thus nutrient content in the experimental site was little different to that in the control site (Wong et al., 1995). Six months after the second sewage discharge, i.e., in July 1995, the sediment NH_4^+-N in the most landward

Table 3. Effect of sewage wastewater discharge on leaf total N and organic C concentrations (% dw) of *K. candel* and *A. corniculatum* in Futian National Nature Reserve (mean and standard deviation values of three replicates are shown).

Distance from land	Year	Total N		Organic C	
		Site A	Site B	Site A	Site B
A. corniculatum					
5 m	12/1994	1.69 ± 0.07	1.57 ± 0.06	33.7 ± 2.7	33.3 ± 1.3
	12/1995	1.68 ± 0.02	1.62 ± 0.08	30.4 ± 3.6	30.4 ± 2.7
10 m	12/1994	1.69 ± 0.07	1.66 ± 0.06	30.5 ± 1.3	33.5 ± 2.1
	12/1995	1.73 ± 0.04	1.63 ± 0.06	34.2 ± 0.8	31.4 ± 2.1
K. candel					
5 m	12/1994	1.78 ± 0.12	1.71 ± 0.04	30.8 ± 2.3	33.3 ± 2.2
	12/1995	1.89 ± 0.07	1.77 ± 0.11	31.6 ± 1.8	30.6 ± 3.4
20 m	12/1994	1.61 ± 0.08	1.68 ± 0.04	34.0 ± 0.6	31.5 ± 2.1
	12/1995	1.69 ± 0.11	1.73 ± 0.05	30.2 ± 2.4	31.8 ± 3.1

region of Site A (2 m away from discharge points) was at least two times higher than that at the same distance in Site B (Figure 3). This difference was even more obvious in samples collected one year after wastewater treatment, i.e. in December 1995. These results indicate that the wastewater-borne N was indeed accumulated in mangrove sediment. However, the increase in ammonium N of Site A was only found in the sampling points close to where wastewater was discharged: differences between Sites A and B became less when sampled towards the sea. As recorded in our previous publication, the nutrient concentrations at both sites decreased from landwards to seaward regions (Tam et al., 1995; Wong et al., 1995). In addition to such spatial variation, the sediment ammonium N content also fluctuated with sampling time. In both sites, July samples tended to have higher NH_4^+-N values than those collected in December (Figure 3). This might relate to the fact that the microbial activity in the sediment was higher in summer due to the high temperature and humidity, which encourages litter decomposition, turnover of the nutrient pool and the release of inorganic nitrogen into the environment. The large spatial and temporal variability further increased the difficulty in examining the effects of sewage discharge on nutrient status of the sediment. The content of nitrite and nitrate nitrogen was low in the mangrove sediment in both sites. Nitrite was below the detection limit (<0.01 mg Kg^{-1}) and nitrate nitrogen was only one tenth of the ammonium nitrogen concentration. No difference in NO_3^--N concentration was found between Sites A and B (Figure 4).

The changes in total N concentration in sediment between Sites A and B were similar to those found in NH_4^+-N, although only a small portion of the total N was ammonium nitrogen. The total N concentration of Site A recorded in December 1994 was comparable with that in Site B, but more N was found in landward samples of Sites A than at Site B when collected in 1995, especially at the end of 1995, i.e. one year after sewage release (Figure 5). Table 4 shows that the differences in total N content between Sites A and B were small and in a random pattern in December 1994 samples. In 1995 samples, higher N was found in Site A than in Site B. Such differences declined with distance away from the landward end, and no obvious accumulation of total N was found in locations 45 m away from the discharge point in Site A (Table 4). This supports our field observation that wastewater from each sewage discharge did not flow more than 50 m away from discharge points in Site A. Most wastewater was infiltrated or absorbed by the sediments which were dry during the low ebb tide period. Sewage was well retained in the sediment before the next rising tide came into the mangrove area.

Sediment phosphorus and organic carbon concentration
The changes in sediment phosphorus and organic C concentrations between Sites A and B were similar to those of the total nitrogen. More P was found in the landward end in Site A than in Site B, and such differences diminished towards the sea (Table 4). The higher P concentrations found in the sediment collected near the discharge points indicated the ability of

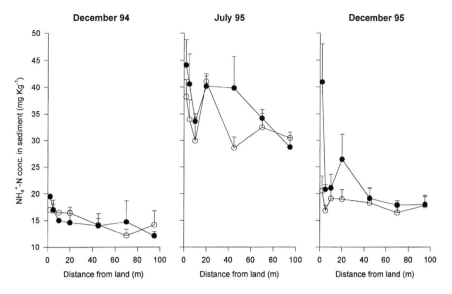

Figure 3. Changes of NH_4^+-N concentrations in sediments of Sites A and B during three sampling periods in 1994–96. (mean and standard deviation of triplicates were shown; Distance from land means distance from the discharge point at the landward edge of the Futian National Nature Reserve; ●: Site A, experimental; ○: Site B, control).

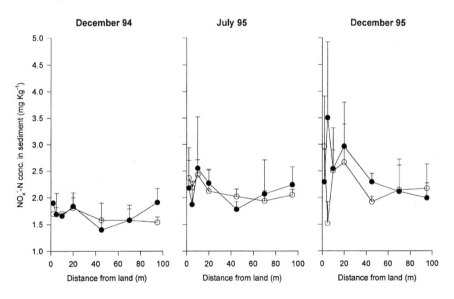

Figure 4. Changes of NO_x^--N concentrations in sediments of Sites A and B during three sampling periods in 1994–96. (Same symbol as Figure 3).

mangrove sediments in retaining P from sewage. The temporal variation in sediment P content was small but the decline of P from land to sea was very clear especially in the first 20 m (Figure 6). On the other hand, the vertical gradient of organic C, from land to sea, was not as obvious as that recorded for P (Figure 7). Addition of sewage also enhanced the organic C level in sediment of Site A.

Percentages of nutrients retained in sediments

A relatively low percentage of the wastewater-borne inorganic nitrogen (less than 30%) was retained and accumulated in the mangrove sediments (Table 5). Boto (1992) reported that inorganic N in mangrove sediments can be quite mobile, owing to the high concentrations of sodium ions which tend to 'swap'

Table 4. Changes in nutrient concentration of sediment between Sites A and B in Futian National Nature Reserve during three sampling times during the year 1995-96.

From	Changes in Total N (%)			Changes in Total P (%)			Changes in Organic C (%)		
land (m)	12/94	07/95	12/95	12/94	07/95	12/95	12/94	07/95	12/95
2	ND	0.060	0.061	ND	0.046	0.07	ND	0.197	0.789
5	0.02	0.031	0.077	0.008	−0.01	0	−0.108	0.231	0.809
10	ND	0.024	0.055	ND	0	0.02	ND	0.438	0.339
20	0.008	0.030	0.045	−0.012	0.02	0	0.273	−0.456	1.150
45	0.01	−0.01	0.021	0.007	0	0.01	−0.219	0.569	0.304
70	−0.003	0.014	−0.01	0	0.015	0.01	0.497	0.498	−0.408
95	0	0	0.018	−0.007	−0.01	0.02	0.263	0.047	ND

From land = distance from the discharge point at the landward edge of the Futian National Nature Reserve;
Changes in soil nutrients = (nutrient conc. in Site A - nutrient conc. in Site B);
Negative value means the conc. in Site A (an experimental site) was lower than that in Site B (the control); ND: not determined.

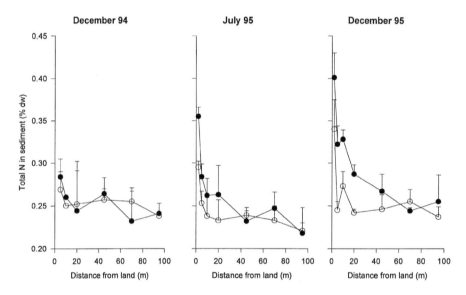

Figure 5. Changes of total N concentrations in sediments of Sites A and B during three sampling periods in 1994–96. (Same symbol as Figure 3).

cation-exchange sites. The high mobility suggests that the ammonium nitrogen added from sewage would be readily lost through leaching and tidal exchange. However, the tidal export of ammonium from the forests as measured by Boto and Wellington (1988) and in our previous study (Wong et al., 1995) showed that the loss of ammonium through tidal export was negligible. The soluble ammonium would have been readily taken up by plants. The very high densities of microbes (10^{11} cells per gram sediment, unpublished data) and their activity in the upper few centimetres of mangrove sediments might also utilize the dissolved inorganic nitrogen (Alongi, 1988). Some nitrogen might also be lost to the air *via* denitrification process (in the form of

nitrogen gas) or *via* ammonia volatilization (in the form of NH_3). Although very little information on the rate of either of these processes in mangrove sediments is available, the anaerobic and reduced condition together with the supply of the organic matter from wastewater would be in favour of denitrification activities in sediments. Corredor and Morell (1994) estimated the mangrove sediment communities were capable of depurating 10 to 15 times the nitrate added in the secondary sewage effluent *via* denitrification. The N loss from ammonia volatilization might be small in this mangrove ecosystem because the mangrove sediment has acidic pH (around 5) and the sediment pH was not altered by wastewater discharge (Wong et al., 1995).

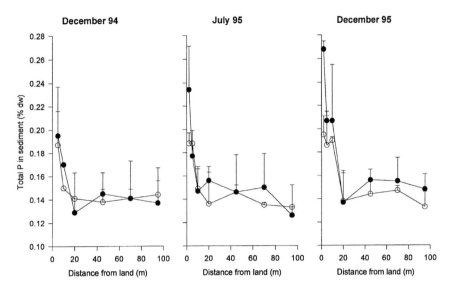

Figure 6. Changes of total P concentrations in sediments of Sites A and B during three sampling periods in 1994–96. (Same symbol as Figure 3).

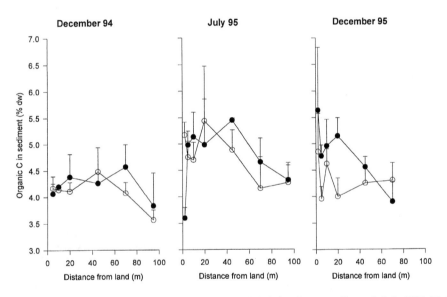

Figure 7. Changes of organic C concentrations in sediments of Sites A and B during three sampling periods in 1994–96. (Same symbol as Figure 3).

For contrast, a much higher portion of the phosphorus from wastewater (85%) was retained in the sediments (Table 5). When soluble reactive phosphate is added to soils, it is usually rapidly immobilized by adsorption reactions depending to a large extent on the soil clay mineralogy, iron content, and redox status (Boto, 1992). Surface Al.OH, Fe.OH, Al.OH$_2$ and Fe.OH$_2$ group that occur at exposed edges of clay minerals, surfaces of Fe and Al oxides on the external surfaces of clay minerals, and surfaces of amorphous Fe and Al hydroxides are important sites for sorption of phosphate anions (Mansell et al., 1985). Our previous studies revealed that mangrove sediments were effective in trapping phosphorus, which had lower mobility and solubility than nitrogen (Tam & Wong, 1993, 1995).

Table 5. Percentage of nutrients retained in mangrove sediments of Futian National Nature Reserve receiving sewage from 1994–1996.

Nutrients	Inputs[a]	Retained in sediment[b]	Balance[c]	Nutrient retention in sediment
	(Kg)	(Kg)	(Kg)	(%)
Nitrogen	195.3	52.84	142.4	27.06
Phosphorus	24.56	20.81	3.74	84.73

[a] Inputs were mainly from sewage and the quantity was calculated according to nutrient conc. × flow (the flow was 5,072 m^3).
[b] Retained in soil was calculated based on the following assumptions and calculation: soil density = 1.02 g cm^{-3} (Tam & Wong, 1995), sewage did not flow more than 50 m from the discharge point (field observation), width of site A was 10 m, soil samples were taken from surface to 20 cm depth, so that total weight of soil considered = 1.02 × 50 × 10 × 0.2 × 1000 Kg = 102,000 Kg, the average increases in total N, P and C in the first 50 m of Site A were 0.0518%, 0.0204% and 0.678%, respectively (Table 4), The quantity of total N accumulated in soil of Site A=102,000 × 0.000518=52.84 Kg, The quantity of total P accumulated in Site A=102,000 × 0.000204=20.81 Kg, The quantity of organic C accumulated in Site A=102,000 × 0.00678=691.56 Kg.
[c] Balance: difference between input and those retained in soil.

Conclusion

The present field trial carried out in Futian National Nature Reserve shows that the discharge of municipal wastewater (with a total hydraulic loading of 5072 m^3), collected from local premises, to the mangrove ecosystem did not cause any significant change in its plant community structure during the period from December 1994 to January 1996. The mangrove site receiving sewage discharge had higher concentrations of NH_4^+-N, total N, total P and organic C in the sediments than that in the control site, especially in the locations close to the discharge points. In spite of an accumulation of nutrients in the sediments, there was no increase in the nutrient concentrations of the young leaves of two dominant plant species. The differences between the treated and control sites in terms of the sediment nutrient content diminished with distance from the discharge points. The nutrient status of the sediments between the two sites became comparable when sediment samples were collected in locations more than 45 m away from the landward edge of the sites. Most of the added P from sewage wastewater (85%) was retained in the sediment but only a relatively small percentage of added N was accumulated in mangrove sediment due to its high mobility and rapid microbial transformation.

Acknowledgments

The authors would like to thank the Research Grant Council of Hong Kong, the Croucher Foundation and the Biotechnology Research Institute (BRI) of Hong Kong for their financial contributions to this project.

References

Alongi, D. M., 1988. Bacterial productivity and microbial biomass in tropical mangrove sediments. Microbiol. Ecol. 15: 59–79.

Boto, K. G. & J. T. Wellington, 1983. Nitrogen and phosphorus nutritional status of a northern Australian mangrove forest. Mar. Ecol. Prog. Ser. 11: 63–69.

Boto, K. G. & J. T. Wellington, 1988. Seasonal variations in concentrations and fluxes of dissolved organic and inorganic materials in a tropical, tidally-dominated, mangrove waterway. Mar. Ecol. Prog. Ser. 50: 151–160.

Boto, K. G., 1992. Nutrients and mangroves. In Connell, D. W. & D. W. Hawker (eds), Pollution in Tropical Aquatic Systems. CRC Press Inc., Ann Arbor, London: 129–145.

Breaux, A., S. Farber & J. Day, 1995. Using natural coastal wetlands systems for wastewater treatment: an economic benefit analysis. J. envir. Mgmt 44: 285–291.

Clough, B. F., K. G. Boto & P. M. Attiwill, 1983. Mangrove and sewage: a re-evaluation. In Teas, H. J. (ed.), Biology and Ecology of Mangroves. Tasks for Vegetation Science Series, Vol. 8, Dr W. Junk Publishers, Lancaster: 151–162.

Corredor, J. E. & J. M. Morell, 1994. Nitrate depuration of secondary sewage effluents in mangrove sediments. Estuaries 17: 295–300.

Dunbabin, J. S. & K. H. Browmer, 1992. Potential use of constructed wetlands for treatment of industrial wastewaters containing metals. Sci. tot. Envir. 111: 151–168.

Dwiedi, S. N. & K. G. Padmakumar, 1983. Ecology of a mangrove swamp near Juhu Beach, Bombay with reference to sewage pollution. In Teas, H. J. (ed.), Biology and Ecology of Mangrove. Tasks for Vegetation Science Series Vol. 8, Dr W. Junk Publishers, Lancaster: 163–170.

Gale, P. M., K. R. Reddy & D. A. Graetz, 1993. Nitrogen removal from reclaimed water applied to constructed and natural wetland microcosms. Wat. envir. Res. 65: 162–168.

Henley, D. A., 1978. An investigation of proposed effluent discharge into a tropical mangrove estuary. In Proceeding of International Conference on Water Pollution Control in Developing Countries. Sept. 1978, Thailand: 43–64.

Kadlec, J. A., 1987. Nutrient dynamics in wetlands. In Reddy, K. R. & W. H. Smith (eds), Aquatic Plants for Water Treatment and Resource Recovery. Magnolia Publishing Inc., Florida: 393–413.

Lin, P. & R. Chen, 1990. Role of mangrove in mercury cycling and removal in the Jiulong estuary. Acta Oceanol. Sinica 9: 622–624.

Munsell, R. S., P. J. McKenna, E. Flaig & M. Hall, 1985. Phosphate movement in columns of sandy soil from a wastewater irrigated site. Soil Sci. 140: 59–68.

Orson, R. A., R. L. Simpson & R. E. Good, 1992. A mechanism for the accumulation and retention of heavy metals in tidal freshwater marshes of the Upper Delaware River Estuary. Estuar. coast. Shelf Sci. 34: 171–186.

Por, F. D., 1984. The ecosystem of the mangrove forests. In Por, F. D. & I. Dor (eds), Hydrobiology of the Mangal. Dr W. Junk Publishers, The Hague: 1–14.

Rai, U. N., S. Sinha, R. D. Tripathi & P. Chandra, 1995. Wastewater treatability potential of some aquatic macrophytes: Removal of heavy metals. Ecol. Eng. 5: 5–12.

Richardson, C. J. & J. A. Davis, 1987. Natural and artificial wetland ecosystems: Ecological opportunities and limitations. In Reddy, K. R. & W. H. Smith (eds), Aquatic Plants for Water Treatment and Resource Recovery. Magnolia Publishing, Florida: 819–854.

Tam, N. F. Y., S. H. Li, C. Y. Lan, G. Z. Chen, M. S. Li & Y. S. Wong, 1995. Nutrients and heavy metal contamination of plants and sediments in Futian mangrove forest. Hydrobiologia 295: 149–158.

Tam, N. F. Y. & Y. S. Wong, 1993. Retention of nutrients and heavy metals in mangrove sediment receiving wastewater of different strengths. Envir. Technol. 14: 719-729.

Tam, N. F. Y. & Y. S. Wong, 1995. Mangrove soils as sinks for wastewater-borne pollutants. Hydrobiologia 295: 231–242.

US EPA (US Environmental Protection Agency), 1987. Report on the Use of Wetlands for Municipal Wastewater Treatment and Disposal. Office of Water, Office of Municipal Pollution Control. Submitted to Senate Quentin N. Burdick, Chairman of Committee on Environmental and Public Works. EPA 430,09-88-005.

Wong, Y. S., C. Y. Lan, G. Z. Chen, S. H. Li, X. R. Chen, Z. P. Liu & N. F. Y. Tam, 1995. Effect of wastewater discharge on nutrient contamination of mangrove soils and plants. Hydrobiologia 295: 243–254.

Zheng, F. Z., P. Lin, W. J. Zheng & Z. X. Zhuang, 1992. Study on the absorption and removal of *Kandelia candel* for pollutant cadmium. Acta Phytoecol. Geobot. Sinica 16: 220–226.

Hydrobiologia **352**: 61–66, 1997.
Y.-S. Wong & N. F.-Y. Tam (eds), Asia–Pacific Conference on Science and Management of Coastal Environment.
©1997 *Kluwer Academic Publishers. Printed in Belgium.*

Wood structure of *Aegiceras corniculatum* and its ecological adaptations to salinities

Qiang Sun & Peng Lin
Department of Biology, Xiamen University, Xiamen, China

Key words: Aegiceras corniculatum, wood structure, quantitative characters, soil salinities

Abstract

We describe the wood structure of *Aegiceras corniculatum* and its differences under various soil salinities. This species had diffuse-porous wood with poorly defined growth rings. Vessels which had single perforations occurred abundantly and in multiples and were storeyed. Intervascular pits between contiguous vessels were alternate bordered ones while half-bordered pit-pairs existed between both vessel-ray and vessel-parenchyma. Homogenous xylem rays were multiseriate and uniseriate. Fiber-tracheids with bordered pits often had thinner walls. Xylem parenchyma cells were scant and distributed diffusely and paratracheally. Differences in the structural and quantitative characters of vessels, xylem rays and fiber-tracheids under diverse soil salinities are described.

Introduction

Mangroves are communities of woody plants distributed in coastal intertidal strips of tropical and subtropical zones. They are special ecosystems appearing in vulnerable coastal strips all over the world. In recent decades, research on mangroves has been focused on the diversity of species, resources, flora, physiological ecology, matter recycling and energy flow in their ecosystems. However, as far as ecological anatomical research is concerned, little work has been done (Janssonius, H. H., 1950; Lin, P., 1988; Panshin, A. J., 1932; Tomlinson, P. B., 1986). *Aegiceras corniculatum* is a widespread mangrove species and one of major species of mangroves in Fujian Province. We studied this species at Jiulongjiang Estuary, Fujian Province, China, concentrating on its wood structure and the differences of wood structure under various salinities, so as to have a better understanding of the characters and ecological adaptations of wood structure in mangrove plants.

Materials and methods

Materials and sample plots

The materials in this paper were all the trunk wood of *Aegiceras corniculatum,* which grew naturally in five different sample plots from the northen coast of Jiulongjiang Estuary to Xiamen Harbour, Fujian Province, China. Details of research materials and sample plots could be got from Figure 1 and Table 1.

Methods

We adopted the following two methods to treat the trunk wood:

a. Sectioning (Miksche, 1976): cut the trunk wood into blocks of 1 cm^3, soften them in hot water, then section them with a sliding microtome in transverse, radial and tangential directions. Keep sections with the thickness of 15 μm.

b. Maceration (Miksche, 1976): divide the trunk wood into slivers thinner than a toothpick, boil in water, then macerate in a solution of glacial acetic acid and hydrogen peroxide. Wash them in water.

Take some of the above treated materials, stain with safranin, dehydrate through an ethyl alcohol series,

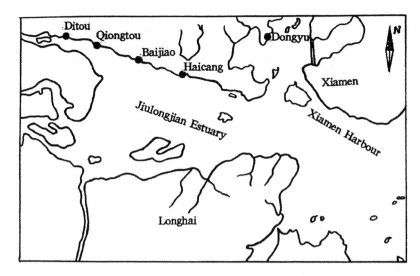

Figure 1. The map illustrating the distribution of sample plots (●).

Table 1. The heights and diameters of plant body of *Aegiceras corniculatum* and soil salinities in different sample plots

Item	Sample plot				
	Ditou	Qiongtou	Baijiao	Haicang	Dongyu
Height of plant body(m)	1.6	1.5	1.8	1.8	1.6
Diameter of trunk(cm)	3.0	2.8	3.2	3.0	2.9
Soil salinity (‰)* (20∼40 cm soil layer)	3.5	12.7	19.2	20.0	23.0

* mean values of monthly soil salinities from January 1988 to December 1992, measured by $AgNO_3$ Methods (Chen, G.Z., 1965) at ebb tide of spring tide.

then clear in xylene and mount for observations under a light microscope. Also take other of the above treated materials, dehydrate through an ethyl alcohol series to absolute alcohol, dry, then coat with gold for observation in the scanning electron microscope (SEM).

Densities of vessels and xylem rays were measured and analysed in 10 fields of view selected randomly. 30 values of other anatomical structures, also selected at random, were measured and analysed.

Results

Wood structure and its differences under various salinities

Aegiceras corniculatum had diffuse-porous wood with poorly defined growth rings. The vessels in transverse section appeared polygonal, round or elliptical, with 273–488 per square millimeter. The length and diameter of vessel elements were 159.6–205.9 μm and 30.7–39.2 μm respectively. In wood which grew in lower salinities (lower than 19‰), almost all vessels formed multiples on a large scale, with scanty solitary vessels (Figure 2); while in higher soil salinities (higher than 19‰), they mainly formed multiples of three to seven cells, but there were some solitary vessels (Figure 3). In this species, vessels exhibited a storeyed arrangement in longitudinal section (Figures 4, 5). Vessel elements had level or oblique end walls with simple perforations (Figure 6). Intervascular pits were alternate bordered ones whose apertures appeared elliptical or extendedly elliptical in transverse direction (Figures 7, 8, 9). Half-bordered pit-pairs, which were of the same size as intervascular pits, existed between both vessel-ray and vessel-parenchyma.

Xylem fibers in this species were fiber-tracheids, whose length and diameter were respectively 234.5–316.1 μm and 17.3–22.3 μm. Their pits were round or elliptical bordered ones (Figure 10). In wood which grew in lower salinities, fiber-tracheids were of thinner

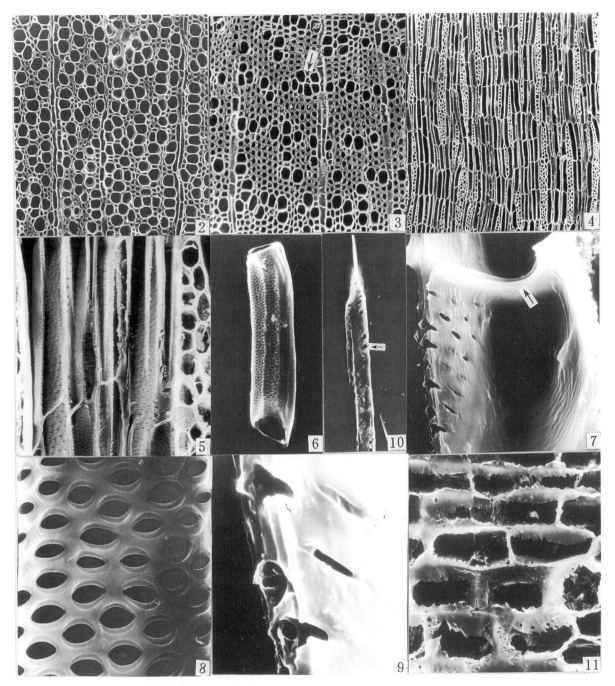

HY4076_2

Figure 2-11. SEM micrographs of wood structure of *Aegiceras corniculatum*. 2. Transverse section of wood grown in lower soil salinity, showing diffuse-porous wood and vessel multiples on a large scale × 150. 3. Transverse section of wood in higher soil salinity (20‰), showing vessel multiples consisting of fewer cells (arrowed) × 150. 4. Tangential section of wood, showing the storeyed arrangement of vessel elements and uniseriate and multiseriate rays × 69. 5. Tangential section of wood, showing the shape of the pits × 300. 6. Vessel element, showing simple perforation and alternate pits × 500. 7. Half of a simple perforation (arrowed) and elliptical bordered pits × 3000. 8. Bordered pits between contiguous vessels × 5000. 9. Surface view of a vessel element, showing alternate elliptical bordered pits × 10 000. 10. Fiber-tracheid, showing round bordered pits (arrowed) × 500. 11. Homogenous xylem ray consisting of procumbent ray cells × 700.

Table 2. Comparison of quantitative characters of secondary xylem of *Aegiceras corniculatum* under various soil salinities (mean value ± standard error)

Item		Soil salinity (‰)					
		3.5	12.7	19.2	20.0	23.0	$r^{(1)}$
Vessel element	Density (/mm²)	488 ± 14.7	413 ± 11.1	345 ± 19.9	310 ± 13.1	273 ± 18.5	−0.9849**
	Diameter (μm)	36.7 ± 1.2	39.2 ± 1.6	39.0 ± 1.2	30.7 ± 0.9	35.9 ± 0.7	−0.2908
	Length (μm)	205.9 ± 5.1	190.6 ± 4.0	172.7 ± 3.7	163.6 ± 3.6	159.6 ± 3.6	−0.9811**
Fiber-tracheid	Diameter (μm)	22.3 ± 0.6	19.9 ± 0.6	17.9 ± 0.7	17.8 ± 0.7	17.3 ± 0.5	−0.9972**
	Length (μm)	267.8 ± 6.5	234.5 ± 5.0	276.2 ± 8.5	241.8 ± 6.1	316.1 ± 4.7	0.3677
Xylem ray	Density (/mm²)	11.0 ± 0.2	9.0 ± 0.6	12.3 ± 0.6	10.5 ± 0.5	13.6 ± 0.5	0.5023
	Height (μm)	366.8 ± 26.9	323.9 ± 23.9	301.2 ± 19.4	292.6 ± 19.6	237.4 ± 21.2	−0.9337*
	Uniseriate ray ratio (%)	45.0 ± 5.1	39.5 ± 6.0	38.8 ± 3.4	36.2 ± 5.7	25.1 ± 3.7	−0.8280

(1) correlation coefficient with soil salinity; ** significant at 1% level; * significant at 5% level.

wall with many pits, while in higher salinities they were of thicker wall with fewer pits.

Xylem parenchyma cells were scanty. They were distributed paratracheally and diffusely in lower salinities, and often paratracheally in higher salinities.

Xylem rays in this species had two types: multiseriate and uniseriate. Both of them were homogenous and consisted of procumbent ray cells (Figure 11). The height and number of rays were 237.4–366.8 μm and 9–13 per square millimeter respectively. Most ray cells contained tannin and resin, but crystals were not observed.

Relationship of quantitative characters of wood structure and soil salinities

We investigated the quantitative characters of vessel, fiber-tracheid and xylem ray of the trunk wood in *Aegiceras corniculatum* which grew naturally under five different soil salinities. We also assessed correlation between them and soil salinity and made a linear regression analysis on those characters with significant correlation. The above data were handled and listed in Table 2 and made into Figure 12.

From Table 2 and Figure 12, it could be seen that these quantitative characters of wood structure varied under various soil salinities. By comparing the diameters, lengths and densities of vessel elements under various soil salinities, it was revealed that, the density (Figure 12A) and length (Figure 12B) of vessel elements were significantly correlated with soil salinity at 1% level and had negative correlation, while between diameter and soil salinity there was no significant correlation and the correlation coefficient was especially low. These meant that, with the increase of soil salinity,

density and length decreased proportionally on a large scale, but diameter changed irregularly.

As for xylem rays, only height and soil salinity had significant negative correlation at 5% level (Table 2, Figure 12D), which indicated that ray height was also in inverse proportion with soil salinity, but no significant correlation occurred between the uniseriate ratio and density of ray and soil salinity (Table 2). Furthermore, density had weaker correlation with soil salinity, showing no regular relationship.

Comparison of the relationship of diameters and lengths of fiber-tracheids and soil salinities showed that the former pair was significantly and negatively correlative at 1% level (Figure 12C), while the latter (Table 2) had no significant correlation with only irregular variation.

Discussion

The relationship between quantitative characters of wood struture and habitat factors has received considerable attention in ecological wood anatomy. However, because of the complicated roles that environment plays upon wood, the data are far from enough to offer an overall understanding of this relationship. How water affects the quantitative characters of vessels is one of the most important problems in this field. It is generally believed that the diameter and length of vessels in the xylem of xerophytes tend to be smaller while their density greater (Bass, et al., 1983; Carlquist, 1975; Fahn, 1964). Mangrove plants grow in salinized soil from which water is scarcely available. Thus, they actually grow in drought habitats which have a physiological influence on the plants. With increase in

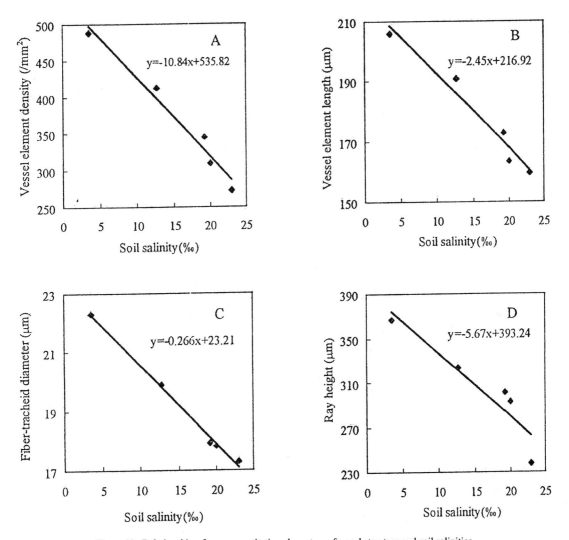

Figure 12. Relationship of some quantitative characters of wood structure and soil salinities.

soil salinity, available water becomes less and drought more serious. Panshin (1932) and Janssonious (1950) who studied the wood structure of mangrove species and non-mangrove species of Rhizophoraceae also discovered that the former had smaller vessels in diameter. But our research showed that, with increase of soil salinity, density and length of vessels decreased proportionally while diameter varied irregularly. So, in our paper, only the variation of vessel length was in accordance with the above trend, and the other two items did not have any connection with it.

Quantitative characters of xylem fibers were often affected by both genetic and habitat factors, though the former was greater than the latter in some plants (Metcalf, et al., 1983). The observation and analysis of fiber-tracheids in this paper revealed that, under lower soil salinities, fiber-tracheids had greater diameters and thinner walls with more pits, while, under higher soil salinities, they had small diameters and, in general, thicker walls with fewer pits. The fact that fiber-tracheids had smaller diameters and thicker walls was conducive to strengthening their mechanical support. It might have resulted from the long-period adaptation of this plant growing on the coast (higher soil salinities) to tidal force. The present study also showed that, there was siginificant and negative correlation between the diameter of fiber-tracheids and soil salinity, but the lengths of fiber-tracheids varied irregularly under various soil salinities, which had no direct connection with soil salinity.

Xylem rays are the radial transporting system in wood. Their quantitative characters are mainly decided by genetic factors, though their density and height are inevitably affected by environment (Metcalf, C. R. et al., 1983). This paper made clear that, with the increase of soil salinity, the height of rays in the wood of *Aegiceras corniculatum* decreased, while there was no significant correlation between the soil salinity and the uniseriate ratio and density of rays.

References

Bass, P., E. Werker & A. Fahn, 1983. Some ecological trends in vessel characters. IAWA Bull. n.s. 4: 141–159.

Carlquist, S., 1975. Ecological strategies of xylem evolution. University of California Press, Barkley, 1–31.

Chen, G. Z., 1965. Analytical chemistry of seawater. China Science Press, Beijing: 11–60.

Fahn, A., 1964. Some anatomical adaptations of desert plants. Phytomorphology 14: 93–102.

Janssonius, H. H., 1950. The vessels in the wood of Javan mangrove trees. Blumea 6: 465–469.

Lin, P., 1988. Mangrove vegetation. China Ocean Press, Beijing: 16–27.

Metcalf, C. R. & L. Chalk, 1983. Anatomy of the dicotyledons (2nd ed.) Clarendon Press, Oxford, Volume II: 1–156.

Miksche, J. P., 1976. Botanical microtechnique cytochemistry. The Iowa State University Press, Iowa: 54–129.

Panshin, A. J., 1932. An anatomical study of the woods of the Philippine mangrove swamps. Philipp. J. Sci. 48: 143–205.

Tomlinson, P. B., 1986. The botany of mangroves. Cambridge University Press, Cambridge: 89–95.

Hydrobiologia **352**: 67–75, 1997.
Y.-S. Wong & N. F.-Y. Tam (eds), Asia–Pacific Conference on Science and Management of Coastal Environment.
©1997 *Kluwer Academic Publishers. Printed in Belgium.*

Accumulation and distribution of heavy metals in a simulated mangrove system treated with sewage

Nora F. Y. Tam[1]* & Yuk-Shan Wong[2]

[1]*Department of Biology and Chemistry, City University of Hong Kong, Hong Kong*
[2]*Research Centre/Biology Department, The Hong Kong University of Science & Technology, Hong Kong*
(*Author for correspondence; tel: (852) 27887793; fax: (852) 27887406; e-mail: bhntam@cityu.edu.hk)*

Key words: heavy metal, accumulation, mangrove, *Kandelia*, sewage

Abstract

Constructed tide tanks were used to examine the accumulation and distribution of heavy metals in various components of a simulated mangrove ecosystem. Young *Kandelia candel* plants grown in mangrove soils were irrigated with wastewater of various strengths twice a week for a period of one year. The amounts of heavy metals released *via* tidal water and leaf litter were monitored at regular time intervals. The quantities of heavy metals retained in mangrove soil and various plant parts were also determined. Results show that most heavy metals from wastewater were retained in soils with little being uptake by plants or released into tidal seawater. However, the amounts of metals retained in plants on a per unit dry weight base were higher than those in soils as the biomass production from the young mangrove plants was much smaller when compared to the vast quantity of soils used in this study. A significantly higher heavy metal content was found in roots than in the aerial parts of the mangrove plant, indicating that the roots act as a barrier for metal translocation and protect the sensitive parts of the plant from metal contamination. In both soil and plant, concentrations of Zn, Cd, Pb and Ni increased with the strengths of wastewater, although the bioaccumulation factors for these metals decreased when wastewater strengths increased. These results suggest that the mangrove soil component has a large capacity to retain heavy metals, and the role of mangrove plants in retaining metals will depend on plant age and their biomass production.

Introduction

Mangrove ecosystems are one of the major types of natural wetlands in tropical and subtropical regions, flooded by fresh river water as well as by salty oceanic water. Similar to other estuarine zones, mangrove ecosystems also receive a large amount of waste from their related drainage and rivers and have become a massive pollution sink. The use of a mangrove ecosystem, the same as other natural wetlands, as an alternative low cost sewage treatment facility has been proposed by a number of researchers (Henley, 1978; Clough et al., 1983; Richardson & Davis, 1987; Tam & Wong, 1993; Breaux & Day, 1994; Corredor & Morell, 1994, Tam et al., 1995; Wong et al., 1995) especially in coastal regions with pressing needs for wastewater treatment. In recent decades, due to the rapid expansion of population and industrial activities of the Shenzhen Special Economic Zone, the People's Republic of China, the amount of pollutant from industrial and municipal waste discharged into the received streams has greatly increased and caused serious pollution problems to surrounding waters. Investigations on the feasibility of using natural mangrove wetland to treat wastewater containing heavy metals and the mechanisms of trapping these pollutants are then of great concern.

The effectiveness of a wetland system to remove the input pollutants is highly dependent on the chemical, biological and/or physical processes, and the entire soil-plant-water system is important in the reduction of pollutants from wastewater (Dunbabin & Bowmer, 1992; Gale et al., 1993). The performance of a natural wetland wastewater treatment system therefore depends very much on the wetland characteristics

Table 1. Characteristics of mangrove soils collected from Futian Mangrove Swamp.

pH	3.65
Particle size (%)	
0.05–1.00 mm	33.64
0.01–0.05 mm	28.09
0.005–0.01 mm	13.93
0.001–0.005 mm	1.04
< 0.001 mm	23.29
Total dissolved salts (%)	1.11
Organic matter (%)	2.51
Total N (%)	0.088
NH_4^+-N (mg Kg^{-1})	11.04
NO_3^--N (mg Kg^{-1})	0.64
Total P (%)	0.026
Total Zn (mg Kg^{-1})	7.26
Total Cd (mg Kg^{-1})	0.77
Total Pb (mg Kg^{-1})	14.97
Total Ni (mg Kg^{-1})	2.52

which are extremely variable. It is difficult, if not impossible, to translate results from one geographical area to another, or from one type of wetland to another (Trattner & Woods, 1989). Despite their significance in purifying wastewater, natural wetlands in many countries including the United States are legally limited to providing only tertiary treatment of secondary waste (Breaux & Day, 1994). Most studies were focused on the removal mechanisms of suspended solids, organic matter and nutrients from domestic or livestock wastewater by wetlands (Corredor & Morell, 1994). Relatively little information is available on the capacity of wetlands in retaining heavy metals from wastewater. The present study aims to understand the retention and distribution of sewage-borne heavy metals in different components of a simulated mangrove ecosystem. The partition of heavy metals in various parts of young *Kandelia candel* plants irrigated with wastewater of different heavy metal content was also determined.

Materials and methods

Experimental setups

One year old *Kandelia candel* plants were transplanted from Futian Mangrove Swamp, Shenzhen Special Economic Zone, the People's Republic of China, and grown in twelve PVC tide tanks (each with a dimen-

sion of $0.7 \times 0.5 \times 0.4$ m^3) filled with mangrove soils collected from the same swamp (Table 1). The tide tanks were flooded with artificial seawater twice a day to simulate the natural tidal regime. A total of 15 young plants was acclimatised in each tank for nine months prior to treatment with wastewater. At the beginning of wastewater irrigation, the plant had an average height of 17.7 ± 1.6 cm, stem diameter of 0.61 ± 0.05 cm, leaf number of 10 ± 2. The total biomass of leaf, stem, hypocotyl and root per tide tank were 23.1, 35.5, 95.1 and 32.1 g dry weight, respectively.

Wastewater with three strengths, namely NW, FW and TW were applied to the tanks twice a week with a hydraulic loading of 1.75 l per tank per irrigation during the low tide (the exposure) period. NW was the synthetic wastewater with nutrient and organic matter concentrations similar to domestic sewage and contained 5 mg l^{-1} Zn, 0.1 mg l^{-1} Cd, 1 mg l^{-1} Pb and 1 mg l^{-1} Ni. FW and TW had five and 10 times of nutrients and heavy metals of the NW, respectively. In between two wastewater irrigations, the tanks were flooded with artificial seawater (prepared by dissolving commercial oceanic salts in deionized water with salinity between 1.2–1.5%) according to the tidal regime, i.e. the tanks were flooded with seawater at 10 am to 2 pm and 10 pm to 2 am which gave 16 hours exposure everyday. Tide tanks irrigated with artificial seawater were used as the control. Artificial seawater from each tank was recycled and reused for two months, then replaced by freshly prepared seawater. All treatments were in triplicates. The experiment lasted for one year under greenhouse conditions.

Analysis of seawater, soil and plant samples

The artificial seawater (acted as flushing tidal water) was collected at the end of 2-months cycle, and the concentrations of total Zn, Cd, Pb and Ni were measured by digesting the samples in nitric acid and hydrogen peroxide, then determined by Inductive Couple Plasma (ICP) technique. Leaf litter samples were collected monthly from each tide tank. Their fresh and oven-dried weight were measured. The samples were digested by conc. nitric acid at 160 °C, followed by analysis for total Zn, Cd, Pb and Ni by ICP. At the end of one-year irrigation, the soils were collected and the plants harvested, then divided into root, stem, leaf and hypocotyl portions. The concentrations of Zn, Cd, Pb and Ni in these samples were determined using the same technique as described above. All data were expressed in terms of 105 °C oven-dried weight base.

Results and discussion

Accumulation of heavy metals in soil component

The mangrove soils collected from Futian Mangrove Swamp were acidic and contained relatively low concentrations of organic matter, nutrients and heavy metals (Table 1). Addition of wastewater to the tide tanks significantly increased the metal levels in the soil component. Figure 1 exhibits that the concentrations of heavy metals in soils increased linearly with that added into the wastewater, and the slopes of the regression lines for the four metals were similar (ranged from 1.59 to 1.69). The linear relationship suggests that the heavy metal binding capacity of this mangrove soil was not exceeded even when it was treated with very strong wastewater. These results, similar to that reported by Banus et al. (1975) and Simpson et al. (1983), reveal that the mangrove soil had a large capacity in retaining heavy metals discharged from wastewater.

Distribution of heavy metals in various plant components

Increased levels of heavy metals in wastewater input generally resulted in higher heavy metal concentrations in mangrove plants, but the degree of increase in different plant parts varied (Figure 2). The heavy metal concentrations in roots were significantly higher than those found in the other plant parts especially for Pb. At low to medium heavy metal inputs (the NW and FW treatments), most added heavy metals were accumulated in roots, with little transportation to the leaf and stem portions. The roots have been known as good absorptive sponge to heavy metal in soil and water. Metals absorbed or adsorbed by roots are often bound with the cell wall material or other macromolecules to prevent them from translocation to the sensitive plant parts. This also explained why growth of the mangrove plants were not inhibited even treated with very strong wastewater (Chen et al., 1995). When the heavy metal concentration in wastewater increased, the defensive mechanism of roots (as a barrier) might have been exhausted and more metals were found in either leaves (e.g. Ni) or stems (e.g. Zn) (Figure 2). The leaf Ni concentrations in plants treated with TW were similar to that in roots, whereas the stem Zn content was comparable to that found in roots when plants were treated with FW and TW. It is clear that under high metal inputs (e.g. TW treatments), the distribution of heavy metals in different plant parts varied between

Table 2. Bioaccumulation factors for Zn, Cd, Ni and Pb in tide tanks by *K. candel* plants (an average of three tide tanks are shown).

Metals	Treatment	Root	Stem	Leaf	Hypocotyl
Zn	Control	2887	1273	896	957
	NW	974	1259	539	326
	FW	466	467	139	216
	TW	237	217	65	59
Cd	Control	695	85	101	136
	NW	748	78	108	127
	FW	466	81	119	103
	TW	346	41	97	57
Ni	Control	308	99	114	113
	NW	230	55	82	42
	FW	129	39	60	39
	TW	118	16	118	16
Pb	Control	251	12	10	11
	NW	241	12	12	9
	FW	287	10	10	54
	TW	199	82	15	71

Bioaccumulation factor = (heavy metal conc. in plant / conc. in soil) $\times$ 100%

metal species. The metal mobility within the plant was in the descending order of Zn>Ni>Cd>Pb, therefore excess Zn and Ni were more likely to move up from roots to the aerial parts of the treated plants than Cd and Pb. In general, the relative ratio of heavy metals in root to leaf decreased as input heavy metals increased (Figure 3), since excess heavy metals might have been translocated to the aboveground portions.

Table 2 shows that in all heavy metals examined, bioaccumulation factors were higher in the roots than those in the aerial parts. Similar findings were reported by Zaranyika et al. (1994) for Ni, Zn and Cd uptake in water hyacinths. Bioaccumulation factors increased in the order of Zn>Cd>Ni>Pb both in roots and the aerial portions, probably because Zn is an essential element for plant growth and metabolism. The bioaccumulation factors decreased significantly when heavy metal concentrations in input wastewater increased (Table 2), indicating that the soil was very effective in binding heavy metals and the uptake by plants decreased with increasing input concentrations.

Concentration of heavy metals in litter

The heavy metal concentrations of leaf litter fluctuated randomly throughout the one-year study, with lower heavy metal content in June (Figure 4). The mean val-

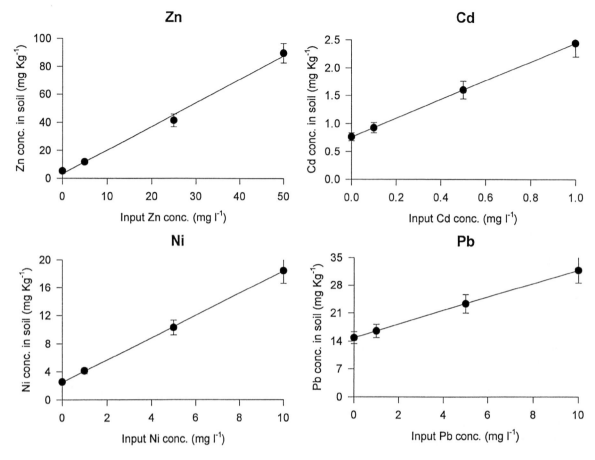

Figure 1. Relationship between concentrations of heavy metals in soil and in wastewater added (average and standard deviation values of three replicate tide tanks are shown; input concentration is the concentration of heavy metals in wastewater added to each tide tank).

Table 3. Concentrations of heavy metals in leaf litter (mg Kg^{-1} dry weight; Mean and standard deviation values of 12 months' data are shown).

Metals	Control	NW	FW	TW
Zn	65.7 ± 12.5	91.6 ± 31.1	76.7 ± 19.5	86.7 ± 18.7
Cd	1.00 ± 0.28	1.29 ± 0.29	1.36 ± 0.51	1.62 ± 0.47
Ni	4.49 ± 1.62	6.49 ± 2.56	7.34 ± 2.53	9.23 ± 3.22
Pb	4.61 ± 3.37	11.2 ± 8.2	15.3 ± 13.4	23.7 ± 19.5

ues over the 12 months of wastewater treatment are shown in Table 3. The concentrations of Cd, Ni and Pb in litter increased with the strengths of wastewater, with highest metal content found in litter collected from tanks treated with TW. This suggests that some of the heavy metals in the wastewater might have been absorbed by litter *via* the soil-plant route. Similar to those reported in the literature (Banus et al., 1975; Simpson et al., 1983; Dubinski et al., 1986), heavy

metal concentrations in litter were generally higher than the aerial plant parts but were significantly lower than those found in the roots (Figure 2).

Distribution of heavy metals in different components of a mangrove ecosystem

The simulated mangrove tide tank system was very effective in removing heavy metals from wastewater as

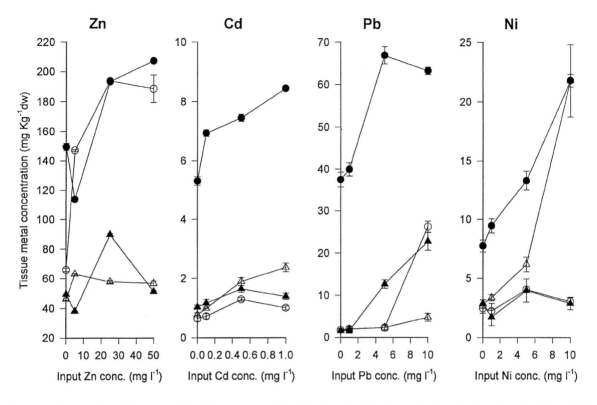

Figure 2. Concentrations of heavy metals in different parts of *Kandelia* plants treated with wastewater of various strengths (●: root; ▲: hypocotyl; ○: stem; △: leaf; average and standard deviation values of three replicate tide tanks with 15 plants each are shown).

Table 4. Distribution of heavy metals in different components of a simulated mangrove ecosystem. (Input was based on one year irrigation of wastewater; while tidal water output was from 6 successive collection of all tidal water, soil and plant output was from all soil and plant samples harvested from each tide tank at the end of one year study, and litter output was from summation of 12 months' data; all these output data are average values of three replicates).

Metal	Treatment	Input from wastewater (g)	Total Output Tidal water (g)	Soil (g)	(g Kg⁻¹)	Plant (g)	(g Kg⁻¹)	Litter (g)	(g Kg⁻¹)
Zn	NW	2.753	1.400	1.261	0.013	0.034	0.136	0.007	0.089
	FW	12.833	1.814	9.786	0.098	0.049	0.196	0.005	0.064
	TW	25.433	1.782	22.750	0.228	0.050	0.200	0.008	0.102
Cd ($\times 10^{-3}$)	NW	50.58	4.27	44.56	0.045	0.56	2.245	0.09	1.145
	FW	252.18	12.29	237.11	2.371	0.86	3.448	0.09	1.145
	TW	504.18	16.77	477.92	4.779	0.81	3.248	0.14	1.781
Pb	NW	0.505	0.044	0.452	0.005	0.002	0.008	0.001	0.013
	FW	2.521	0.102	2.376	0.024	0.008	0.032	0.001	0.013
	TW	5.041	0.124	4.815	0.048	0.016	0.064	0.002	0.025
Ni	NW	0.508	0.050	0.453	0.005	0.001	0.004	0.001	0.013
	FW	2.524	0.192	2.224	0.022	0.003	0.012	0.001	0.013
	TW	5.043	0.266	4.559	0.046	0.005	0.020	0.001	0.013

72

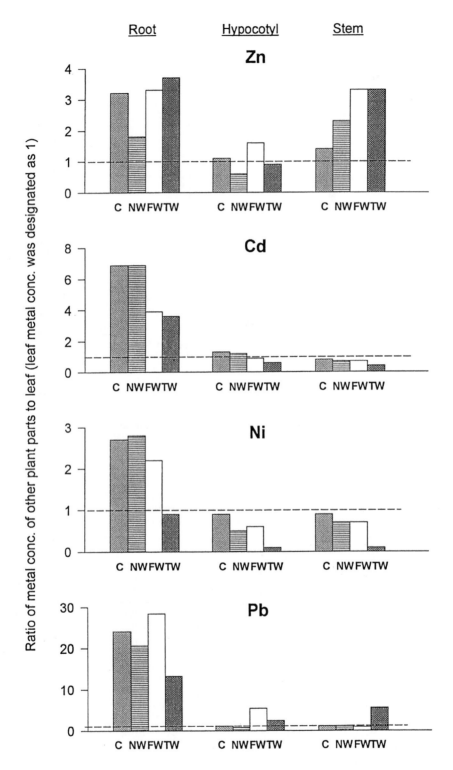

Figure 3. Ratio of heavy metal concentrations in different parts of *K. candel* plants (C: control; NW: normal municipal wastewater; FW: 5 times of NW; TW: 10 times of NW; average values of three replicate tide tanks with 15 plants each are shown; the horizontal line represents the leaf heavy metal concentrations being set to 1).

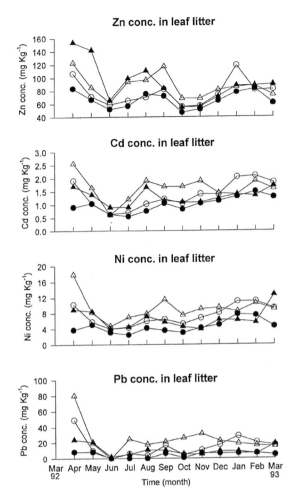

Figure 4. Changes of heavy metal concentrations in leaf litter throughout the one-year study (●: control; ▲: NW; ○: FW; △: TW; average values of three replicate tide tanks are shown).

only a very small amount of heavy metal was released into the flushing seawater (Table 4). Most heavy metals discharged from wastewater were retained in the soil component of this simulated mangrove ecosystem (Table 4), and in most treatments (Table 5), more than 85% removal of heavy metals (except Zn) by soil was found. Although a slightly higher amount of heavy metal was found in tidal water collected from tanks treated with strong wastewater (TW) (Table 4), the overall percentages of heavy metals released into tidal water were lower in TW than those in FW and NW (Table 5). It is clear that even in tide tanks treated with very strong wastewater (TW), heavy metals were still trapped by mangrove soils and the problem of released back to the surrounding water *via* tidal flushing was insignificant. The removal of heavy metals by plant

uptake was small and often less than 2% (Table 5), although plants are the most obvious biological component of the wetland ecosystem. The present study shows that the return of heavy metal *via* litter decomposition was also found to be insignificant as the total quantity of heavy metals in the litter was quite small. These findings, similar to that reported by Dubinski et al. (1986), suggest that the vegetation and litter only played a minor role in the retention of heavy metals present in wastewater.

The role of marsh plants in wetland wastewater treatment has always been debatable and controversial. Many study reported that marsh plants are capable of absorbing heavy metals from the environment (Simpson et al., 1983; Orson et al., 1992; Rai et al., 1995). The plants normally remove heavy metals by surface adsorption or absorption plus incorporation into their own system, or store it in a bound form. However, there are reports in the literature stating that uptake of pollutants by wetland vegetation cannot by itself account for the high pollutant removal efficiencies often observed at the high loading rates (Nichols, 1983). The metal uptake by plants depends upon many factors, including pH and redox changes, species of the plants and changes in seasonal growth rates, as well as the age of the plant. It has been reported that live plants accumulate metals throughout the growing season in tidal marshes, and the litter acts as a short-term sink (Whigham et al., 1989). In the present study, the mangrove plants were relatively small, the average tree height at the end of the 1-year study was around 30 cm and the biomass of roots, hypocotyl, stem and leaves per tide tank were 43.6, 98.4, 65.5 and 41.9 g dried weight, respectively. Similarly, the total biomass of litter produced per tide tank was 78.6 g dry weight. Such weight, compared to the amount of soils in each tank (100 kg), were indeed insignificant. However, the quantities of heavy metal retained per unit weight of plant materials were higher than those retained in soil on a per unit weight base especially in NW and FW treatments (Table 4). These findings suggest that the mangrove plants might play a more significant role in retaining heavy metals under field conditions as the mangrove plants would be taller and produced larger biomass than the young plants used in this study. Moreover, the plant growth and uptake of heavy metals are affected by various environmental conditions, in particular, light intensity and duration. The light intensity in natural tropical mangrove forests is generally higher than the greenhouse conditions employing in this study, i.e. the field condition will support faster plant

Table 5. Percentage distribution of wastewater-borne heavy metals in different components of the simulated mangrove ecosystem.

Treatment	Component	Heavy Metals (% retention or release)			
		Zn	Cd	Pb	Ni
NW	Tidal water	50.9	9.8	8.7	9.8
	Plant	1.2	2.0	0.4	0.2
	Soil	45.8	88.2	89.5	89.2
FW	Tidal water	14.1	4.8	4.1	7.6
	Plant	0.4	0.4	0.3	0.1
	Soil	76.3	94.1	94.3	88.1
TW	Tidal water	7.0	3.4	2.5	5.3
	Plant	0.2	0.2	0.3	0.1
	Soil	89.5	94.8	95.5	90.4

growth with more biomass production and more metal uptake. Therefore, the actual role of mangrove plants in taking up heavy metals from wastewater discharge under a natural mangrove forest situation required a more in-depth study.

Moreover, the metal solubility and bio-availability are directly affected by the pH and redox potential of the soils. Gambrell et al. (1980) reported that cadmium uptake by marsh plants was higher under acidic and oxidizing condition. In this study, the mangrove soils were found to have a very acidic pH (around 3.5). Treatment with wastewater or seawater did not cause any significant change in soil pH, and soil pH only fluctuated between 3.7 to 4.4 throughout the study. The low pH could increase the solubility and bio-availability of heavy metals in soil and enhance the plant uptake. On the other hand, the reduced condition of the mangrove soil (always at negative redox potential) might reduce the solubility and limit the bio-availability of heavy metals to plant uptake. Tam & Wong (1993 & 1995) found that most of the heavy metals added to mangrove soil from discharged sewage were bound either by adsorption to ion-exchange sites, incorporation into lattice structure or by precipitation as insoluble sulphide, and only a very small portion was bio-available. This may also explain why the amounts of heavy metals retained in plant tissues of this simulated mangrove ecosystem was relatively low.

Conclusion

The present greenhouse experiment shows that the mangrove wetland ecosystem has a large capacity to remove heavy metals from wastewater. Soil is the most effective component in retaining heavy metals from wastewater. Although the total amounts of heavy metals retained in mangrove plants were lower than those in soil, heavy metals did accumulate in plants, especially in roots. Excess metals were translocated from root to stem, then to leaf, and the degree of upward movement depended on the mobility of heavy metals. The role of mangrove plants in immobilizing wastewater-borne heavy metals related to plant age, growth and biomass production. The heavy metal concentrations in various plant parts and leaf litter increased with the strengths of wastewater.

Acknowledgments

The authors would like to thank the technicians of Futian Nature Reserve, Shenzhen Special Economic Zone and Zhongshan University, Guangzhou, the PRC for their assistance in field sampling and laboratory analyses. We would also like to express our gratitude to the Croucher Foundation, Research Grant Council and the Hong Kong University of Science and Technology for their financial support.

References

Banus, M., I. Valiela & J. M. Teal, 1975. Pb, Zn and Cd budgets in experimentally enriched salt marsh ecosystems. Estuar. coast. mar. Sci. 3: 421–430.

Breaux, A. M. & J. W. Day Jr, 1994. Policy considerations for wetland wastewater treatment in the coastal zone: a case study for Louisiana. Coast. Mgmt 22: 285–307.

Chen, G. Z., S. Y. Miao, N. F. Y. Tam, Y. S. Wong, S. H. Li & C. Y. Lan, 1995. Effect of synthetic wastewater on young *Kandelia candel* plants growing under greenhouse conditions. Hydrobiologia 295: 263–274.

Clough, B. F., K. G. Boto & P. M. Attiwill, 1983. Mangrove and sewage: a re-evaluation. In Teas, H. J. (ed.), Biology and Ecology of Mangroves. Tasks for Vegetation Science Series, Vol. 8, Dr W. Junk Publishers, Lancaster: 151–162.

Corredor, J. E. & J. M. Morell, 1994. Nitrate depuration of secondary sewage effluents in mangrove sediments. Estuaries 17: 295–300.

Dubinski, B. J., R. L. Simpson & R. E. Good, 1986. The retention of heavy metals in sewage sludge applied to a freshwater tidal wetland. Estuaries 9: 102–111.

Dunbabin, J. S. & K. H. Bowmer, 1992. Potential use of constructed wetland for treatment of industrial wastewaters containing metals. Sci. tot. Envir. 111: 151–168.

Gale, P. M., K. R. Reddy & D. A. Graetz, 1993. Nitrogen removal from reclaimed water applied to constructed and natural wetland microcosms. Wat. envir. Res. 65: 162–168.

Gambrell, R. P., V. Collard & W. H. Patrick Jr., 1980. Cadmium uptake by marsh plants as affected by sediment physicochemical conditions. In Baker, R. A. (ed.), Contaminants and Sediments, Vol. 2. Ann Arbor Science Publishers, Inc., Michigan: 425–443.

Henley, D. A., 1978. An investigation of proposed effluent discharge into a tropical mangrove estuary. In Proceedings of International Conference on Water Pollution Control in Developing Countries. Sept. 1978, Thailand: 43–64.

Nichols, S., 1983. Capacity of natural wetlands to remove nutrients from wastewater. J. Wat. Pollut. Cont. Fed. 55: 495–505.

Orson, R. A., R. L. Simpson & R. E. Good, 1992. A mechanism for the accumulation and retention of heavy metals in tidal freshwater marshes of the Upper Delaware River Estuary. Estuar. coast. Shelf Sci. 34: 171–186.

Rai, U. N., S. Sinha, R. D. Tripathi & P. Chandra, 1995. Wastewater treatability potential of some aquatic macrophytes: Removal of heavy metals. Ecol. Eng. 5: 5–12.

Richardson, C. J. & J. A. Davis, 1987. Natural and artificial wetland ecosystems: Ecological opportunities and limitations. In Reddy, K. R. & W. H. Smith (eds), Aquatic Plants for Water Treatment and Resource Recovery. Magnolia Publishing, Florida: 819–854.

Simpson, R. L., R. E. Good, R. Walker & B. R. Frasco, 1983. The role of Delaware River freshwater tidal wetlands in the retention of nutrients and heavy metals. J. envir. Qual. 12: 41–48.

Tam, N. F. Y. & Y. S. Wong, 1993. Retention of nutrients and heavy metals in mangrove sediment receiving wastewater of different strengths. Envir. Technol. 14: 719–729.

Tam, N. F. Y. & Y. S. Wong, 1995. Mangrove soils as sinks for wastewater-borne pollutants. Hydrobiologia 295: 231–242.

Trattner, R. B. & S. J. E. Woods, 1989. The use of wetlands systems for municipal wastewater treatment. In Cheremisinoff, P. N. (ed.), Encyclopedia of Environmental Control Technology Vol. 3, Wastewater Treatment Technology. Gulf Publishing Co., Houston: 591–622.

Whigham, D. F., R. L. Simpson, R. E. Good & F. A. Sickels, 1989. Decomposition and nutrient-metal dynamics of litter in freshwater tidal wetlands. In Sharitz, R. & J. W. Gibbons (eds), Freshwater Wetlands and Wildlife. Office of Science and Technical Information DOE-CONS. 860326 USDOE, Washington, D.C.: 167–188.

Wong, Y. S., C. Y. Lan, G. Z. Chen, S. H. Li, X. R. Chen, Z. P. Liu & N. F. Y. Tam, 1995. Effect of wastewater discharge on nutrient contamination of mangrove soils and plants. Hydrobiologia 295: 243–254.

Zaranyika, M. F., F. Mutoko & H. Murahwa, 1994. Uptake of Zn, Co, Fe and Cr by water hyacinth (*Eichhornia crassipes*) in Lake Chivero, Zimbabwe. Sci. tot. Envir. 153: 117–121.

Hydrobiologia **352**: 77–87, 1997.
Y.-S. Wong & N. F.-Y. Tam (eds), Asia–Pacific Conference on Science and Management of Coastal Environment.
©1997 *Kluwer Academic Publishers. Printed in Belgium.*

Incidence of heavy metals in the mangrove flora and sediments in Kerala, India

George Thomas & Tresa V. Fernandez
Department of Aquatic Biology & Fisheries, University of Kerala, Trivandrum, 695007, India

Key words: Heavy metals, mangrove flora, sediments, Kerala coast

Abstract

The mangroves of Kerala on the south-west coast of India are fast disappearing due to land reclamation and other anthropogenic disturbances. There are very few ecosystem level studies made in these much threatened biotopes in Kerala. The present study involves the measurement of heavy metals in the mangrove flora and sediments of three mangrove habitats along the Kerala coast. Sampling was carried out for a period of one year at bi-monthly intervals, with concentrations of Fe, Mn, Cu, Zn, Pb and Co analysed using atomic absorption spectrophotometry. An appreciable variation was observed in metal concentrations in different mangrove species. Cu, Zn and Pb were found to be in higher concentrations in *Avicennia officinalis* whereas higher levels of Fe, Mn and Co were observed in the species *Barringtonia racemosa*. The analysis of heavy metals indicated a high level of metal pollutants such as Fe, Cu, Zn and Pb in the mangrove habitats of Quilon and Veli compared to the relatively uncontaminated areas of Kumarakom.

Introduction

In recent years it has become important to estimate the heavy metal concentrations in organisms which act as bioaccumulators, in sediment and in overlying water in freshwater, estuarine and marine environments (Fernandez & Jones, 1987) because many of these organisms are indicators of metal contamination. Many marine organisms are known to accumulate and concentrate metals in relatively higher levels than the surrounding environment. The seaweeds are reported to concentrate heavy metals several thousand times more than their concentrations in seawater (Preston et al., 1972; Yamamoto, 1972). Red and brown algae are more efficient in metal accumulation than the green algae (Fernandez & George Thomas, 1995; Fernandez et al., 1995).

Many mangrove species are reported to contain very high concentrations of certain heavy metals such as Fe and Mn (Untawale et al., 1980). The toxic effects of these metals are reduced due to the action of chelating substances present in these plants (Untawale et al., 1980). On the other hand, mangrove sediments are considered as both a sink and source for trace metals (Harbison, 1986). Therefore, high metal levels may be expected in mangrove areas, and the study of metal contamination in these systems is of great relevance in pollution assessment in the tropics (Lacerda et al., 1987).

Few studies have been conducted in India regarding the heavy metal pollution in the mangrove environment (Untawale et al., 1980; Seralathan, 1987; Chakrabarti et al., 1993). The present study is aimed at assessing heavy metal incidence and distribution in the flora and sediments of three selected mangrove ecosystems of Kerala and to find out possible relationships between metal concentrations in plants and sediments.

Study sites

The three mangrove areas studied are located along the south-west coast of India in Kerala district (Figure 1). Kumarakom (9°37′ N Lat., 76°26′E. Long.) is a famous bird sanctuary located on the banks of Vembanad estuary, the largest estuarine system of Kerala. This area can be considered as uncontaminated because no major industrial units are located in the near vicinity of this location. The mangroves studied at Quilon (8°54′ N Lat., 76°36′ E Long.) were a small patch in one of the smaller creeks of the Ashtamudi estuary. The area receives considerable amounts of metal inputs from nearby industries such as steel-wire plant, aluminium industries and automobile work shop which discharge their effluents directly into the estuary. The southern most site was at Veli (08°28′ N Lat., 76°57′ E Long.), one of the smaller estuaries of Kerala (∼ 3 ha). Veli estuarine system is under the threat of effluents from nearby titanium factory and frequent sewage inflow from the urban areas. In all the three ecosystems, mangroves were found fringing the banks of backwaters.

Materials and methods

Analysis of Fe, Mn, Cu, Zn, Pb and Co in the mangrove plants and sediment were carried out adopting standard methods (EPA, 1972 and APHA, 1980) using atomic absorption spectrophotometer (Model 551, Instrumentation Laboratory, USA). The mangrove species analysed were *Avicennia officinalis*, *Acanthus ilicifolius*, *Bruguiera gymnorrhiza*, *Sonneratia caseolaris* and *Barringtonia racemosa*. Twigs from the different mangrove species were collected from Kumarakom, Quilon and Veli mangrove regions at bi-monthly intervals from October 1988 to September 1989. They were washed carefully, using fresh followed by distilled water to remove any adhering dirt or dust particles. Small twigs (including leaves and small stems upto 15 cm from the tip) of various plant species were oven-dried at 70 ± 0.5 °C for constant weight. Pre-weighed samples (in triplicate) were digested with a mixture of nitric acid, hydrochloric acid and perchloric acid. Digested samples were filtered using Whatman No. 42 filter paper and made up to 25 ml using double distilled water for further analysis using AAS. Reagent blanks were also prepared for background corrections.

Sediment samples were collected using a PVC corer of 5 cm diameter and 15 cm depth from all the mangrove locations. Three replicate samples were pooled and oven-dried at 70 ± 0.5°C and digested with nitric acid-perchloric acid mixture over low heat. Other procedures are same as for plant samples described above. Sediment texture was determined by pipette method (Carver, 1971) and organic carbon was estimated by the method of Walkley & Black (1934).

Results

Heavy metal concentrations in the mangrove flora

The annual ranges, mean and standard deviation of heavy metal concentrations in the mangrove flora and sediments are given in Table 1. Figures 2 to 7 represents the seasonal (bi-monthly) variation of heavy metals in various species of mangrove plants.

Iron
Iron content was found to be very high in *Barringtonia racemosa* and *Sonneratia caseolaris*. Overall concentrations of Fe in five species tested ranged from 140 to 6050 μg g^{-1} air dry wt (Table 1). Respective ranges in each species were 160 to 840 in *Avicennia officinalis*, 140 to 1640 in *Acanthus ilicifolius*, 440 to 1460 in *Bruguiera gymnorrhiza*, 1200 to 5940 in *Sonneratia caseolaris* and 2140 to 6050 in *Barringtonia racemosa*. Although seasonal variation was not much pronounced, a slight increase in concentrations could be observed during the post-monson months (Figure 2).

Manganese
Comparatively high concentrations of manganese were observed in *B. racemosa* and *S. caseolaris*. Mn concentrations were found to be lowest in *A. ilicifolius*. The overall range of Mn concentration was between 46 and 2472 μg g^{-1} air dry wt. The individual ranges were 475 to 680 in *A. officinalis*, 46 to 163 in *A. ilicifolius*, 339 to 599 in *B. gymnorrhiza*, 258 to 1793 in *S. caseolaris* and 628 to 2472 in *B. racemosa*. The concentrations were slightly higher during the monsoon season in *Barringtonia* and *Sonneratia* (Figure 3).

Copper
Copper was found to be higher in *Avicennia officinalis* than the other mangrove species studied with a peak value of 190.2 μg g^{-1} air dry wt. In general the concentrations of Cu ranged from 6.1 to 190.2 μg g^{-1} air dry wt in the different mangrove species tested. Individual ranges were from 29.8 to 190.2 in *A. officinalis*, 6.1 to 62.8 in *A. ilicifolius*, 6.2 to 38.0 in *B. gymnorrhiza*, 16.1 to 45.2 in *S. caseolaris* and 12.7 to 103.5 in *B. racemosa*. No seasonal pattern was discernible in the case of Cu content (Figure 4).

Zinc
High concentrations of zinc were observed in *A. officinalis*, *A. ilicifolius* and *B. racemosa*. Comparatively

79

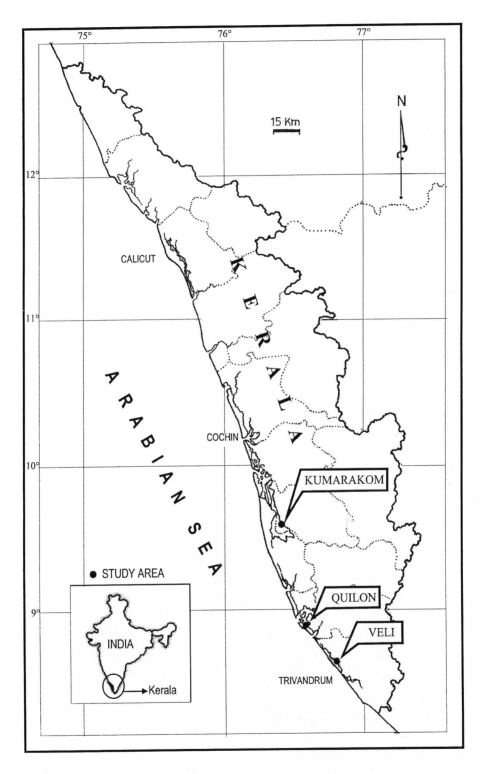

Figure 1. Map of Kerala showing the study sites.

Table 1. Heavy metal concentrations (annual range, mean, standard deviation) in the mangrove flora and sediments

Mangrove species	Metal concentrations (μg g^{-1} air dry wt)								
	Fe			Mn			Cu		
	Range	Mean	S.D	Range	Mean	S. D	Range	Mean	S. D
Avicennia officinalis	160–840	678	243	475–680	579	80	29.8–190.2	101.7	53.5
Acanthus ilicifolius	140–1640	703	479	46–163	104	42	6.1–62.8	74.4	54.2
Bruguiera gymnorrhiza	440–1460	890	367	339–599	526	87	6.2–38.0	22.8	11.7
Sonneratia caseolaris	1200–5940	2850	1766	258–1793	925	534	16.1–45.2	32.6	9.1
Barringtonia racemosa	2140–6050	4593	1535	628–2472	1281	689	12.7–103.5	41.9	31.8
Sediments									
Kumarakom	46909–121091	77603	25206	318–1089	629	262	10–44	28	13
Quilon	116727–146182	133818	9377	698–1124	881	171	652–845	758	58
Veli	99272–149455	123636	16720	158–456	348	114	62–256	125	92

Mangrove species	Metal concentrations (μg g^{-1} air dry wt)								
	Zn			Pb			Co		
	Range	Mean	S.D	Range	Mean	S. D	Range	Mean	S.D
Avicennia officinalis	70.5–212.0	119.5	44.3	75–225	146	51	5.7–22.8	14.2	6.3
Acanthus ilicifolius	93.3–160.0	135.7	28.3	25–125	63	35	5.7–45.5	25.6	12.2
Bruguiera gymnorrhiza	18.0–55.0	35.3	11.6	50–75	63	13	11.4–22.8	17.1	5.7
Sonneratia caseolaris	41.0–67.0	56.7	8.7	25–125	67	31	34.1–68.2	47.4	12.1
Barringtonia racemosa	61.5–248.0	112.3	64.9	25–200	88	66	90.0–125.0	100.1	12.1
Sediments									
Kumarakom	31–155	100	44	1100–1300	1217	79	159–261	205	36
Quilon	1550–2372	1880	250	1800–1950	1888	64	148–248	168	36
Veli	223–370	313	51	1250–1475	1346	80	91–114	104	8

Iron

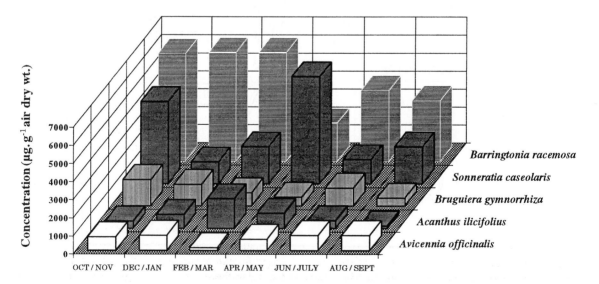

Figure 2. Bi-monthly variation of iron in various mangrove species.

Table 2. Bi–monthly variation of particle size distribution and organic carbon (%) in the three mangrove areas

Months	Site	Organic carbon %	Sand %	Silt %	Clay %
OCT – NOV		0.57	88.37	9.10	2.53
DEC – JAN		0.99	65.23	31.38	3.39
FEB – MAR	KUMARAKOM	1.41	60.92	34.83	4.25
APR – MAY		0.51	61.71	36.79	1.50
JUN – JUL		0.36	78.74	18.23	3.03
AUG – SEP		1.29	56.31	39.41	4.28
OCT – NOV		0.60	69.25	17.94	12.81
DEC – JAN		3.45	50.49	36.43	13.08
FEB – MAR	QUILON	7.65	78.42	16.56	5.08
APR – MAY		0.90	84.70	7.65	7.65
JUN – JUL		1.65	63.87	26.57	9.56
AUG – SEP		2.40	50.50	32.65	16.85
OCT – NOV		1.02	79.58	7.66	12.76
DEC – JAN		0.57	80.68	14.28	5.04
FEB – MAR	VELI	0.57	86.23	3.19	10.58
APR – MAY		0.78	76.29	8.47	15.24
JUN – JUL		0.90	80.63	6.74	12.63
AUG – SEP		0.84	83.09	7.61	9.30

Table 3. Correlation coefficients between metal concentrations in mangrove flora and respective concentrations in sediments

Sl. No.	Concentration in mangrove flora	Concentration in Sediments					
		Fe	Mn	Cu	Zn	Pb	Co
1.	*Avicennia officinalis*	0.7319	0.3604	0.4578	–0.2426	–0.4949	0.6650
2.	*Acanthus ilicifolius*	–0.8706*	–0.6671	0.3398	–0.7038	0.3988	0.6566
3.	*Bruguiera gymnorrhiza*	–0.3100	–0.3955	–0.5791	–0.4406	–0.1060	0.5232
4.	*Sonneratia caseolaris*	0.0888	–0.1463	–0.3744	–0.5160	–0.6015	–0.4924
5.	*Barringtonia racemosa*	–0.0887	–0.5941	–0.5304	0.1941	0.3682	–0.1794

* Significant at 5% level

lesser concentrations were found in *B. gymnorrhiza* and *S. caseolaris*. The overall concentrations of zinc ranged from 18.0 to 248.0 μg g^{-1} air dry wt. The individual ranges were from 70.5 to 212.0 in *A. officinalis*, 93.3 to 160.0 in *A. ilicifolius*, 18.0 to 55.0 in *B. gymnorrhiza*, 41.0 to 67.0 in *S. caseolaris* and 61.5 to 248.0 in *B. racemosa*. Concentrations of Zn were slightly higher during the post-monsoon season than the other two seasons (Figure 5).

Lead
Concentrations of lead were comparatively high in *Avicennia officinalis* and during certain months in *Barringtonia racemosa*. The overall concentrations of Pb

ranged from 25 to 225 μg g^{-1} air dry wt. The concentrations (μg g^{-1} air dry wt) in the different mangrove species ranged from 75 to 225 in *A. officinalis*, 25 to 125 in *A. ilicifolius*, 50 to 75 in *B. gymnorrhiza*, 25 to 125 in *S. caseolaris* and 25 to 200 in *B. racemosa*. Seasonal variation was not prominent in the Pb content in different mangrove species (Figure 6).

Cobalt
Fairly high concentrations of cobalt were observed in the mangrove species such as *Barringtonia racemosa* and *Sonneratia caseolaris*. The concentrations were similar in the other three mangrove species analysed. Considering all the species, the concentrations of Co

82

Manganese

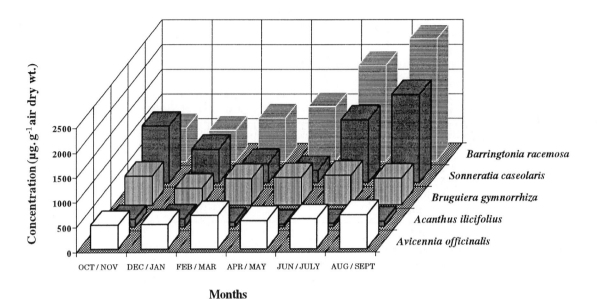

Figure 3. Bi-monthly variation of manganese in various mangrove species.

Copper

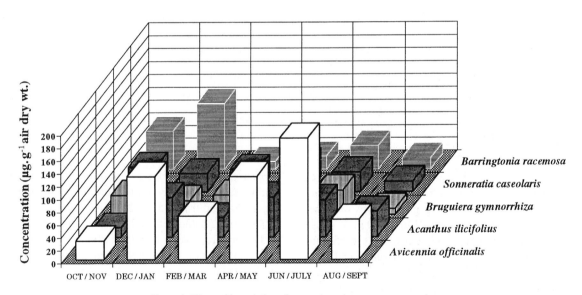

Figure 4. Bi-monthly variation of copper in various mangrove species.

ranged from 5.7 to 125.0 μg g^{-1} air dry wt. The corresponding ranges in the different mangrove species were from 5.7 to 22.8 in *A. officinalis*, 5.7 to 45.5 in *A. ilicifolius*, 11.4 to 22.8 in *B. gymnorrhiza*, 34.1 to 68.2 in *S. caseolaris* and 90.0 to 125.0 in *B. racemosa*. The Co concentrations were generally higher during

Zinc

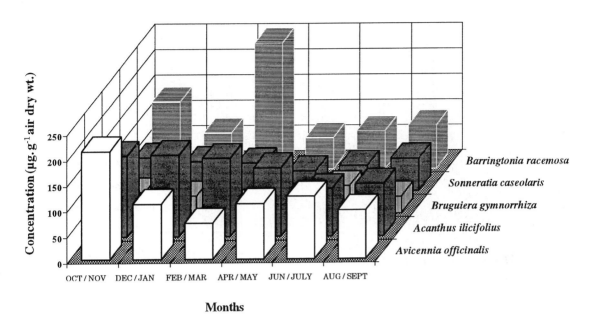

Figure 5. Bi-monthly variation of zinc in various mangrove species.

Lead

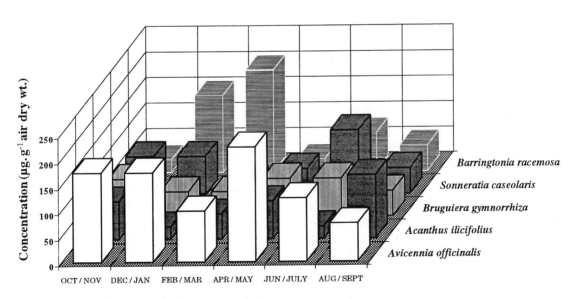

Figure 6. Bi-monthly variation of lead in various mangrove species.

the post-monsoon season and lower during the monsoon season (Figure 7).

Heavy metal concentration in sediments

Iron
Extremely high concentrations of Fe were observed in all the three mangrove systems. Concentrations

Cobalt

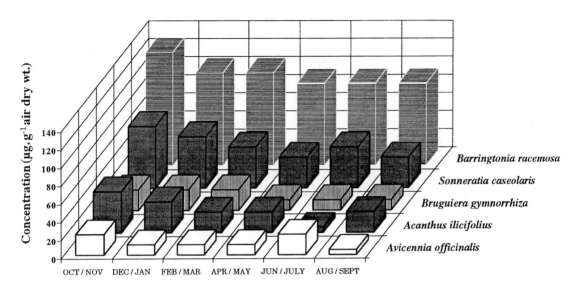

Figure 7. Bi-monthly variation of cobalt in various mangrove species.

of Fe (μg g^{-1} air dry wt) ranged from 46 909 to 121 091, 116 727 to 146 182 and 99 272 to 149 455 in Kumarakom, Quilon and Veli respectively (Table 1).

Manganese

Concentrations of Mn were comparatively higher in Kumarakom and lower in Veli mangrove sediments. In Kumarakom Mn ranged from 318 to 1089 μg g^{-1} air dry wt. The respective ranges obtained in Quilon were from 698 to 1124 μg g^{-1} air dry wt and in Veli Mn ranged from 158 to 456 μg g^{-1} air dry wt.

Copper

Very high concentrations of Cu were observed in Quilon region, while the least values were recorded at Kumarakom. The respective ranges (μg g^{-1} air dry wt) were from 10 to 44 in Kumarakom, 652 to 845 in Quilon and 62 to 256 in Veli.

Zinc

Very high concentrations of Zn were observed in the sediments of Quilon mangrove region, which ranged from 1550 to 2372 μg g^{-1} air dry wt. Whereas, the respective ranges in Kumarakom and Veli were 31 to 155 and 223 to 370 μg g^{-1} air dry wt.

Lead

The concentrations of lead were very high in the sediments of all the three mangrove ecosystems studied. The concentrations (μg g^{-1} air dry wt) ranged from 1100 to 1300 in Kumarakom, 1800 to 1950 in Quilon and 1250 to 1475 in Veli.

Cobalt

Slightly higher concentrations of Co were observed in Kumarakom, than the other two systems, with a range from 159 to 261 μg g^{-1} air dry wt. The concentrations in Quilon ranged from 148 to 248 μg g^{-1} air dry wt. Whereas in Veli Co ranged from 91 to 114 μg g^{-1} air dry wt.

Sediment texture and organic carbon

Particle size distribution and percentage of organic carbon are given in Table 2. Particle size variations showed that the Veli region is more sandy in nature, whereas the sediments of other two regions were composed of silty-sand.

Organic carbon content was very high in Quilon region ranging from 0.45 to 7.65% whereas the respective ranges in Kumarakom and Veli were from 0.36 to 2.37% and 0.33 to 1.44%.

Discussion

Differences in elemental composition in plants can arise from numerous sources including environmental variation, biological and environmental changes, as well as environment-organism interactions (Garten et al., 1977). The mineral metabolism of mangroves is known to be influenced by many factors, such as soil, water and climatic conditions (Kotmire & Bhosale, 1979).

In the present study, high concentrations of Fe, Mn and Co were detected in *Barringtonia racemosa* and *Sonneratia caseolaris*. The highest value of Fe recorded in the species *Sonneratia alba* in Goa mangroves was 8111.8 μ g^{-1} air dry wt (Untawale et al., 1980), which is comparable with the results obtained in the present study for *Sonneratia caseolaris*. However, the species, *Acanthus ilicifolius* in Goa mangroves has been reported to contain very high levels of Fe with a maximum value of 13 000 μg g^{-1} air dry wt (Untawale et al., 1980) which is substantially higher compared to the values obtained in the present study in the same species.

Copper content was found to be higher in *A. officinalis* than the other mangrove species analysed. In *A. officinalis*, Untawale et al. (1980) have reported Cu concentrations ranging from 36.81 to 119.41 μg g^{-1} air dry wt which agrees well with the present values. However, Bhosale (1979) have obtained a Cu concentration of 9.2 μg g^{-1} air dry wt in the leaves of *A. officinalis*. They have also reported Cu concentration 6.0 and 9.0 μg g^{-1} air dry wt in *B. gymnorrhiza* and *A. ilicifolius* which are very low compared to the present values. Low concentrations of Cu and Zn have also been reported in the leaves of *Kandelia candel* in Taiwan (Chiu & Chou, 1991).

Zn and Cu followed a similar pattern of seasonal variation in the present study. This shows that the efficacy of accumulation and regulation of these two metals are somewhat similar in the mangrove plants tested. Zn concentrations of 12.8, 17.5 and 24.7 μg g^{-1} air dry wt have been reported in the mangrove species *B. gymnorrhiza*, *A. Officinalis* and *A. ilicifolius* respectively (Bhosale, 1979). Their values are much lower compared to the present values obtained for these species.

Lead, though toxic to many marine animals and plants, were found in high quantities in the mangrove species studied. High values of Pb have been reported in *A. officinalis* in Goa regions (Untawale et al., 1980) with a peak value of 66.08 μg g^{-1}. In the present study also concentrations of Pb were comparatively higher in *A. officinalis* than the other species studied.

Rhizophora mangle plants are reported to have higher affinities for elements such as cobalt and cerium which may or may not be essential to them (Jayasekhera, 1991). Cobalt concentrations in the present study varied considerably among the different mangrove species. In Goa mangroves, cobalt content in *B. gymnorrhiza* ranged from 6.03 to 30.11 μg g^{-1} (Untawale et al., 1980) which is in accordance with the present concentrations obtained in this species.

Mangrove sediments are considered as a sink and source for trace metals (Harbison, 1986). A number of authors have shown that mangrove sediments operate as a biogeochemical sink for heavy metals, mainly due to the high concentrations of organic matter and sulphides under permanently reducing conditions (Harbison, 1981; Lacerda & Abrao, 1984; Lacerda et al., 1988; Silva, 1988). According to Lacerda et al. (1988) marine suspended matter with elevated metalic load is actively trapped within the mangrove environment. Immobilization mechanisms might include the binding of settled suspended matter by the tangled roots abundant in the mangrove environment (Harbison, 1986), thus elevating the metal levels in the sediments.

In the present study very high concentrations of heavy metals were obtained in the mangrove sediments. Among the three ecosystems examined, Quilon mangroves rank highest with respect to the concentrations of most of the heavy metals studied. The effluents from the neighbouring industries might have contributed substantially, metals such as Pb, Zn and Fe. In all the three systems iron content was exceptionally high ranging from 46 909 to 149 455 μg g^{-1} air dry wt. These high concentrations might be due to the precipitation of iron as iron sulphides which is common in mangrove ecosystems. Iron is generally described as the principal metal that precipitates with sulphidic compounds in anaerobic sediments (Howarth, 1979; Berner, 1984) and these sulphides form a major sink for heavy metals in the mangrove area. In the Ashtamudi estuary Nair et al. (1987) have reported Fe concentrations in the sediments ranging from 1100 to 3900 μg g^{-1}. In the mangrove sediments of eastern Australia, Hirner et al. (1990) have reported an iron concentration of 80 000 mg kg^{-1} which is within the range of values found in mangrove biotopes in the present study.

Mn and Co were found to be in comparatively higher concentrations in the sediments of Kumarakom mangrove region than the other two systems. In Cochin backwater Venugopal et al. (1982) have reported concentrations of Mn and Co in the sediments with peak values of 317.5 ppm and 43.1 ppm respectively.

Lacerda et al. (1987) have obtained a concentration of Mn ranging from 5 to 1260 ppm in the sediments of Brazilian mangrove region which is comparable with the present values in the three mangrove biotopes.

Very high concentrations of Cu, Zn and Pb were characteristics for the Quilon mangrove region. According to Elderfield et al. (1979) trace metals precipitate with Fe forming polysulphide minerals in particular Cu, Zn and Pb. Similar formation of polysulphides can also be expected in the Quilon mangrove region which receives a high load of organic matter (Table 2). In the mangrove environments of Thane creek (Bargaonkar & Gokhale, 1992) maximum concentrations of Cu, Zn and Pb of 51.83 μg g^{-1}, 90.55 μg g^{-1} and 10.10 μg g^{-1} respectively were recorded. The present values for these metals are much higher than the above reported values. In New Zealand mangrove areas, Zauke et al. (1992) have reported sediment Cu and Zn concentrations ranging from 198 to 226 mg kg^{-1} and 4460–6530 mg kg^{-1} respectively, which is comparable with the high value recorded in the present study.

Many authors have made attempts to link the sediment metal concentrations to that present in the mangrove plants. In the present study positive correlations were obtained between the concentrations of Fe and Co in *A. officinalis* and the respective concentrations in sediments (Table 3). However, the other mangrove species studied have not shown any significant positive correlations and some species showed a negative correlation to the concentration of metals in the sediment. In the mangrove environment though the metals are found in excess quantity in the sediments, it may not be available to the plants. This is due to the precipitation of these metals as sulphides or due to the effect of chelating substances such as humic acids. Subramanian & Pragatheeswaran (1985) have also observed the absence of any relationship of Fe in plant parts with the concentration in the sediments.

Particle size analysis showed more sandy substrates in the Veli region compared to silty sands in Quilon and Kumarakom. However, the metal distributions showed little relationships to the textural characteristics of the sediments in the present study areas.

In conclusion, the results of the present study revealed high concentrations of heavy metals especially, Cu, Zn, Pb and Co in above ground tissues of the mangrove species analysed. Certain metals such as Cu, Zn and Pb were found in very high concentrations in the sediments of Quilon and Veli regions compared to the relatively uncontaminated, areas of Kumarakom. Levels in plant tissues generally showed little correlation with total concentrations in the sediments, suggesting a high proportion of the heavy metals in the sediments were unavailable or effectively excluded by mangrove plants growing in them. Similarly, it seems that mangroves can accumulate certain metals such as Zn and Pb without much harmful effects for their growth.

Acknowledgments

One of the authors (G.T) is grateful to Council of Scientific and Industrial Research (C.S.I.R), New Delhi for the financial assistance in the form of a Senior Research Fellowship.

References

APHA (American Public Health Association), 1980. Standard methods for the examination of water and waste water. 15th edn., APHA-AWWA-WPCF, 1134 pp.

Berner, R. A., 1984. Sedimentary pyrite formation: an update. Geochim. Cosmochim. Acta 48: 605–615.

Bhosale, L. J., 1979. Distribution of trace elements in the leaves of mangroves. Indian J. Mar. Sci. 8: 58–59.

Borgaonkar, S. S. & K. S. Gokhle, 1992. Distribution of copper, zinc, cadmium and lead in Thane Creek and its relation with textural types and organic carbon. J. envir. Biol. 13: 39–45.

Carver, R. E., 1971. Sedimentary petrology. John Wiley & Sons Inc., London, 653 pp.

Chakrabarti, C., S. K. Kundu, P. B. Ghosh & A. Choudhury, 1993. A preliminary study on certain trace metals in some plant and animal organisms from mangroves of Sunderbans, India. Mahasagar 26: 17–20.

Chiu, C. Y. & C. H. Chou, 1991. The distribution and influence of heavy metals in mangrove forests of the Tamshui estuary in Taiwan. Soil. Sci. Plant Nutr. 37: 659–669.

Elderfield, H., A. Hepworth, P. N. Edwards & L. M. Holiday, 1979. Zinc in the Conway river and estuary. Estuar. coast. mar. Sci. 9: 403–422.

EPA, 1983. Methods for Chemical Analysis of Water and Wastes. United States Environmental Protection Agency, EPA-6004-79-020, USA. Methods: 110.1–430.2.

Fernandez, T. V. & N. V. Jones, 1987. Some studies on the effect of zinc on *Nereis diversicolor* (Polychaete – Annelida). Trop. Ecol. 28: 9–21.

Fernandez, T. V. & George Thomas, 1995. Distribution of essential elements in seaweeds along the South West coast of India. In Majumdar, S. K., E. W. Miller & F. J. Brenner (eds), Environmental Contaminants, Ecosystems and Human Health. Pennsylvania Academy of Science, USA: 112–125.

Fernandez, T. V., George Thomas, T. Jamila, Lekha Mammachan, S. Deepthi, K. Krishnakumar & T. Anil Kumar, 1995. Elemental composition of the marine algae occurring along the South-West coast of India. Envir. Conserv. 22: 359–361.

Garten, C. T. Jr, J. B. Gentry & R. R. Sharitz, 1977. An analysis of elemental concentrations in vegetation bordering a South-Eastern United States Coastal Plain Stream. Ecology, 58: 972–992.

Harbison, P., 1981. The case for the protection of mangrove swamps: geochemical considerations. Search, 12: 273–276.

Harbison, P., 1986. Mangrove muds – A sink and a source for trace metals. Mar. Pollut. Bull., 17: 246–250.

Hirner, A. V., K. Kritsotakis & H. J. Tobschall, 1990. Metal organic associations in sediments-1. Comparison of unpolluted recent and ancient sediments and sediments affected by anthropogenic pollution. Appl. Geochem., 5: 491–505.

Howarth, R. H., 1979. Pyrite: its rapid formation in a salt marsh and its importance in ecosystem metabolism. Science, 203: 49–51.

Jayasekhera, R., 1991. Chemical composition of the mangrove *Rhizophora mangle* L. J. Plant Physiol. 138: 119–121.

Kotmire, S. Y. & L. J. Bhosale, 1979. Some aspects of chemical composition of mangrove leaves and sediments. Mahasagar 12: 149–154.

Lacerda, L. D. & J. J. Abrao, 1984. Heavy metal accumulation by mangrove and salt marsh intertidal sediments. Rev. Bras. Biol., 7: 49–52.

Lacerda, L. D., L. A. Martinelli, C. E. Rezende, A. A. Mozeto, A. R. C. Ovalle, R. L. Victoria, C. A. R. Silva & F. B. Nogueira, 1988. The fate of trace metals in suspended matter in a mangrove creek during a tidal cycle. Sci. tot. Envir. 75: 169–180.

Lacerda, L. D., C. E. Rezende, C. A. R. Silva & J. C. Wasserman, 1987. Metallic composition of sediments from mangroves of the SE. Brazilian Coast. In Lindberg, S. E. & T. C. Hutchinson (eds), Proc. int. Conf. Heavy Metals in the Environment, New Orleans 2: 464–466.

Nair, N. B., P. K. A. Azis, H. Suryanarayanan, M. Arunachalam, K. Krishnakumar & Tresa V. Fernandez, 1987. Distribution of heavy metals in the sediments of the Ashtamudi estuary, S.W. Coast of India. Contr. Mar. Sci. 269–289.

Preston, A., D. F. Jefferies, J. W. R. Dutton, B. R. Harvey & A. K. Steele, 1972. British isle coastal waters. The concentrations of selected heavy metals in sea water, suspended matter and biological indicators – a pilot survey. Envir. Pollut. 3: 69–82.

Seralathan, P., 1987. Trace element geochemistry of modern deltaic sediments of the Cauvery River, east coast of India. Indian J. Mar. Sci. 16: 235–239.

Silva, C. A. R., 1988. Distribuicao e ciclagem interna de metais pesados em um ecossistema de manguezal dominado por Rhizophora mangle, Baia de Sepetiba, Rio de Jeneiro. T. Mest. Inst. Quimica, Univ. Fed. Fluminense.

Subramanian, A. N. & V. Pragatheeswaran, 1985. Accumulation of iron by some mangrove plants and seaweeds. In Krishnamurthy, V. & A. G. Untawale (eds), Proc. All India Symp. Marine Plants, their biology, chemistry and utilization, Dona Paula, Goa: 133–136.

Untawale, A. G., S. Wafer & N. B. Bhosale, 1980. Seasonal variation in heavy metal concentration in mangrove foliage. Mahasagar 13: 215–223.

Venugopal, P., K. Sarala Devi, K. V. Remani & R. V. Unnithan, 1982. Trace metal levels in the sediments of the Cochin backwaters. Mahasagar 15: 205–214.

Walkley, A. & T. A. Black, 1934. An estimation Degitijaraff method for determining soil organic matter and proposed modification of the chromic acid titration method. Soil Sci. 37: 23–38.

Yamamoto, T., 1972. The relation between concentration factors in seaweeds and residence time of some elements in seawater. Res. Oceanogr. Works, Japan 11: 65–72.

Zauke, G. P., J. Harms & B. A. Foster. 1992. Cadmium, lead, copper and zinc in *Eleminius modestus* Darwin (Crustacea, Cirripedia) from Waitemata and Manukau Harbours, Auckland, New Zealand. N.Z.J. Mar. Freshwat. Res. 26: 405–415.

Hydrobiologia **352**: 89–96, 1997.
Y.-S. Wong & N. F.-Y. Tam (eds), Asia–Pacific Conference on Science and Management of Coastal Environment.
©1997 *Kluwer Academic Publishers. Printed in Belgium.*

Response of *Aegiceras corniculatum* to synthetic sewage under simulated tidal conditions

Yuk-Shan Wong[1], Nora F. Y. Tam[2]*, Gui-Zhu Chen[3] & Hua Ma[3]
[1]*Research Centre/Biology Department, The Hong Kong University of Science & Technology, Hong Kong*
[2]*Department of Biology and Chemistry, City University of Hong Kong, Hong Kong*
[3]*Institute of Environmental Science, Zhongshan University, Guangzhou, PRC*
(*Author for correspondence; tel: (852) 27887793; fax: (852) 27887406; e-mail: bhntam@cityu.edu.hk.)*

Key words: Mangrove plant, pollution, wastewater, tolerance, *Aegiceras*

Abstract

Young plants of *Aegiceras corniculatum*, a dominant mangrove species, were collected from Futian Mangrove Swamp in Shenzhen, The People's Republic of China, and grown in simulated tide tanks containing mangrove sediments. After acclimatisation in the greenhouse for 6 months, the plants were irrigated with either synthetic sewage of various strengths (NW, FW and TW) or artificial seawater (as control). NW had the characteristics and strength equivalent to municipal wastewater, while FW and TW contained 5 and 10 times the nutrient and heavy metal concentrations of the NW, respectively. Results showed that the young plants of *A. corniculatum* were able to tolerate the wastewater (TW) with highest concentration of nutrients and heavy metals after one year treatment. The growth of TW treated plants, measured in terms of stem height, basal diameter and biomass, was comparable to that found in the control. The plants treated with NW and FW had significantly greater growth than the control, indicating that the nutrients contained in sewage are beneficial to mangrove plants. Physiological parameters such as chlorophyll content, the ratio of chlorophyll *a* and *b*, proline concentration and root activity did not show any significant changes among plants treated with wastewater of various strengths and the control, suggesting that sewage addition did not cause any apparent physiological impact on growth of *A. corniculatum*. Nevertheless, the plants which received sewage with highest levels of heavy metals (TW treatment) appeared to have lower content of free water but higher amount of bound water than FW, NW and the control. Higher electric conductance was also found in plants treated with TW.

Introduction

Natural wetlands such as mangrove ecosystems have been found to be capable of removing pollutants from domestic and industrial wastewater and could be used as a low cost sewage treatment facility (Clough et al., 1983; Tam & Wong, 1993; Breaux & Day, 1994; Corredor & Morell, 1994; Wong et al., 1995). Many studies have been carried out to understand the mechanism of pollutant reduction and to improve the performance of mangroves in treating wastewater. In wetland systems, higher plants are the most obvious biological components and are considered to be of great importance in removing pollutants from sewage (Gam-

brell et al., 1987; Reed et al., 1988; Rogers et al., 1991). However, the actual effects of sewage discharge, especially those containing toxic metals, onto wetland plants are not clear. It has been suggested that the discharge of wastewater may be beneficial to wetlands and should be incorporated as a component of coastal management (Breaux & Day, 1994). Effluents serve as vital sediment and nutrient sources, which can stimulate plant growth and biomass production and increase accretion and subsequent soil formation in mangrove wetlands (Onuf et al., 1977; Boto & Wellington, 1983; Boto, 1992; Breaux et al., 1995; Chen et al., 1995). In contrast, a number of studies have reported that addition of sewage can cause sub-lethal damages in

Table 1. Characteristics of the standard artificial wastewater (NW) used in experiments.

Pollutants	Concentration (mg l^{-1})	Pollutants	Concentration (mg l^{-1})
NH_4^+-N	40	Ni^{2+}	1
NO_3^--N	1	Cu^{2+}	2
Organic N	10	Zn^{2+}	5
Total P	10	Mn^{2+}	5
COD	500	Pb^{2+}	1
K^+	50	Cr^{6+}	0.5
		Cd^{2+}	0.1

Table 2. Growth of *Aegiceras* plants at the end of one year treatment with sewage, measured in terms of annual increment.

Growth parameters		Control	NW	FW	TW
Stem Height (cm)	Mean ± S.D.	39.04 ± 12.97	45.55 ± 10.59	43.42 ± 16.43	38.88 ± 14.99
	(min–max)	(16.50–71.00)	(23.60–61.70)	(18.00–81.00)	(11.00–82.00)
Basal Diameter (cm)	Mean ± S.D.	0.63 ± 0.19	0.80 ± 0.27	0.78 ± 0.23	0.64 ± 0.21
	(min–max)	(0.33–1.46)	(0.33–1.46)	(0.32–1.25)	(0.25–1.07)
Biomass (g)	Mean ± S.D.	17.61 ± 6.65	22.83 ± 8.81	21.30 ± 8.66	17.59 ± 7.55
	(min–max)	(4.03–32.17)	(8.42–44.52)	(3.09–37.09)	(5.89–32.43)

plant such as changes in photosynthetic rates, chlorophyll concentrations and enzymatic activities (Culic, 1984; Peng, 1990) and in one case has contributed to the gradual disappearance of mangrove vegetation near Bombay, India (Navalkar, 1951). In many countries, natural wetlands are subject to legal limitation for providing only tertiary sewage treatment or to treat domestic sewage with low levels of toxic pollutants (Breaux & Day, 1994).

The impacts of wastewater discharge vary significantly from wetland to wetland, and it is not possible to make predictions about the impact of wastewater on a specific wetland without first examining its characteristics (Trattner & Woods, 1989). Furthermore, even within the same type of wetland, different plant species with varied growth rates and physiological adaptation can have varying responses to sewage discharge. In addition, the response of plants to wastewater may also depend on the concentration and composition of the sewage effluent. The present study is therefore aimed (1) to study the effects of sewage discharge on the growth and physiology of young plants of *Aegiceras corniculatum*, a dominant mangal species, and (2) to compare the impact of three concentrations of synthetic sewage on this plant, growing under controlled greenhouse conditions.

Materials and methods

Tide tank operation under greenhouse conditions

One year old young *Aegiceras corniculatum* plants were collected from Futian Mangrove Swamp, Shenzhen, the People's Republic of China (114°05'E, 22°32'N), and transplanted in 12 simulated tide tanks situated in a greenhouse condition. Fifteen plants were grown in each tide tank (made of PVC with a dimension of $1.2 \times 0.6 \times 0.5$ m^3) containing 100 kg (dry weight equivalent) fresh mangrove sediments collected from the same swamp. The tide tanks were flooded with artificial seawater (1.2–1.6% salinity, prepared from commercially available ocean salts) twice a day, to simulate the natural tidal regime. The plants were allowed to acclimatise in these tide tanks for six months prior to treatment with wastewater. Three types of wastewater, namely NW, FW and TW, were applied to the tanks twice a week for a year with a hydraulic loading of 1.75 l per tank per irrigation during the low tide (exposure) period. NW represented the synthetic wastewater with a composition similar to that of municipal sewage discharged in Hong Kong (Table 1). FW and TW represented five and 10 fold higher in loadings of nutrients and heavy metals of the NW, respectively. In between two wastewater irrigations, all tanks were flooded with

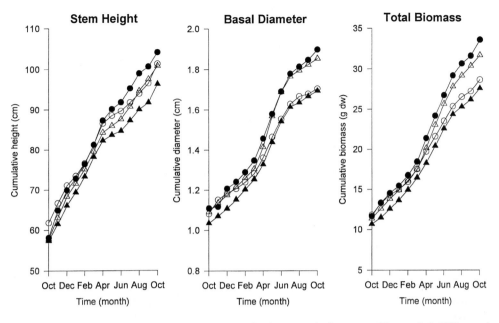

Figure 1. Growth of *A. corniculatum* under treatments with wastewater of various strengths for one year (C: control, ○; NW: normal wastewater, ●; FW: 5 × NW, △; TW: 10 × NW, ▲).

artificial seawater according to the tidal regime. Tide tanks treated with artificial seawater alone were used as the control. All experimental treatments were carried out using three replicates.

Determination of plant growth and physiological changes

The growth of *Aegiceras* in terms of stem height (H) and the basal diameter (D) was measured at monthly intervals. Biomass was estimated by the non-destructive allometric equation: log biomass = a + b log (D^2H) suggested by Snedaker & Snedaker (1984). The constant value and slope of this equation were determined by measuring the height, stem basal diameter, and oven dried weight (70 °C) of 20 young *Aegiceras* plants (about 18 months old) at the beginning of the experiment. The regression equation for total biomass was: log biomass (g) = −0.126 + 0.642 log (D^2H) with $r = 0.80$ (significant at $p < 0.01$).

At the beginning of the experiment, and at intervals of approximately three months, 5–8 mature leaves, from the third pair of the leaves below the apical part of the stem of every plant in each tank, were collected, pooled together and analysed for physiological parameters. These included: (1) concentrations of chlorophyll *a* and *b*, extracted by acetone and measured at a wavelength of 663 and 645 nm (Zhu et al., 1990);

(2) free proline, extracted by glacial acetic acid, triketohydrindene and benzene, absorbance measured at 515 nm (Zhu, 1983); (3) total, free and bound water content. Total water content was measured by weight loss after drying at 105 °C; free water content was determined by Abbe refractometry and bound water was calculated as the difference between total and free water content (Zhu et al., 1990); (4) electric conductance, by shaking and heating leaf discs (0.5 cm^2) in deionized water for 10 min to destroy the cell membranes and the amount of electrolytes released measured by an electrical conductivity meter. At the beginning and the end of the experiment, roots were collected from each tank and root activity was assayed by reduction of TTC (2, 3, 5 triphenyl tetrazolium chloride) to triphenyl formazan (Zhu et al., 1990). The mean and standard deviation of the triplicated samples were calculated. The results collected were analysed by a two-way ANOVA, with treatment and time as the main effects. The least significant difference (LSD) values at 5% probability level was calculated if the results of the ANOVA indicated significant difference. All statistical analyses were performed by means of an IBM-compatible computing package 'SPSS' (SPSS/PC Plus for IBM PC, SPSS Inc., 1985).

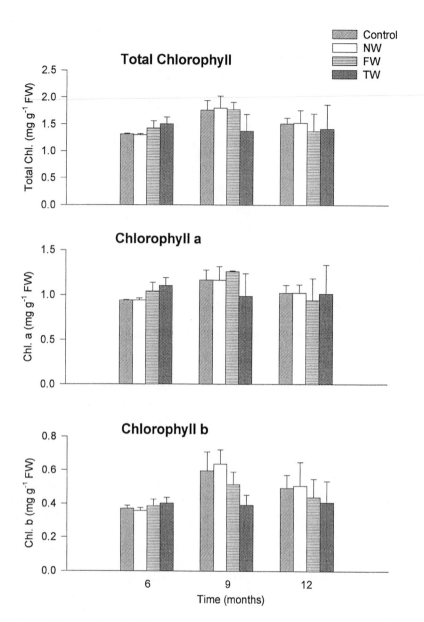

Figure 2. Chlorophyll content of *A. corniculatum* leaf during one year irrigation of wastewater of various strengths. (The mean and standard deviation values of three replicates from 45 plants were shown).

Results and discussion

Effect of sewage treatment on growth of Aegiceras *plants*

The initial mean stem height, basal diameter and total biomass of the plants were 58.76 ± 2.04 cm, 1.08 ± 0.03 cm and 11.49 ± 0.39 g, respectively. Application of normal wastewater (NW) and FW

($5 \times$ NW) enhanced plant growth (Figure 1). Both the basal diameter and biomass of plants treated with NW and FW were higher than the control especially towards the latter part of the study (from March onwards). The annual increment of height, diameter and biomass was significantly larger (Table 2) in these two treatments than those found in the control and TW (concentrated wastewater, $10 \times$ NW). It indicates that addition of wastewater containing nutrients and organic mat-

Table 3. Physiological and biochemical properties of *A. corniculatum* plants after receiving wastewater of various strengths for one year.

Parameter	Control	NW	FW	TW	2-ways ANOVA Results
Root activity					
(μg g^{-1}dw h^{-1})	4.35 (1.21)	3.78 (0.33)	3.20 (1.71)	3.38 (1.49)	Not significant
Bound: Free water ratio	0.25^a (0.16)	0.63^b (0.48)	1.22^c (0.32)	1.49^c (0.16)	Significantly different at probability level of 0.05
Chl. a: Chl b ratio	2.11 (0.40)	2.11 (0.43)	2.18 (0.46)	2.49 (0.04)	Not significant
Electric conductance					
(ms m^{-1})	964^a (59)	1011^a (77)	1076^a (59)	1211^b (39)	Significantly different at probability level of 0.05
Proline content					
(μg^{-1} g^{-1} fw)	30.69 (8.59)	34.75 (10.08)	32.01 (17.45)	38.47 (5.89)	Not significant

The mean and standard deviation (in brackets) values of three parallel groups (15 plants in each group) were calculated. The mean values followed by different superscripts within each row indicated that they were significantly different at a probability level of 0.05.

Table 4. Electric conductance (ms m^{-1}) of *A. corniculatum* leaf during one year irrigation of wastewater of various strengths.

Time (month)	Control	NW	FW	TW
0 (initial)		653 $\pm$ 103		
6	596 $\pm$ 12^a	771 $\pm$ 11^b	658 $\pm$ 8^b	771 $\pm$ 34^b
9	745 $\pm$ 37^a	672 $\pm$ 36^a	865 $\pm$ 36^b	836 $\pm$ 37^b
12	964 $\pm$ 59^a	1011 $\pm$ 77^a	1076 $\pm$ 59^a	1211 $\pm$ 39^b

The mean and standard deviation values of three parallel groups (15 plants in each group) were calculated. The values followed by different superscripts within each row indicated that they were significantly different at a probability level of 0.05 according to ANOVA tests.

ter were beneficial to mangrove plants by improving mangrove soils, which are often deficient in nutrients (Clough et al., 1983; Boto, 1992). In this respect, Boto & Wellington (1983) showed that addition of N and P to the mangrove communities resulted in 30% increase in plant growth coupled with a higher nutrient concentration in plant tissues. Similar findings were also reported by Naidoo (1987) and Chen et al. (1995).

Plants treated with TW showed similar growth response as the control, indicating that strong sewage had no obvious harmful effect on *Aegiceras* plants. Mangrove plants colonizing the intertidal habitat are well known to be resistant to environmental stresses such as high salinity and waterlogging. These plants are specially adapted to an unstable substratum and the alternating aerobic and anaerobic conditions (Lugo & Snedaker, 1974; Clough et al., 1983). Lin & Chen (1990) reported that mangrove plants are able to tolerate high concentrations of heavy metals, and the growth of *Avicennia alba*, *Rhizophora muoronata* and *R. mangle* was not inhibited by Zn, Pb, Cd and Hg ions in the soil (Chen & Lin, 1988). Our previous

study (Chen et al., 1995) also found that *Kandelia candel*, another dominant plant species in the Futian Mangrove Swamp, was capable of withstanding strong wastewater under similar greenhouse tide tank condition. Moreover, most of the added heavy metals from wastewater are likely to have been retained by soil *via* adsorption on ion-exchange sites, incorporation into lattice structures and by precipitation as insoluble sulphide, with only a small portion remained bio-available (Cooke et al., 1990; Tam & Wong, 1993, 1995). This may explain why mangrove plants are able to grow in wastewater containing very high concentrations of heavy metals.

Physiological responses of Aegiceras *plants to sewage of various strengths*

The concentrations of total chlorophyll, chlorophyll *a* and *b* (Figure 2), and ratios of chlorophyll *a* to chlorophyll *b* (Table 3) in leaves treated with wastewater were similar to that found in the control. The plants treated with TW for 9 months appeared to have the

Table 5. Proline concentrations (μg g^{-1} fw) of *A. corniculatum* leaf during one year irrigation of wastewater of various strengths.

Time (month)	Control	NW	FW	TW
0 (initial)		29.2 $\pm$ 2.9		
6	32.1 $\pm$ 5.2	25.6 $\pm$ 8.2	37.6 $\pm$ 5.1	41.9 $\pm$ 3.5
9	34.1 $\pm$ 3.1	30.6 $\pm$ 1.3	40.0 $\pm$ 7.9	32.1 $\pm$ 10.9
12	30.6 $\pm$ 8.6	34.8 $\pm$ 10.1	32.1 $\pm$ 17.5	38.5 $\pm$ 5.9

The mean and standard deviation values of three parallel groups (15 plants in each group) were calculated. There was no significant difference between wastewater treatment and control according to ANOVA tests.

lowest chlorophyll *b* content (Figure 2) although it was not statistically different to the other treatments. The concentrations of chlorophyll *a* and *b* as well as the chlorophyll *a:b* ratio often change when the plants are under stress. Lin et al. (1984) found that under high soil salinity, the synthesis and accumulation of chlorophyll in *K. candel* were reduced and the ratio of chlorophyll *a:b* rose. Our previous study on *K. candel* (Chen et al., 1995) also reported that addition of sewage caused a significant drop in chlorophyll *a* and *b* content (but a rise in chlorophyll *a:b* ratio). The minimal changes in chlorophyll content recorded in sewage treated *A. corniculatum* plant suggest that this mangrove species might be more adaptable to strong wastewater than *K. candel*.

The free and bound water contents in plants treated with wastewater were not significantly different from the control in the first 9 months of treatment (Figure 3). However, at the end of one-year treatment, the plants treated with stronger wastewaters (FW and TW) had significantly lower content of free water but a higher amount of bound water in their leaves compared to the control plants (Figure 3). The ratio of bound to free water content was highest in TW and decreased in the order of FW>NW>control (Table 3). It has been proposed that amounts of free and bound water could reflect the activity of the plants; higher amounts of free water indicated that the plants possessed a high metabolic rate and a rapid growth, but low resistance to stress (Peng, 1990). Under stress environments such as drought conditions, plants will usually lose their free water first, resulting in an increase in bound:free water ratios. The high bound:free water ratios recorded in TW treated plants in this study also suggested that strong sewage did cause some degree of stress to *A. corniculatum,* if the treatment was prolonged for more than one year.

The leaf electric conductance increased with treatment time and sewage strength (Table 4). The highest conductance, which was found in TW treated leaves recorded at the end of the experiment, may have been due to the damage of plasma membranes by this strong wastewater. Wang et al. (1984) similarly found that total leaf electric conductance was directly proportional to the concentration of pollutants in wastewater applied to aquatic plants. Other physiological and biochemical properties such as root activity (Table 3) and proline content (Table 5) did not show any significant difference between plants treated with wastewaters and the control; similar results have been recorded in *K. candel* treated with same types of wastewater (Chen et al., 1995). Proline has been shown to accumulate in tissues/organs of plants subject to environmental stresses such as drought, salt or temperature stress (Aspinall & Paleg, 1981; Chandler & Thorpe, 1987). Saradhi & Saradhi (1981) also reported that Cd was the most potent inducer, among four tested heavy metals, for proline accumulation. A recent study by Chen & Kao (1995) revealed that Cd induced an accumulation of proline in roots, and the amount of proline found in roots was closely associated with the reduction in root growth of rice seedlings by Cd. In the present study, proline levels fluctuated with time, and plants treated with strong wastewater (TW) appeared to have slightly higher proline content. However, due to the large variability found between replicates of the same treatment and the fluctuation with treatment time, the differences between TW, FW, NW treatments and the control were not significant according to ANOVA.

Conclusion

The present greenhouse experiment demonstrates that the application of sewage of normal (NW) or medium

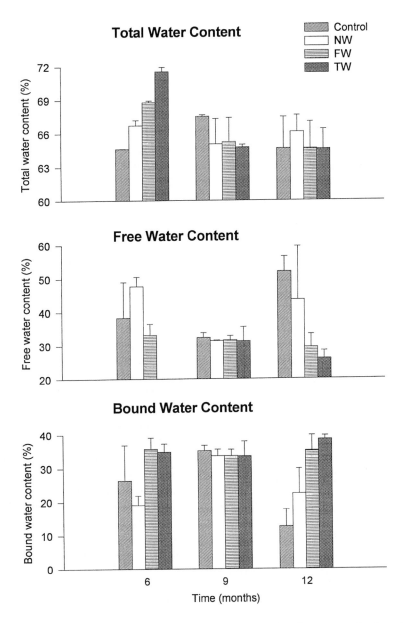

Figure 3. Bound and free water content of *A. corniculatum* leaf during one year irrigation of wastewater of various strengths. (The mean and standard deviation values of three replicates from 45 plants were shown).

(FW) strength stimulated growth of young *A. corniculatum* plants, probably by supplying nutrients to the mangrove soils. Even very strong wastewater (TW) did not reduce the plant growth in terms of both diameter, height and biomass when compared with the control. The concentrations of total chlorophyll, chlorophyll *a* and *b*, proline content and root activity of plants treated with wastewater of various strengths were all similar to the control, indicating no apparent physiological damage in *A. corniculatum* plants treated with sewage. This plant species appeared to be capable of tolerating strong wastewater containing excess nutrients and heavy metals. However, strong wastewater (TW) did cause some degree of stress to the plants, reflected by the subtle changes in the more sensitive physiological parameters, such as the ratio of bound and free water content and the electric conductance of leaves. The long-term effect of treating mangrove plants with

96

heavily polluted wastewater should be fully explored before employing this natural wetland system as wastewater treatment facility on a large scale basis.

Acknowledgments

This research was supported by grants from Research Grant Council of Hong Kong and the Croucher Foundation. The authors wish to thank Mr S. H. Li, Zhongshan University, Guangzhou, and the officers in Futian Nature Reserve, Shenzhen Special Economic Zone, the PRC for their assistance in the field, laboratory and greenhouse work.

References

Aspinall, D. & L. G. Paleg, 1981. Proline accumulation: physiological aspects. In Paleg, L. G. & D. Aspinall (eds), The Physiological and Biochemistry of Drought Resistance in Plants. Academic Press, Sydney: 280–295.

Boto, K. G., 1992. Nutrients and mangroves. In Connel, D. W. & D. W. Hawker (eds), Pollution in Tropical Aquatic Systems. CRC Press, Inc., Ann Arbor, London: 129–145.

Boto, K. G. & J. T. Wellington, 1983. Nitrogen and phosphorus nutritional status of a northern Australian mangrove forest. Mar. Ecol. Press Ser. 11: 63–69.

Breaux, A. M. & J. W. Day Jr, 1994. Policy considerations for wetland wastewater treatment in the coastal zone: a case study for Louisiana. Coast. Mgmt 22: 285–307.

Breaux, A., S. Farber & J. Day, 1995. Using natural coastal wetlands systems for wastewater treatment: an economic benefit analysis. J. envir. Mgmt 44: 285–291.

Chandler, S. F. & T. A. Thorpe, 1987. Characterization of growth, water relations, and proline accumulation in sodium sulfate tolerant callus of Brassica napus L. cv Westar (Canola). Pl. Physiol. 84: 106–111.

Chen, S. L. & C. H. Kao, 1995. Cd induced changes in proline level and peroxidase activity in roots of rice seedlings. Pl. Growth Reg. 17: 67–71.

Chen, G. Z., S. Y. Miao, N. F. Y. Tam, Y. S. Wong, S. H. Li & C. Y. Lan, 1995. Effect of synthetic wastewater on young Kandelia candel plants growing under greenhouse conditions. Hydrobiologia 295: 263–274.

Chen, R. H. & P. Lin, 1988. Influence of mercury and salinity on the growth of seedlings of three mangrove species. Universitatis Amoiensis Acta Scientiarum Naturalium 27: 110–115.

Clough, B. F., K. G. Boto & P. M. Attiwill, 1983. Mangrove and sewage: a reevaluation. In Teas, H. J. (ed.), Biology and Ecology of Mangroves, Vol. 8. Dr. W. Junk Publishers, The Hague, Lancaster: 151–161.

Cooke, J. G., A. B. Cooper & N. M. U. Clunie, 1990. Changes in the water, soil, and vegetation of a wetland after receiving a sewage effluent. New Zealand J. Ecol. 14: 37–47.

Corredor, J. E. & J. M. Morell, 1994. Nitrate depuration of secondary sewage effluents in mangrove sediments. Estuaries 17: 295–300.

Culic, P., 1984. The effects of 2,4-D on the growth of Rhizophora stylosa Griff. seedlings. In Teas, H. J. (ed.), Physiology and Management of Mangroves, Vol. 9. Dr. W. Junk Publishers, The Hague, Lancaster: 57–63.

Gambrell, R. P., R. A. Khalid & W. H. Jr. Patrick, 1987. Capacity of a swamp forest to assimilate the TOC loading from a sugar refinery wastewater stream. J. Wat. Pollut. Cont. Fed. 59: 897–904,

Lin, P., D. H. Chen & W. J. Li, 1984. A preliminary study of the interrelationship between the physiological characteristics of certain enzymes in the leaves of mangrove plants and soil salinity in tidal swamp. Univ. Amoiensis Acta Sci. 27: 110–115.

Lin, P. & R. Chen, 1990. Role of mangrove in mercury cycling and removal in the Jiulong estuary. Acta Oceanol. Sinica 9: 622–624.

Lugo, A. E. & S. C. Snedaker, 1974. The ecology of mangroves. Annu. Rev. Ecol. Syst. 5: 39–64.

Naidoo, G., 1987. Effects of salinity and nitrogen on growth and water relations in the mangrove, Avicennia marina (Forsk.) Vierh. New Phytol. 107: 317–325.

Navalkar, B. S., 1951. Succession of the mangrove vegetation of Bombay and Salsette Islands. J. Bombay Nat. Hist. Soc. 50: 157–161.

Onuf, C. P., J. M. Teal & I. Valiela, 1977. Interactions of nutrients, plant growth and herbivory in a mangrove ecosystem. Ecology 58: 514–526.

Peng, S. Q., 1990. The Stress Physiological Fundamentals of Plants. University of Northeast Forestry Press, Harbin, China.

Reed, S. C., E. J. Middlebrooks & R. W. Crites, 1988. Natural Systems for Waste Management and Treatment. McGraw-Hill Book Co., New York, N.Y.

Rogers, K. H., P. F. Breen & A. J. Chick, 1991. Nitrogen removal in experimental wetland treatment systems: evidence for the role of aquatic plants. J. Wat. Pollut. Cont. Fed. 63: 934–941.

Saradhi, A. & P. P. Saradhi, 1981. Proline accumulation under heavy metal stress. J. Pl. Physiol. 138: 554–558.

Snedaker, S. C. & J. G. Snedaker, 1984. The Mangrove Ecosystem: Research Methods. United Nations Educational, Scientific and Cultural Organization Publisher, Richard Clay (The Chaucer Press) Ltd., U.K.

Tam, N. F. Y. & Y. S. Wong, 1993. Retention of nutrients and heavy metals in mangrove sediment receiving wastewater of different strengths. Envir. Technol. 14: 719–729.

Tam, N. F. Y. & Y. S. Wong, 1995. Mangrove soils as sinks for wastewater-borne pollutants. Hydrobiologia 295: 231–242.

Trattner, R. B. & S. J. W. Woods, 1989. The use of wetlands systems for municipal wastewater treatment. In Cheremisinoff, P. N. (ed.), Encyclopedia of Environmental Control Technology Vol. 3, Wastewater Treatment Technology. Gulf Publishing Co., Houston TX: 591–622.

Wang, W. C., Z. C. Sun & P. Kuan, 1984. Accumulation, distribution and toxicity of Cd to aquatic higher plants. Envir. Sci. 4: 248–257.

Wong, Y. S., C. Y. Lan, G. Z. Chen, S. H. Li, X. R. Chen, Z. P. Liu & N. F. Y. Tam, 1995. Effect of wastewater discharge on nutrient contamination of mangrove soils and plants. Hydrobiologia 295: 243–254.

Zhu, G. L., 1983. Determination of free proline in plants. Pl. Physiol. Commun. 1: 35–37.

Zhu, G. L., H. W. Zhang & A. Q. Zhang, 1990. Plant Physiological Experiments. Beijing University Press, Beijing: 122–124.

Hydrobiologia **352**: 97–106, 1997.
Y.-S. Wong & N. F.-Y. Tam (eds), Asia–Pacific Conference on Science and Management of Coastal Environment.
©1997 *Kluwer Academic Publishers. Printed in Belgium.*

Size-fractionated productivity and nutrient dynamics of phytoplankton in subtropical coastal environments

Haili Wang, Bangqin Huang & Huasheng Hong
Research Laboratory of SECD on Marine Ecological Environment, Environmental Science Research Center, Xiamen University, Xiamen 361005, China

Key words: size-fractionation, phosphate uptake, productivity, phytoplankton, coastal environment

Abstract

It is now well established that the size distribution of phytoplankton plays an important role in primary production processes and nutrient dynamics of coastal environment. *In situ* observations showed that nanophytoplankton (3~20 μm) contributed 72.08% and 58.18% of phytoplankton biomass and 58.32% and 41.14% of primary productivity to Xiamen Western Waters and the northern Taiwan Strait, respectively; picophytoplankton (0.2~3 μm) dominated the biomass (64.70%) and productivity (66.09%) in the southern Taiwan Strait. Furthermore, nanophytoplankton accounted for 75% of phosphate uptake with the highest rate constant (8.3×10^{-5} s^{-1}) and uptake rate in unit water volume (5.4×10^{-5} mmol dm^{-3} s^{-1}); picophytoplankton had the highest uptake rate in unit biomass (5.4×10^{-5} mmol mg^{-1} s^{-1}) and photosynthetic index (3.8 mgC mgChl a^{-1} h^{-1}). All the results highlighted the remarkable characteristics of small size ranged (0.2~20 μm) phytoplankton in subtropical coastal environments: main contributor to phytoplankton biomass and production, high efficiency on organic carbon production and nutrient recycling. The far reaching environmental and ecological implications were discussed.

Introduction

It has been recognized for a long time that rates of many processes occurring in planktonic ecosystems are body-size dependent. As a consequence, the size structure is of great importance in determining the ecology of aquatic plankton communities. Data on size-fractionated biomass (mainly chlorophyll *a*) and production of phytoplankton are documented in more and more studies. Recently, the substantial contribution of the smaller phytoplankton to total chlorophyll, and total primary productivity in oceanic ecosystems has been demonstrated (Malone, 1980; Davis et al., 1985; Larry & El-Sayed, 1987; Chavez, 1989). Nanophytoplankton (3~20 μm) accounted for a high proportion of the biomass and primary production of most phytoplankton communities. With the application of techniques such as epifluorescence microscopy and flow cytometry, an even smaller portion of picophytoplankton (<3 μm) was found to be of considerable quantitative significance in many parts of the world's oceans

(Platt et al., 1983; Stockner & Antia, 1986; Ray et al., 1989; Ishizaka et al., 1994; Iriate & Purdie, 1995).

This study focused on the size-fractionated phosphate uptake kinetics and the contribution of different size fractions (micro-phytoplankton, nanophytoplankton and pico-phytoplankton) to primary productivity. The study areas are Western Xiamen Bay, Jiulong River Estuary, northern and southern Taiwan Strait, which represent subtropical coastal region of China, and possess distinctive ecological features of phosphorus limitation and high primary production.

Description of sites studied

The investigations were carried out in Western Xiamen Bay (WXB), Jiulong River Estuary (JRE) and the Taiwan Strait.

WXB is located to the west of Xiamen Island (Figure 1). It became a semi-enclosed water body after the construction of the dike which links the mainland and

Xiamen Island. WXB has an area of about 50 km² with a depth range of 6~25 m. Characterized by typical subtropical oceanic climate, this region is highly influenced by monsoon. It has semi-diurnal tide and the tide range is about 4 m. Sea water in WXB is well mixed so that the temperature and salinity are nearly homogeneous, without remarkable horizontal and vertical gradients.

JRE is to the southwest of WXB. Being the second longest river in Fujian, Jiulong river has an average annual runoff of about 1.48×10^8 m³. The flood season and dry season of this river is in May–June and December–January, respectively. The runoff affects the hydro-chemical feature of WXB significantly.

The Taiwan Strait is a dynamic channel on the board continental shelf between the East China Sea and the South China Sea (Figure 2). With irregular and sophisticated bottom topography, water in this region is mostly not deeper than 100 m. The climate is affected by subtropical monsoon. The 'Narrow Pipe Effect' is quite distinct due to the corresponding NE-SW trend of the strait and the mountains along two sides. Upwelling appears to be common in this region owing to the effect of monsoon and its topography.

Materials and methods

Sample collection

Size-fractionated kinetics of phytoplankton was studied at Station XM5 of WXB and Station XM9 of JRE (Figure 1), during five expeditions in this region, which were carried out in August and November 1993 and March, May and October 1994. Water samples were collected from the surface and 4 m depth at high-tide, and were brought immediately back to the laboratory for incubation and uptake experiment. The size structure of chl *a* biomass and primary productivity were investigated in WXB and JRE in November 1993, March, May and October 1994 and May and August 1995, and in the Taiwan Strait (21 °–27 °N, 116.5 °–122.5 °E) during two cruises carried out in August 1994 and February 1995. Water samples of WXB and JRE were collected from the surface layer and 4 m depth; those of Taiwan Strait were collected from depths of 0, 10, 20, 30, 50m of each stations studied (Figure 2).

Table 1. Size-fractionated (shown in percentage) phosphate uptake, chl.a biomass and primary productivity in sites studied (WXB: Western Xiamen Bay; JRE: Jiulong River Estuary; NTS: northern Taiwan Strait; STS: southern Taiwan Strait; MICRO: micro-phytoplankton; NANO: nano-phytoplankton; PICO: pico-phytoplankton)

Sites	MICRO	NANO	PICO
	a.Phosphate uptake		
WXB	23.4	70.0	6.6
JRE	11.3	79.7	9.0
	b. Chl.a biomass		
NTS	22.77	58.18	19.05
STS	10.32	24.98	64.70
WXB	23.82	72.08	4.10
	c. Primary productivity		
NTS	28.44	41.14	30.42
STS	14.22	19.69	66.09
WXB	30.99	58.32	10.69

Size-fractionation

The procedure of size-fractionation are illustrated in Figure 3. Before incubation, 200 μm mesh was used to remove zooplankton. Then radioisotopes (^{14}C or ^{32}P) were added and the sample was incubated. After incubation, the sample was divided into three replicates. Two of them were respectively filtered onto 0.2, 3.0 μm Nucleopore filters; another one, after the filtration with 20 μm mesh, was filtered onto 3.0 μm Nucleopore filter. These three filters were termed as fractions A, B and C. Therefore, micro-fraction (20~200 μm) = B–C; nano-fraction (3~20 μm) = C; pico-fraction (0.2~3 μm) = A–B.

Uptake rate constant

The kinetic constant of phosphate uptake was measured by isotope dilution method using carrier-free ^{32}P-labeled PO$_4$-P (Lean & White, 1983). At time zero, about 10 μCi carrier-free ^{32}P-orthophosphate was added into 100 ml water sample. Aliquots of 10 ml were removed at timed interval of 80, 120, 160, 220, 300, 400 s after the addition of radioisotope. The aliquots of 10 ml was filtered through phosphate-soaked Millipore HA filters with different pore size (see procedure of Figure 3). The filters were air-dried for 48 h and placed into 20 ml scintillation vials, each added with 1 ml of ethyl acetate to dissolve the filter, and then 9 ml of isobutyl alcohol was added. The radio-activity was determined by a liquid scintillation

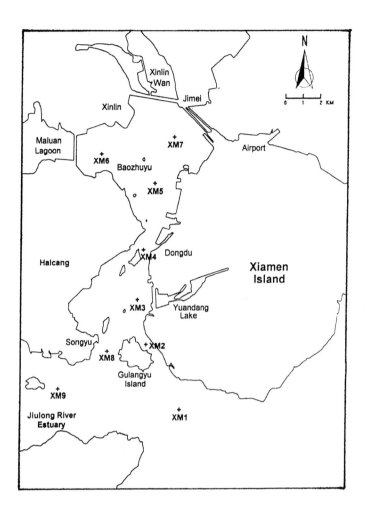

Figure 1. Sampling stations in Western Xiamen Bay and Jiulong River Estuary.

counter (Packard Tri-carb). The uptake rate constant can be calculated by the following equation (Heath, 1986):

$$U = dM/dt = K * P \qquad (1)$$

where U is the uptake rate, M is the amount taken up by algal cells at time t, P is the amount of dissolved phosphate available at that time, and K is the uptake rate constant. Equation (1) can be expressed as:

$$dX/dt = K(P_0 - X) \qquad (2)$$

where P_0 is the radioactivity at time zero, X is the radioactivity of algae cells at time t. The integrated form of equation (2) is:

$$\ln(P_0/(P_0 - X)) = K * t$$

It was reported that algae uptake phosphate very quickly at the initial (several hundred seconds). We plotted $\ln(P_0/(P_0 - X))$ against t of the early stage (linear relationship), and the slope gives the uptake rate constant, K.

Uptake rate

The uptake rates, expressed in unit volume (U) and unit chl a biomass (U'), can be estimated as: $U = K * P$; $U' = U/\text{chl } a$, where

K is the uptake rate constant (s^{-1});

U is the uptake rate per unit volume (mol l^{-1} s^{-1});

chl a is the concentration of chlorophyll a (mg l^{-1} or mg m^{-3});

U' is the uptake rate per unit chl a biomass (mol μg chl a^{-1} s^{-1})

100

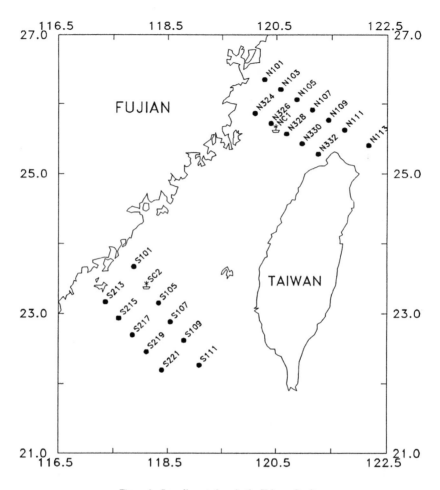

Figure 2. Sampling stations in the Taiwan Strait.

Estimation of phosphate uptake percentage for different size fractions

Phosphate uptake proportion of one size fraction is the average of uptake proportion measured according to the time intervals described above. The uptake percentage at time t is equal to the ratio (%) of ^{32}P taken by that size fraction with that of total phytoplankton.

Chlorophyll a

The determination of chl a was performed by fluorescent analysis, followed the method described by Parsons et al. (1984), with a slight modification that the volume of seawater sample was 150 ml, and *in vitro* measurements were conducted in a Hitachi 850 Flurospectrometer with the excitation and emission wave length set at 430 nm and 670 nm.

Photosynthesis

Phytoplankton photosynthesis rate was measured by $H^{14}CO_3^-$ assimilation method (Parsons et al., 1984).

Statistical analysis

A one-way ANOVA was used to test the differences of contribution percentage of chl a biomass and primary productivity among sites studied for each size-fraction (after arcsine transformation of percentage data).

Results

Size-fractionated phosphate uptake and its seasonal variation

In WXB, about 70% of the phosphate uptake was attributed to nano-phytoplankton (NANO), while the smallest size fraction, pico-phytoplankton (PICO),

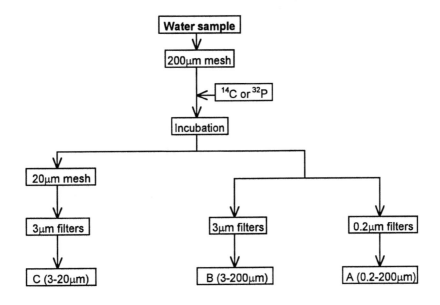

micro-fraction = B - C;   nano-fraction = C;   pico-fraction = A - B

Figure 3. Procedures for size fractionation.

Table 2. Size-fractionated phosphate uptake rate (U) and uptake rate constant (K) of phytoplankton (per unit water volume) in Western Xiamen Bay

Time (m)	Station	Depth	$U(10^{-12} \text{mol } l^{-1} \text{ s}^{-1})$			$K(10^{-6} \text{ s}^{-1})$		
			MICRO	NANO	PICO	MICRO	NANO	PICO
Nov. 1993	XM5	0	63.8	120	4.52	85.0	160	6.03
Nov. 1993	XM5	4	8.82	87.4	10.3	10.5	104	12.3
May 1994	XM5	0	7.19	48.1	7.55	14.1	94.4	14.8
May 1994	XM9	0	9.02	24.5	0.35	33.4	90.7	1.28
Oct. 1994	XM5	0	30.2	36.3	19.5	27.2	32.7	17.6
Oct. 1994	XM9	0	4.54	6.22	0	10.8	14.8	0
Average			20.6	53.8	7.04	30.2	82.8	8.67

only accounted for less than 10% (Table 1a). Similar results were obtained in JRE. All the three size-fractions appeared to be significantly variable with seasons. PICO always consisted a negligible proportion to phosphate uptake, except in summer and early autumn, when it accounted for nearly 1/5 of the total phytoplanktonic phosphate uptake. Micro-phytoplankton (MICRO) contributed over 1/2 of the uptake in May, in contrast to the fact that it contributed little in other seasons (Figure 4). The seasonal variation of phosphate uptake by different size fraction in JRE followed to the same pattern of WXB.

Uptake kinetics parameters of different size fractions
Similar to proportional contribution to phosphate uptake, when compared among different phytoplankton size fractions, the uptake rate constant (K), and uptake rate in unit water volume (U) also followed the same order of NANO>MICRO>PICO (Table 2).

Moreover, they varied sharply in different seasons. U and K values of NANO and MICRO in late autumn were much higher than that of other seasons. PICO also took phosphate quicker in autumn than in other seasons, whereas the seasonal variation was relatively unremarkable. It should be pointed out that if compared

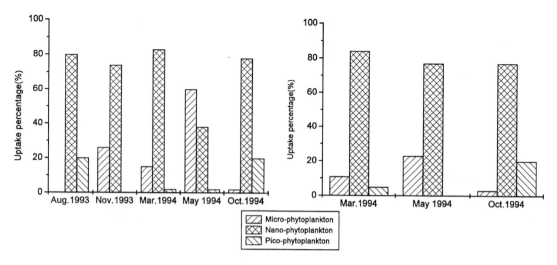

Figure 4. Seasonal variation of size-fractionated phosphate uptake by phytoplankton (left: Western Xiamen Bay; right: Jiulong River Estuary).

Table 3. Size-fractionated phosphate uptake rate of photyplankton (per unit chl.a biomass, U') is Western Xiamen Bay

Time	Station	Depth (m)	$U'(10^{-12}\text{mol }\mu\text{gChl.a}^{-1}\text{ s}^{-1})$		
			MICRO	NANO	PICO
Nov. 1993	XM5	0	/	36.8	126
Nov. 1993	XM5	4	16.8	36.1	35.3
May 1994	XM5	0	1.63	3.72	/
Oct. 1994	XM5	0	26.2	11.4	/
Oct. 1994	XM9	0	1.09	1.72	0
Average			11.4	17.9	53.8

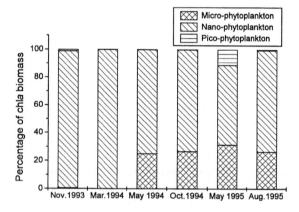

Figure 5. Seasonal variation of the contribution to chl *a* biomass by different size fraction at the surface water of Station XM5 in WXB.

in unit chl *a* biomass, things changed dramatically. The smallest size fraction showed an overwhelming advantage in uptaking phosphate. U' of PICO was about 2 to 4 times higher than that of NANO and MICRO (Table 3).

Size structure of chl a biomass with seasonal pattern
The smaller size fractions of phytoplankton ($<20\ \mu$m) played an important role in the regions we studied. NANO dominated chl *a* biomass in northern Taiwan Strait (NTS) and WXB (Table 1b), and the contribution percentage of NANO to WXB was significantly ($p<0.001$) higher than that to NTS. PICO accounted for nearly 2/3 of the chl *a* biomass in southern Taiwan Strait (STS). In contrast, the percentage of PICO in NTS and WXB were significantly ($p<0.001$) lower. It should be pointed out that the means for percentage of MICRO in NTS, STS and WXB were not significantly different ($p=0.14$), implying that its contribution

to chla biomass was of high variation with stations in each region (data not shown). Size structure of chl *a* biomass at the surface water of Station XM5 in WXB was determined in six expeditions in two years of field research. The results indicated that size structure of phytoplankton varied with seasons. NANO dominated during cold season (November and March), but in warm season, MICRO contributed about 1/4 of chl *a* biomass (Figure 5). A contribution of more than 10% by PICO was observed in May 1995, indicating the fact that PICO fraction occasionally played an important role in phytoplankton assemblage.

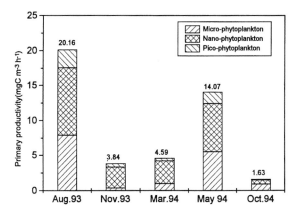

Figure 6. Seasonal variation of size-fractionated primary productivity in WXB (data shown on the stacked column represent the sum of three size-fractions).

Size structure of primary productivity with seasonal pattern

The relative importance of the different phytoplankton size fractions to primary productivity were, for all three regions we studied, similar to those described for chl *a* biomass (Table 1c). The means of NANO and PICO in three regions were of highly significant difference ($p<0.001$). Those for MICRO were not significantly different ($p = 0.26$).

In WXB, the total primary productivity and the contribution of three size fractions were of significant seasonal variation (Figure 6). WXB is most productive in August, and the second largest organic carbon yield by phytoplankton appears in May. MICRO played an important role in productive season, which could be indicated that it contributed over 1/3 of the total primary productivity. However, MICRO was the dominator in October 1994, the least productive period of the five expeditions we conducted. NANO accounted more than 2/3 of total primary productivity in winter. For PICO, seasonal variation of contribution percentage was not so remarkable as that of other two size fractions. It consisted about 10% of total primary productivity all the time.

Discussion

Factors controlling the size distribution of phytoplankton

It has been well recognized that the size structure of phytoplankton populations is latitudinally determined.

Larger size fraction, especially MICRO, tends to be associated negatively with temperature. In contrast, picoplankton are more important in subtropical and tropical oceanic waters (Takahashi & Bienfang, 1983; Le Bouteiller et al., 1992). Besides, size distribution patterns of phytoplankton are also highly associated with trophic conditions. Picoplankton are more important in oligotrophic ecosystems, while large phytoplankton cells dominate in eutrophic waters. However, trophic state alone (oligotrophic *vs* eutrophic) was not a good indicator of the importance of picoplankton (Wehr, 1991). Smaller cells were relatively abundant when phosphorus was limiting other phytoplankton, but also when nitrogen was in surplus. This relationship was strongly affected by light and grazer levels.

Nutrient uptake kinetics is another main controlling factor. Picoplankton have higher light harvesting ability and growth rate than larger cells. In WXB, picoplankton also have higher phosphate uptake rate per unit chl *a* biomass. Therefore, their contribution to phytoplankton biomass was highest in May, when the concentration of soluble phosphate was the lowest (Song & Wang, 1987). Generally, however, the proportions of phosphate uptake by three size fractions were positively associated with corresponding proportions of chl.a biomass and photosynthesis (Huang et al., 1995). Having the highest phosphate uptake rate per unit water volume, in addition to the fact that phosphate turnovers very quickly in WXB (Hong et al., 1994), NANO was not limited by low P concentration. As a result, they became the dominant size fraction of phytoplankton communities in this region (Tables 4 & 5). The remarkable difference of phytoplankton size structure between NTS and STS was mainly attributed to their different latitude (water temperature).

The role of grazing in scaling the size structure of phytoplankton populations is still controversial. On one hand, it has been suggested that picophytoplankton are possibly more tightly controlled by grazing than larger phytoplankton (Thingstad & Sakshang, 1990); on the other hand, Legendre et al. (1993) found that in the Arctic Domain, primary production was generally dominated by cells $>5~\mu$m, but the standing stock was dominated by the $<5~\mu$m fraction. This difference between production and standing stock was potentially due to strong grazing on the large cells. We have no data in the present study on concentrations of possible consumers of each phytoplankton size fraction and grazing rates, but it could be implied from the coupling between chl *a* biomass and productivity

104

Table 4. Comparison of size-fractionated standing crop (chlorophyll *a*) of phytoplankton in different regions (shown in percentage)

Region	MICRO	NANO	PICO	Nutrient Status	Reference
Hawaii	2	18	80	/	Takahashi & Bienfang, 1983
Gulf of Maine	/	/	60~97	oligotrophic	Stockner & Antia, 1986
Mediterrancean	/	/	60~80	oligotrophic	Stockner & Antia, 1986
Baltic Sea	/	/	10~25	eutrophic	Stockner & Antia, 1986
Tsingtao	30	60	10	/	Chen & Qian, 1992
WXB	23.8	72.1	4.1	mesotrophic	This study
NTS	22.8	58.2	19.0	mesotrophic	This study
STS	10.3	25.0	64.7	mesotrophic	This study

Table 5. Comparison of size-fractionated primary productivity of different regions (shown in percentage)

Region	MICRO	NANO	PICO	Nutrient Status	Reference
Baltic Sea	/	/	3~43	eutrophic	Stockner & Antia, 1986
Gulf of Maine	/	/	60~80	oligotrophic	Stockner & Antia, 1986
Mediterrancean	/	/	25~90	oligotrophic	Stockner & Antia, 1986
WXB	31.0	58.3	10.7	mesotrophic	This study
NTS	28.5	41.1	30.4	mesotrophic	This study
STS	14.2	19.7	66.1	mesotrophic	This study

that the grazing pressure on all size fraction was nearly equivalent.

The relationship between phytoplankton size distribution and dynamics of organic carbon

The size distribution of phytoplankton populations is important in determining the direction and magnitude of carbon and energy fluxes within the euphotic zone of pelagic ecosystems and via sedimentation to below the euphotic zone and sediments. Traditional viewpoints of oceanic size distribution has been increasingly challenged. Using 0.2 μm Nucleopore filters instead of Whatman GF/F, Dickson & Wheeler (1993) demonstrated that surface chl *a* concentrations in North Pacific (28 °–48 °N along 152 °W) in March 1991 were up to 4 fold higher than those measured with Whatman GF/F filters. The largest difference between the two filter types was found in subtropical waters, where picoplankton were a major constituent of the phytoplankton assemblage. Integrated chl.a concentrations based on measurements with 0.2 μm Nucleopore filters were nearly constant. As a consequence, the authors questioned the existence of previously reported latitudinal gradients in integrated chl *a* concentrations in the North Pacific. The emergence of newly developed par-

adigm of size-structured primary producer also questions the traditional oceanic organic carbon dynamics since there is a rapid cycling of carbon and relevant biogenic elements in the open sea that is driven by synergistic picoplankton interactions between functional compartments within microbial loop (Glover, 1991). The microbial loop is thought to enhance ecosystem efficiency through rapid recycling and reduced sinking rates, therefore reducing the loss of nutrients contained in organisms remaining within the euphotic zone. Systems with greater relative importance of autotrophic picoplankton had significantly reduced loss rates (Wehr et al., 1994). We suggest that this can well explain the fact that picoplankton-dominated phytoplankton assemblage support an productive pelagic coastal upwelling fishing ground in STS.

Ecological implications of size-fractionated phosphate uptake

Our results in WXB are consistent with the view that large-celled phytoplankton control the major fluctuations in biomass and primary productivity, while picoplankton account for a comparatively stable background productivity. The intrinsic property of phosphate uptake kinetics for different phytoplankton size fractions, coupling with the effects of nutrient status in ambient water body, therefore have far-reaching ecological implications. It was suggested that the examination on the relationships between the dominant phytoplankton and the factors of water quality is one of the useful methods to clarify the distinctive features of phytoplankton and to forecast the occurrence of algal bloom (Yoshida & Makahara, 1995). Smaller cells of phytoplankton dominate primary producers in WXB and larger cells only occur in seasons of high standing stock. The sudden loading of nutrients such as phosphate in this area may be liable to cause a shift from the dominance of NANO and picoplankton toward an increasing contribution from size fraction of >20 μm, thus having the potential in triggering blooms of diatom or dinoflagellates. The red tide event occurred during mid-June to early July 1986 in the inner WXB (Zhang, 1988) could be served as an convincing example for revealing the environmental implications of the interactions between phytoplankton size structure and environmental factors.

Conclusion

Size-fractionated phosphate uptake kinetics and contribution of different size fractions to primary productivity were studied in Western Xiamen Bay (WXB), Jiulong River Estuary (JRS) and the Taiwan Strait. Over 2/3 of phosphate uptake was attributed to NANO. The proportions of phosphate uptake by MICRO and PICO were of great seasonal fluctuation. The uptake rate constants and uptake rates per unit volume of three size fractions were in the sequence of NANO>MICRO>PICO, while the sequence for the uptake rates per unit chl.a biomass were PICO>NANO>MICRO.

NANO and PICO dominated the standing stock and production of phytoplankton in WXB, JRE, northern Taiwan Strait and in southern Taiwan Strait, respectively. Larger cells of phytoplankton tended to act as significant components of chl a biomass only in warm seasons. In WXB and JRE, PICO accounted for a comparatively stable background productivity, while NANO and MICRO controlled the major seasonal fluctuation in primary productivity. Size distribution of phytoplankton depended on water temperature, nutrient supply and the intrinsic nutrient uptake kinetics. The size structure of primary producers was tightly related to pelagic organic carbon dynamics. The kinetics of phosphate uptake in WXB was of fundamental

importance in triggering blooms of larger phytoplankton such as red tide species.

References

Chavez, F. P., 1989. Size distribution of phytoplankton in the central and eastern tropical Pacific. Glob. Biogeochem. Cycles 3: 27–35.
Chen, H. & S. Qian, 1992. The study on nanoplankton and picoplankton in nearshore waters of Tsingtao. Acta Oceanol. Sinica 14: 105–113 (in Chinese).
Davis, P. G., D. A. Caron, P. W. Johnson & J. McN. Sieburth, 1985. Phototrophic and apochlorotic components of picoplankton and nanoplankton in the North Atlantic: geographic, vertical, seasonal and diel distributions. Mar. Ecol. Prog. Ser. 21: 15–26.
Dickson, M. L. & P. A. Wheeler, 1993. Chlorophyll-a concentrations in the North Pacific – Does a latitudinal gradient exist? Limnol. Oceanogr. 38: 1813–1818.
Glover, H. E., 1991. Oceanic phytoplankton communities-our changing perception. Revue aquat. Sci. 5: 307–331.
Heath, R. T., 1986. Dissolved organic phosphorus compounds: Do they satisfy planktonic phosphate demand in summer? Can. J. Fish. aquat. Sci. 43: 343–350.
Hong, H., M. Dai, B. Huang & W. Li, 1994. Studies on rate of phosphate uptake by phytoplankton in Xiamen Harbour. Oceanol. Limnol. Sinica 25: 54–59 (in Chinese with English abstract).
Huang, B., H. Hong, H. Wang & L. Hong, 1995. Size-fractionated phosphate uptake by phytoplankton in West Xiamen Harbour. J. Oceanogr. Taiwan Strait 14: 269–273 (in Chinese with English abstract).
Iriarte, A. & D. A. Purdie, 1995. Size distribution of chlorophyll a biomass and primary production in a temperate estuary (Southampton Water): the contribution of photosynthetic picoplankton. Mar. Ecol. Prog. Ser. 115: 283–297.
Ishizaka, J., H. Kiyosawa, K. Ishida, K. Ishikawa & M. Takahashi, 1994. Meridional distribution and carbon biomass of autotrophic picoplankton in the Central North Pacific Ocean during Late Northern Summer 1990. Deep Sea Res. 41: 1745–1766.
Larry, H. W. & S. Z. El-Sayed, 1987. Contributions of the net-, nano and picoplankton to the phytoplankton standing crop and primary productivity in the Southern Ocean. J. Plankton Res. 9: 973–994.
Lean, D. R. S. & E. White, 1983. Chemical and radiotracer measurements of phosphorus uptake by lake plankton. Can. J. Fish. aquat. Sci. 40: 147–155.
Le Bouteiller, A., J. Blanchot & M. Podier, 1992. Size distribution patterns of phytoplankton in the western Pacific: towards a generalization for the tropical open ocean. Deep Sea Res. 39: 805–823.
Legendre, L., M. Gosseline, H. J. Hirche, G. Katner & G. Rosenberg, 1993. Environmental control and potential fate of size-fractionated phytoplankton production in the Greenland Sea (75N). Mar. Ecol. Prog. Ser. 98: 297–313.
Malone, T. C., 1980. Size fractionated primary productivity of marine phytoplankton. In Falkowski, P. G. (eds), Primary productivity in the sea. Plenum Press, New York, 301–319.
Parsons, T. R., Y. Maita & C. M. Lalli, 1984. A manual of chemical and biological method for seawater analysis. Pergamon Press, Oxford.
Platt, T., D. V. Subba Rao & B. Irwin, 1983. Photosynthesis of picosynthesis in the oligotrophic ocean. Nature 301: 702–704.

Ray, R. T., L. W. Haas & M. E. Sieracki, 1989. Autotrophic picoplankton dynamics in a Chesapeake Bay sub-estuary. Mar. Ecol. Prog. Ser. 52: 273–285.

Song, J. & W. Wang, 1987. Soluble phosphate in the Xiamen Harbour waters. J. Oceanogr. Taiwan Strait 6: 308–326 (in Chinese with English abstract).

Stockner, J. G. & N. J. Antia, 1986. Algal picoplankton from marine and freshwater ecosystems: A multidisciplinary perspective. Can. J. Fish. aquat. Sci. 43: 2472–2503.

Takahashi, M. & P. K. Biengang, 1983. Size structure of phytoplankton biomass and photosynthesis in subtropical Hawaiian waters. Mar. Biol. 76: 203–211.

Thingstad, T. F. & E. Sakshang, 1990. Control of phytoplankton growth in nutrient recycling ecosystems: Theory and terminology. Mar. Ecol. Prog. Ser. 63: 261–272.

Wehr, J. D., 1991. Nutrient and grazing-mediated effects on picoplankton and size structure in phytoplankton communities. Int. Revue ges. Hydrobiol. 76: 643–656.

Wehr, J. D., J. Le & L. Campbell, 1994. Does microbial biomass affect pelagic ecosystem efficiency – an experimental study. Microb. Ecol. 27: 1–17.

Yoshida, Y. & H. Nakahara, 1995. Relationships between the dominant phytoplankton and DIN:DIP ratios in Lake Biwa. Nippon Suisan Gakkaishi 61: 561–565 (in Japanese with English abstract).

Zhang, S., 1988. An observation of red tide event in Western Xiamen Harbour. Acta Oceanol. Sinica. 10: 602–608 (in Chinese).

Hydrobiologia **352**: 107–115, 1997.
Y.-S. Wong & N. F.-Y. Tam (eds), Asia–Pacific Conference on Science and Management of Coastal Environment.
©1997 *Kluwer Academic Publishers. Printed in Belgium.*

Long-term changes in hydrography, nutrients and phytoplankton in Tolo Harbour, Hong Kong

Y.-K. Yung [1,*], C. K. Wong [2], M. J. Broom[1], J. A. Ogden[1], S. C. M. Chan [3] & Y. Leung[3]
[1]*Water Policy and Planning Group, Hong Kong Government Environmental Protection Department, 130 Hennessy Road, Wanchai, Hong Kong*
[2]*Department of Biology & The Centre for Environmental Studies, The Chinese University of Hong Kong, Shatin, Hong Kong*
[3]*The Centre for Environmental Studies, The Chinese University of Hong Kong, Shatin, Hong Kong*
(*Author for correspondence)

Key words: coastal waters, nutrients, phytoplankton

Abstract

This paper presents the results of a long-term survey of the hydrography, nutrients and phytoplankton in Tolo Harbour carried out between 1982 and 1992. Some nutrients such as total inorganic nitrogen, ammonia and total phosphorus increased during the 10 year period, but chlorophyll *a*, which indicated algal biomass, did not show an increasing trend. The phytoplankton of Tolo Harbour consisted largely of diatoms. Dinoflagellates and minor algal groups such as cryptomonads and small flagellates constituted a smaller fraction of the phytoplankton population. Densities of diatoms and minor algal groups increased in some stations, but the density of dinoflagellates remained relatively unchanged during the study period. Most nutrient variables were negatively correlated with densities of diatom and total phytoplankton, and positively correlated with densities of minor algal groups. While dinoflagellate densities were positively correlated with total nitrogen in some stations, no correlation existed between dinoflagellate density and most of the nutrient variables. Our results show that there is a gradual change in phytoplankton community in Tolo Harbour, most notably in the nutrient-rich inner harbour waters, with the smaller algae assuming increasing abundance. Thus there was a net increase in density of total phytoplankton even though chlorophyll *a* concentrations did not increase. No evidence was found in this study to show that increased nutrient loading would inevitably lead to increase in densities of dinoflagellates in Tolo Harbour. Instead, dinoflagellate densities showed stronger correlations with physical variables such as temperature, pH and salinity.

Introduction

Tolo Harbour is a semi-enclosed inlet in the northeastern part of Hong Kong. The bay comprises of an inner harbour and a narrow channel that opens into the Pacific Ocean. Total area is about 50 km^2 and mean depth is 15 m. Tidal exchange in the inner parts of the harbour is severely restricted because of the 'bottleneck' topography. Tolo Harbour received secondary treated sewage from two waste treatment plants that serve over one million people. Studies conducted in the early 1970's have shown that Tolo Harbour was organically polluted (Wear et al., 1984). Increased

nutrient loading and extensive coastal land reclamation have resulted in reduction of benthic fauna (Wu, 1988) and changes in the phytoplankton community structure (Hodgkiss & Chan, 1983). Beginning in the late 1970's the incidence of red tides in Tolo Harbour has increased dramatically (Holmes & Lam, 1985; Holmes 1988; Lam & Ho, 1989) and Hodgkiss and Chan (1983) reported that there was a trend of gradual replacement of diatoms by dinoflagellates in the Tolo Harbour Channel. Recognising the seriousness of the organic pollution problem, the Environmental Protection Department of the Hong Kong Government established a programme to monitor the water qual-

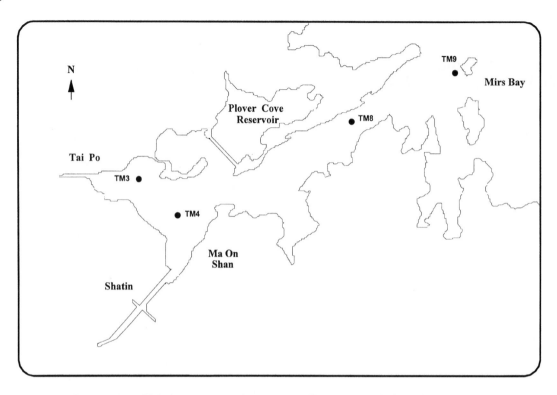

Figure 1. Map of Tolo Harbour showing location of sampling stations TM3, TM4, TM8 and TM9.

ity and phytoplankton of Tolo Harbour in the early 1980's. The programme has generated over ten years of hydrographical, nutrient and biological data. This paper describes long-term changes in the abundance of major phytoplankton groups in relation to nutrient concentrations and some hydrographic factors during the ten years period from 1982 to 1992. The main objectives are to detect long term trends in hydrographical and nutrient factors and to identify those that show significant correlation with phytoplankton densities in Tolo Harbour.

Materials and methods

Field sampling

Marine water was sampled monthly at 4 fixed stations (TM3, TM4, TM8 and TM9) in Tolo Harbour from 1982 to 1992 (Figure 1). Hydrographical parameters such as temperature, salinity, pH, turbidity and dissolved oxygen concentration were measured *in situ* using a Sea-Bird SBE 19 Seacat electronic profiler. Seawater was collected at 1 m below the surface with a

3-l Van Dorn water sampler. An aliquot of at least 200-ml was removed from each sample and preserved in Lugol's iodine solution (1: 100 ml) for phytoplankton identification and enumeration. The remaining sample was poured into a polyethylene bottle, kept in a cool and dark container, and transported to the Hong Kong Government Laboratory for nutrient analysis. All analytical methods are listed in the technical memorandum by the Hong Kong Environmental Protection Department (1991).

Phytoplankton enumeration

In the laboratory, a 1-ml aliquot from each preserved phytoplankton sample was settled in a Sedgwick-Rafter counting cell. The sample was identified and enumerated using a Nikon Optiphot compound microscope at $100 \times$ magnification. Taxonomic references consulted included Dodge (1982), Yamaji (1984) and Larsen & Moestrup (1989).

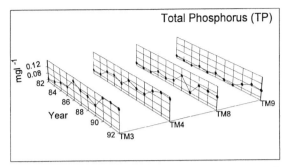

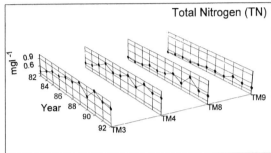

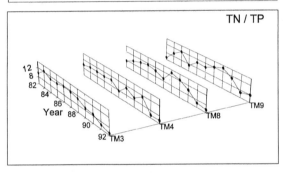

Figure 2. Changes in total phosphorus (TP), total nitrogen (TN) and the ratio of TN/TP in 4 sampling stations in Tolo Harbour, 1982–1992.

Statistical analysis

One-way analysis of variance and Tukey's test was used to detect spatial differences in the hydrographical, nutrient and phytoplankton variables. Farrell's aligned rank method (van Belle & Hughes, 1984) was used to detect the presence of temporal trends in the hydrographical, nutrient and phytoplankton data. Kendall's coefficient of rank correlation (Sokal & Rohlf, 1981) was used to measure the degree of association between phytoplankton variables and each of the hydrographical and nutrient parameters.

Results

The hydrographical, nutrient and biological data of Tolo Harbour are summarized in Table 1. During the study period, the annual average surface water temperature in the 4 sampling stations ranged between 21.9 °C and 25.7 °C. The highest and lowest surface water temperature recorded during the study period were 33.2 °C and 11.4 °C respectively. Thermal stratification of the water column occurred each year during the summer months from April and October. The annual average surface salinity in the 4 sampling stations ranged from 28.8 to 31.7‰ and was slightly less than the 33 to 38‰ generally recorded for open sea waters. As expected, the 10 year mean of annual average surface salinity increased from the inner harbour (30.1‰ in TM3 and 30.0‰ in TM4) to the channel (31.0‰ in TM8) and the outer harbour (31.5‰ in TM9).

In general, the surface water concentration of all nutrient variables was higher in the inner harbour compared with the channel and outer harbour. The annual average concentrations of orthophosphate phosphorus (PO_4-P), nitrate nitrogen (NO_3-N) and ammonia nitrogen (NH_4-N) were significantly higher in TM3 and TM4 than in TM8 and TM9 ($P<0.05$). The annual average biochemical oxygen demand (BOD), which was an indicator of organic pollution, was also significantly higher in TM3 and TM4 than in TM8 and TM9.

The 10-year mean of annual average surface water chlorophyll *a* concentration in TM3 and TM4 (14.6 mg m^{-3} and 15.1 mg m^{-3} respectively) were significantly higher than those in TM8 (5.5 mg m^{-3}) and TM9 (1.8 mg m^{-3}) ($P<0.01$). The annual average densities of total phytoplankton, diatoms and dinoflagellates were also significantly higher in TM3 and TM4 than in TM9 ($P<0.05$). The annual average Secchi disk transparency in the 4 sampling stations ranged from 1.79 to 4.92 m and the annual average values were significantly lower in TM3 and TM4 compared with TM8 and TM9 ($P<0.05$).

Results of Farrell's test for trends in the various hydrographical, nutrient and phytoplankton factors are summarized in Table 2. In general, all nutrient factors showed increasing trend from 1982 to 1992. The trends for total phosphorus (TP) was significant in all 4 sampling stations and those for total nitrogen (TN) was significant at TM9 (Figure 2). The ratio of TN/TP showed a statistically significant decreasing trend in TM3 and TM4 (Figure 2).

Table 1. Ranges and means of some hydrographical, nutrient and biological factors in Tolo Harbour, 1982–1992. (All measurements were taken from the surface water except for Secchi disk depth and bottom dissolved oxygen.)

		TM3	TM4	TM8	TM9
Temperature	Min-Max	22.83–25.73	21.99–25.32	21.93–23.68	21.91–23.46
(°C)	(Mean)	(23.70)	(23.34)	(23.05)	(22.87)
pH	Min-Max	8.27–8.76	8.25–8.85	8.20–8.67	8.16–5.57
	(Mean)	(8.43)	(8.43)	(8.35)	(8.29)
Turbidity	Min-Max	2.39–4.97	2.46–4.93	1.42–8.47	0.70–2.60
(NTU)	(Mean)	(3.79)	(3.56)	(2.80)	(1.81)
Suspended solids	Min-Max	2.11–7.40	2.08–6.70	1.32–11.33	0.84–4.15
($mg\,l^{-1}$)	(Mean)	(3.99)	(3.90)	(3.34)	(2.04)
Secchi disk depth	Min-Max	1.53–2.10	1.39–2.09	2.99–5.03	3.43–5.81
(m)	(Mean)	(1.79)	(1.81)	(3.77)	(4.92)
Salinity	Min-Max	29.40–30.82	28.77–31.03	30.38–31.67	30.61–32.26
(‰)	(Mean)	(30.10)	(29.99)	(31.03)	(31.49)
D.O. (Surface)	Min-Max	7.29–9.43	8.05–10.09	7.08–8.37	6.84–8.14
($mg\,l^{-1}$)	(Mean)	(8.38)	(8.65)	(7.80)	(7.39)
D.O. (Bottom)	Min-Max	3.38–5.21	3.00–5.44	3.89–6.62	4.69–7.21
($mg\,l^{-1}$)	(Mean)	(4.09)	(3.99)	(4.99)	(5.48)
NO_2-N	Min-Max	0.004–0.037	0.004–0.020	0.002–0.008	0.002–0.008
($mg\,l^{-1}$)	(Mean)	(0.011)	(0.008)	(0.004)	(0.004)
NO_3-N	Min-Max	0.016–0.085	0.011–0.109	0.004–0.036	0.005–0.020
($mg\,l^{-1}$)	(Mean)	(0.047)	(0.049)	(0.014)	(0.011)
NH_4-N	Min-Max	0.046–0.229	0.040–0.182	0.019–0.072	0.014–0.051
($mg\,l^{-1}$)	(Mean)	(0.134)	(0.093)	(0.039)	(0.024)
TIN	Min-Max	0.071–0.350	0.061–0.266	0.027–0.111	0.024–0.066
($mg\,l^{-1}$)	(Mean)	(0.191)	(0.150)	(0.057)	(0.039)
TKN	Min-Max	0.344–0.971	0.354–0.862	0.239–0.787	0.172–0.469
($mg\,l^{-1}$)	(Mean)	(0.710)	(0.652)	(0.437)	(0.313)
TN	Min-Max	0.370–1.064	0.373–0.939	0.250–0.829	0.179–0.498
($mg\,l^{-1}$)	(Mean)	(0.768)	(0.708)	(0.454)	(0.327)
PO_4-P	Min-Max	0.006–0.069	0.010–0.066	0.005–0.025	0.005–0.019
($mg\,l^{-1}$)	(Mean)	(0.033)	(0.032)	(0.012)	(0.008)
TP	Min-Max	0.044–0.160	0.045–0.155	0.033–0.114	0.024–0.088
($mg\,l^{-1}$)	(Mean)	(0.102)	(0.098)	(0.068)	(0.047)
TSi	Min-Max	0.450–0.913	0.408–1.055	0.402–0.694	0.492–0.828
($mg\,l^{-1}$)	(Mean)	(0.582)	(0.592)	(0.565)	(0.627)
TSi/TN	Min-Max	0.53–1.19	0.68–1.65	0.83–2.47	1.49–4.94
	(Mean)	(0.71)	(0.96)	(1.71)	(2.76)
TSi/TP	Min-Max	3.56–7.03	3.61–15.23	6.15–21.64	7.36–26.09
	(Mean)	(5.82)	(7.98)	(14.01)	(19.63)
TN/TP	Min-Max	6.36–12.27	6.05–11.38	6.02–13.44	5.72–13.60
	(Mean)	(9.78)	(9.22)	(9.72)	(9.69)
BOD_5	Min-Max	2.79–5.13	2.43–4.49	1.19–3.30	0.84–1.67
($mg\,l^{-1}$)	(Mean)	(3.70)	(3.40)	(1.94)	(1.13)
Chlorophyll *a*	Min-Max	5.51–22.54	5.81–31.91	1.35–15.28	0.79–3.17
($mg\,m^{-3}$)	(Mean)	(14.63)	(15.06)	(5.46)	(1.83)
E. coli	Min-Max	80–473	141–460	2–34	1–4
CFU/100 ml	(Mean)	(298)	(239)	(7)	(2)

Table 2. Farrel's test for trends of some hydrographical, nutrient and biological factors in Tolo Harbour, 1982–1992. (* statistically significant at $p<0.05$; ** statistically significant at $p<0.01$).

Parameters	TM3	TM4	TM8	TM9
Temperature	−0.5631	−0.2697	−0.1995	−0.201
pH	−0.0874	0.3053	0.1882	0.2359
Turbidity	−0.7022	0.0999	0.3845	1.6727
Suspended solids	−1.5185	−1.1290	0.5158	1.2544
Dissolved Oxygen	−0.0986	0.9943	1.4511	1.1433
Sicchi disk depth	0.4130	−1.0782	−1.8449	−1.9969*
Salinity	−0.8239	−0.8118	0.3071	0.6050
NO_2-N	1.1868	1.1522	1.1930	1.8065
NO_3-N	1.6670	1.2616	1.6592	1.7114
NH_4-N	1.9683*	2.2230*	2.2157*	1.9666*
TIN	2.0453**	2.2568**	2.3897*	2.4212*
TKN	1.2493	1.5419	1.7925	3.1306*
TN	1.5600	1.8286	1.8206	3.2731**
PO_4-P	3.3817**	3.5583**	2.9381**	1.1833
TP	3.0649**	3.4744**	2.7066**	2.3568*
TSi	N/A	0.4602	0.1414	0.3075
TN/TP	−2.5072*	−1.9825*	−0.7146	0.2842
BOD	0.4645	0.9498	1.4713	1.0295
Chlorophyll *a*	−1.5472	−0.4165	0.0276	−0.4237
Total phytoplankton density	1.3204	1.9877*	0.9861	0.5862
Diatom density	1.4034	1.9700*	0.8748	0.5472
Dinoflagellate density	−1.3399	−1.6714	−0.7615	−1.1594
Minor algal groups density	2.2352*	2.4743*	2.2780*	1.7775

Table 3. Kendall correlations for phytoplankton and hydrogeographical factors in Tolo Harbour, 1982–92. (+/- Statistically significant at $p<0.05$; ++/– Statistically significant at $p<0.01$).

		Temp	pH	Turb	SS	DO	SD	Salin
	Chl *a*		++	++	++	++	–	
	Total	+	++	++	++	++	–	
TM3	Diatom	++	++	+	++	++	–	
	Dinophyte	-	++	++	++	++	–	–
	Others	–	+					
	Chl *a*		++	++	++	++	++	–
	Total		++	+		++	–	-
TM4	Diatom	++	++			++	–	
	Dinophyte	-	++	++		++	–	-
	Others	-	++			++	–	
	Chl *a*		++	++	++	++	–	–
	Total	++	++	++		++	–	-
TM8	Diatom	++	++			++	–	
	Dinophyte		++	++	++	++	–	
	Others						–	
	Chl *a*			++	++	+	–	-
	Total	+					–	–
TM9	Diatom	++	+				–	–
	Dinophyte	–				++	–	
	Others					++	–	

The phytoplankton community of Tolo Harbour consisted largely of diatoms. Dinoflagellates and other minor algal groups such as cryptomonads and other small flagellates constituted a smaller fraction of the phytoplankton population (Figure 3). Although dinoflagellates such as *Prorocentrum* and *Noctiluca* tended to bloom frequently and irregularly to form red tides in the harbour, no regular spring or autumn blooms of diatoms and dinoflagellates were observed. Algal biomass, as inferred by chlorophyll *a* concentrations, appeared to decrease in TM3, TM4 and TM8 after 1989 even though no statistically significant trends were found (Figure 4). Statistically significant increasing trends were observed for total phytoplankton and diatoms in TM4 and for minor algal groups in TM3, TM4 and TM8 (Figure 4). No statistically significant temporal trends were found for dinoflagellates, but the group as a whole was decreasing in all sampling stations (Figure 4).

Kendall's correlations between phytoplankton variables and each of the hydrographical factors in stations TM3, TM4, TM8 and TM9 are summarized in Table 3. Chlorophyll *a* concentrations correlated positively with turbidity, suspended solids and dissolved oxygen in all 4 stations, and with pH in TM3, TM4 and TM8. Densities of total phytoplankton, diatoms and dinoflagellates tended to show positive correlation with temperature, pH, turbidity, suspended solids and dissolved oxygen. As expected, Secchi disk transparency correlated negatively with most of the phytoplankton variables in all 4 stations. Densities of minor algal groups tended not to correlate with hydrographical factors.

Table 4 summarizes Kendall's correlations between phytoplankton variables and nutrient factors. With few exceptions, chlorophyll *a* concentrations and densities of total phytoplankton and diatoms were negatively correlated with nutrient factors. Positive correlation between chlorophyll *a* concentrations and total nitrogen (TN) occurred in stations TM4 and TM8. There was also positive correlation between chlorophyll *a*

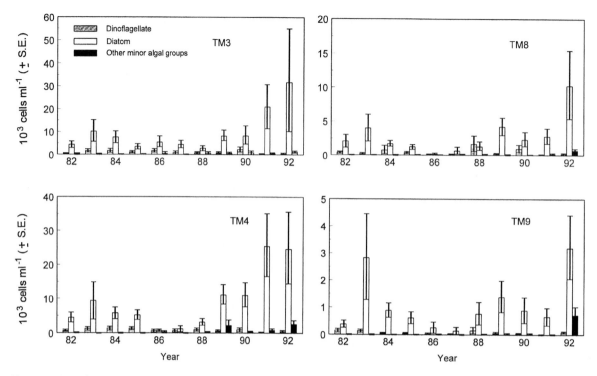

Figure 3. Annual average concentration of diatom, dinoflagellate and other minor algal groups in 4 sampling stations in Tolo Harbour, 1982–1992.

concentrations and total Kjeldahl nitrogen (TKN) in TM3, TM4 and TM8. No significant correlation was found between chlorophyll *a* concentration and TP in any of the station. It should be noted that most nutrient variables (except for TKN and TN) were negatively correlated with the densities of diatoms and dinoflagellates, while positively correlated with the density of the other minor algal groups.

Discussion

Formerly rural, the catchment area of Tolo Harbour has been under intense urbanization since the early 1970's. Increased sewage discharge from the human populations in urbanized areas around the inner harbour has resulted in serious organic pollution problems. The daily input of total nitrogen into Tolo Harbour has been estimated to be over 6000 kg in 1986 and around 4000 kg in 1994 (Hong Kong Government Environmental Protection Department, 1994). Because Tolo Harbour is a shallow, confined body of water with low flushing rate (Chan & Wong, 1993), nutrient concentrations were higher in the inner harbour compared with

the channel and the outer harbour. At the same time, organic pollution, as inferred by BOD level, chlorophyll *a* concentration and Secchi disk transparency was also higher in the inner harbour and lower in the channel and the outer harbour.

Hodgkiss & Chan (1983) measured the concentrations of selected nutrients in Tolo Harbour. Their results were generally comparable to those reported in this study. Despite of the efforts by the Hong Kong Government to reduce the pollution load into Tolo Harbour in recent years, the concentrations of many of the nutrients continue to increase in the surface water (Table 2). It is speculated that the bottom sediments might be a source of the observed nutrients increase. It is known that the sediments could function as reservoirs of the excess nutrients accumulated in the system from previous years. These excess nutrients could gradually be released back into the water column long after the external nutrient supplies had ceased or reduced. It is worth noting that the ratio of TN/TP in TM3 and TM4 had decreased significantly over the 10 year period. This decreasing trend was likely to be the result of the nitrogen removal scheme being implemented at the Tai Po sewage treatment works. Phosphorus removal was

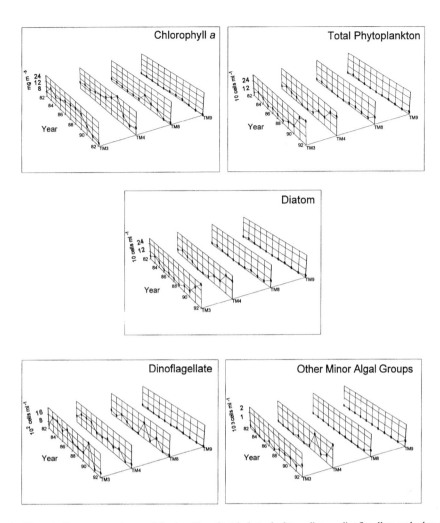

Figure 4. Changes in chlorophyll *a* concentrations and the densities of total phytoplankton, diatom, dinoflagellate and other minor algal groups in 4 sampling stations in Tolo Harbour, 1982–1992.

not carried out because phosphorus was not regarded as a limiting nutrient for phytoplankton growth in Tolo Harbour (Holmes, 1988).

The chlorophyll *a* concentrations reported in this study were also comparable to those measured by Chan and Hodgkiss (1987). However, it should be noted that the annual average chlorophyll *a* concentrations in the 4 sampling stations did not increase during the 10 year period when the levels of nutrients were increasing. A casual inspection of the data suggests that chlorophyll *a* concentrations in TM3, TM4 and TM8 reached a peak in around 1988 and began to decline after 1989 (Figure 4), but the statistical analysis employed in this study does not allow us to recognize the presence of such a 'trend reversal' in an objective manner. We do not wish to speculate on the long-term trends in

chlorophyll *a* concentrations until further analysis is performed to incorporate the more recent data.

The phytoplankton in Tolo Harbour was studied by Hodgkiss & Chan (1983), Chan & Hodgkiss (1987) and Lam & Ho (1989). These authors speculated that the eutrophication of Tolo Harbour had led to a shift in algal species composition, with dinoflagellates replacing diatoms as the dominant phytoplankton in the water column. These authors further argued that increasing dominance of dinoflagellates was the major cause of the dramatic increase in red tide occurrence in Tolo Harbour during the 1980's and early 1990's.

Contrary to the above reports, our results show that diatom was the most important algal group in Tolo Harbour during the study period, and there was no evidence of diatoms being replaced by dinoflagellates

114

Table 4. Kendall correlations for phytoplankton and nutrient factors in Tolo Harbour, 1982–192. (+/- Statistically significant at $p<0.05$; ++/– Statistically significant at $p<0.01$)

		NO$_2$-N	NO$_3$-N	NH$_4$-N	TIN	TKN	TN	PO$_4$-P	TP	TSi	TSi/TN	TSi/TP	TN/TP
TM3	Chl *a*	−	−	−	−	+		−		−	−	-	+
	Total	−	−	−	−			-		−	-	-	-
	Diatom	−	−	−	−	−	−	−		−	−	−	
	Dinophyte												
	Others	++	++	++	++	++	++	++	++				−
TM4	Chl *a*	-		−	−	++	++	−		−	−	−	
	Total	−	−	−	−			-		−	−	−	
	Diatom	−	−	−	−			−		−	−	−	++
	Dinophyte					++	++						
	Others		+			++	++	++	++			−	−
TM8	Chl *a*		-	−	−	++	++	-		−	−	−	++
	Total	-	−	-	−	+				−	−	−	
	Diatom	−	−	−	−			−		−	−	−	
	Dinophyte	−	−			+	+			+			
	Others	++	++		+	++	++	++	++				
TM3	Chl *a*												
	Total									−	−	−	
	Diatom									−	-		
	Dinophyte	−	−						-				
	Others	+	+	++	++	++	++		++		−	−	

(Figure 3). Although dinoflagellates were more abundant than diatoms on some occasions, these were isolated incidents. In fact, while dinoflagellates did not show long-term trend, the densities of diatoms had increased significantly in the inner harbour.

The density of the other minor algal groups showed increasing trend in the inner harbour and channel stations. The major components of the minor algal groups consisted of smaller algae such as cryptomonads and other small flagellates, and their increase correlated strongly and positively with nutrient factors. On the other hand, the nutrient factors correlated either weakly with density of dinoflagellates or negatively with densities of total phytoplankton and diatoms. It is speculated that the positive correlations between the density of minor algal groups and nutrients were due to the faster growth rates of the smaller algae and their abilities to take advantage of the higher nutrients situation (r-strategists). The negative correlations between the densities of total phytoplankton and diatoms were probably due to the slower growth rates of the larger algae (k-strategists) where nutrients were being removed from

the water column by the organisms. Further study should be carried out to investigate the potential of using members of the minor algal groups as indicators of water quality in Tolo Harbour.

Our results suggest that the structure of the phytoplankton community in Tolo Harbour might be changing. The larger algae with a slower growth rate, such as many large diatoms and dinoflagellates, were gradually being replaced by the smaller and faster growing algae, including cryptomonads, small flagellates and diatoms. Thus there was a net increase in densities of total phytoplankton and diatoms in TM4 and other minor algal groups in TM3, TM4 and TM8, even though the overall algal biomass, as inferred by chlorophyll *a* concentrations, had not increased in any of the stations.

Acknowledgments

We thank the staff of the Environmental Protection Department for assistance in collecting and analysing the samples. We are also grateful to Joanne Leung and John Chui for preparing the graphics, the Government Chemist for analysing the nutrient samples and the Director of Environmental Protection for his permission to publish the data. The opinions in this paper are those of the authors and do not necessarily reflect the views or policies of the Hong Kong Government.

References

Chan, B. S. S. & I. J. Hodgkiss, 1987. Phytoplankton productivity in Tolo Harbour. Asian mar. Biol. 4: 79–90.

Chan, A. L. C. & C. K. Wong, 1993. Impact of eutrophication on marine plankton in Tolo Harbour, 1988–89. In Morton, B. (ed.), The Marine Biology of the South China Sea, Hong Kong University Press, Hong Kong: 543–558.

Dodge, J. D., 1982. Marine dinoflagellates of the British Isles. Her Majesty's Stationery Office, London, 303 pp.

Hodgkiss, I. J. & B. S. S. Chan, 1983. Pollution studies in Tolo Harbour, Hong Kong. Mar. envir. Res. 10: 1–44.

Holmes, P. R., 1988. Tolo Harbour – the case for integrated water quality management in a coastal environment. J. Inst. Wat. envir. Mgmt 2: 171–179.

Holmes, P. R. & C. W. Y. Lam, 1985. Red tides in Hong Kong waters – response to a growing problem. Asian mar. Biol. 2: 1–10.

Hong Kong Government Environmental Protection Department, 1991. Standards for effluents discharged into drainage and sewerage systems, inland and coastal waters. HKEPD Technical Memorandum, Hong Kong Government Printer, Hong Kong, 26 pp.

Hong Kong Government Environmental Protection Department, 1994. Marine water quality in Hong Kong for 1994. Hong Kong Government Printer, Hong Kong, 121 pp.

Lam, C. W. Y. & K. C. Ho, 1989. Phytoplankton characteristics of Tolo Harbour. Asian mar. Biol. 6: 5–18.

Larsen, J. & Ø. Moestrup, 1989. Guide to toxic and potentially toxic marine algae. The Fish Inspection Service, Danish Ministry of Fisheries, 61 pp.

Sokal, R. R. & F. J. Rohlf, 1981. Biometry. W. H. Freeman & Company, San Francisco, California, 859 pp.

Wear, R. G., G. B. Thompson & H. P. Stirling, 1984. Hydrography, nutrients and plankton in Tolo Harbour, Hong Kong. Asian mar. Biol. 1: 59–75.

Wu, R. S. S., 1988. Marine pollution in Hong Kong: a review. Asian mar. Biol. 5: 1–23.

van Belle, G. & J. P. Hughes, 1984. Nonparametric tests for trend in water quality. Wat. Resour. Res. 20: 127–136.

Yamaji, I., 1984. Illustrations of the marine plankton of Japan. Hoikusha Publ. Ltd. Osaka, 538 pp.

Hydrobiologia **352**: 117–140, 1997.
Y.-S. Wong & N. F.-Y. Tam (eds), Asia–Pacific Conference on Science and Management of Coastal Environment.

Environmental and nutritional factors which regulate population dynamics and toxin production in the dinoflagellate *Alexandrium catenella*

Gavin K. Y. Siu, Maria L. C. Young & D. K. O. Chan
Department of Zoology, The University of Hong Kong, Pokfulam Road, Hong Kong

Key words: toxin production, cell cycle, population dynamics, nutritional factors, environmental factors, paralytic shellfish poisons, toxin composition, *Alexandrium catenella*, dinoflagellate, toxic red tide

Abstract

The effects of environmental and nutritional factors on population dynamics and toxin production were examined in *Alexandrium catenella*, maintained in enriched K media in laboratory cultures. Starting with a density of 50 cell ml^{-1}, the dinoflagellate population typically showed a lag phase and an exponential growth phase which lasted 14 days each, and then entered the stationary phase, with a maximal capacity of 12–18,000 cell ml^{-1}. Population densities showed distinct diurnal patterns, with population growth beginning 2–4 hours in darkness. The optimal physical conditions for growth were pH 8.5, salinity of 30–35‰, temperature of 20–25°C, and photoperiod of 14//10D to 16L/8D. The cell cycle was determined by flow cytometry on synchronized batch cultures maintained at optimal pH, salinity, temperature and under 5 different photoperiod regimes. It was found that the G_1 phase was timed to end at approximately 3 h after onset of darkness, and the G_2/M phase had begun at 4 hours. Nutrient supply markedly affected population growth. Under optimal physical conditions, the optimal concentrations for macronutrients and micronutrients were: NH_4^+ 0.025–0.2 mM, NO_3^- 0.22–8.83 mM, glycerophosphate 0.04–0.06 mM, silicate 0.1–0.54 mM; FeEDTA 0.07–0.11 mM; Co 0.1 μM, Cu 0.005–0.04 μM; Mn 0.22–7.2 μM; Mo 0.03–0.6 μM; Se 0.02–0.1 μM; Zn 0.04–1.6 μM; thiamin 0.075–6 μM; vitamin B_{12} 0.0004–0.004 μM; biotin 0.007–0.015 μM; EDTA 5–40 μM.

The toxin profile of A. catenella was determined by HPLC and found to include in descending order: GTX-4, GTX-3, GTX-1, B2, neosaxitoxin, saxitoxin. Toxin content per cell was highest in cell populations in the early exponential phase. The highest toxin per litre medium was recorded at 20°C at the beginning of the stationary phase, when cell density was highest and toxin/cell was still relatively high. At 10°C, the cell density was low while the amount of toxin/cell was high; while at 30°C, the population at full capacity was low and the toxin/cell was also low. The population and toxin data thus provided an explanation for the peak level of PSP contamination in shellfish during the months of March–April around the eastern and southern side of Hong Kong and a minor peak extending to the western side in September–October, when the physical conditions of the seawater provided the right environment for toxin accumulation.

Toxin content in the dinoflagellate reached its maximum during the S-phase of the cell cycle. Nitrogen restriction in the medium reduced population growth and toxin production, while phosphorus restriction reduced only population growth but enhanced toxin accumulation in the cells.

Introduction

Toxic red tides and paralytic shellfish poisoning (PSP) resulting from eating PSP-contaminated shellfish have been reported all over the world, particularly in temperate and subtropical regions (Steidinger, 1975). In

Hong Kong, the occurrence of toxic red tides and PSP-shellfish has been shown to be highly seasonal, with higher incidences were recorded when the seawater temperature was around 20–23 °C (Chan & Liu, 1991). The major peak occurred during the spring months of March–April and PSP-contaminated shellfish were

118

found mainly in the eastern and southern parts of the territory which received oceanic waters. A minor peak was observed during September–October, when the flow of freshwater down the Pearl River was reduced and contamined shellfish could be found in the western side of the harbour as well. Crabs and finfish could also be contaminated by PSP, and their peak seemed to lag behind the shellfish peak by about one month (Chan & Siu, 1995).

Many factors affect algal blooms, e.g. nutrients, light, wave action, physical and hydrographic conditions, etc. Dinoflagellates usually represent a very minor component of the microplankton. However, when conditions were right, they could replicate quickly and became the dominant species within 7–10 days. To elucidate the role of these factors which favour (excystment and replication) or terminate (encystment or death) a bloom, there is a need to study the biology of the dinoflagellate in the laboratory.

In March–April 1989, a highly toxic redtide bloom occurred in Tai Tam Bay on the south side of Hong Kong Island. The level of toxin in the water was raised to detectable levels (2–10 mouse-units l^{-1}) for about two weeks and serious PSP-contamination of shellfish followed. The causative organism was found to be the dinoflagellate *Alexandrium catenella* (Whedon & Kofoid) Balech. At the peak of the bloom, cell count in the seawater reached 15 000 cells ml^{-1}, and the PSP level in shellfish harvested from the Bay reached 18 000 mouse-units kg^{-1}, rendering them hazardous for human consumption. Officers from the Environmental Protection Department successfully isolated vegetative cells from this species and cultured them under axenic conditions. A sub-culture was maintained in excellent condition in our laboratory since then. Although this species can be found in the oceanic waters around Hong Kong throughout the year, an increase in cell numbers did not necessarily lead to high PSP-contamination in shellfish. Indeed, on several occassions, particularly during November and May, a complete dissociation between the two events had been recorded. Thus environmental and nutritional factors might affect the population dynamics and toxin production in a complex manner. The present study was undertaken to examine this interrelationship using the stock culture of *A. catenella* under controlled laboratory culture conditions.

Materials and methods

Alexandrium catenella cultures

The stock culture of the dinoflagellate *A. catenella* used in this study was maintained in triplicate in 1-litre bottles containing the 'K medium' (Keller & Guillard, 1985, Keller et al., 1987), which was made up with artificial seawater (salinity = 33 ± 1 ‰ pH = 8.30 ± 0.20). Enrichment nutrient (final concentration: $NaNO_3$ 8.83 × 10^{-4} M; NH_4Cl 5 × 10^{-5} M; Na_2 glycerophosphate 1 × 10^{-5} M; Na_2SiO_3 $9H_2O$ 5.4 × 10^{-5} M; thiamin-HCl 3 × 10^{-7} M; biotin 2.1 × 10^{-9} M; B12 3.7 × 10^{-10} M; Fe-EDTA 1.17 × 10^{-5} M; $MnSO_4$ 9 × 10^{-7} M; $ZnSO_4$ 8 × 10^{-8} M; $CoCl_2$ 5 × 10^{-8} M; Na_2MoO_4 3 × 10^{-8} M; $CuSO_4$ 1 × 10^{-8} M; SeO_2 1 × 10^{-8} M; $EDTA(Na_2)$ 1 × 10^{-4} M; and Trizma base (pH = 7) 1 × 10^{-3} M) was added aseptically to the sterilized and filtered seawater (0.22 µm GS membrane filter, Millipore). Batches were kept at 25 ± 1 °C and illuminated with 'tropical day light' fluorescent tubes (20W, 38 mm × 600 mm, Thorn EMI Lighting Co.) at a light intensity of 120 ± 10 µE m^{-2} s^{-1} and a 14/10 L/D photoperiodic cycle. The medium in the stock culture was changed every 7 days to ensure that the cells were maintained in the exponential growth phase. Possible contamination was monitored periodically by microscopic inspection. Any contaminated bottles were discarded immediately.

Population growth

Cellular density

For the experimental batch cultures, the dinoflagellate inoculum was obtained from the stock culture during the mid-exponential growth phase (density ≈ 4000 cells ml^{-1}). The exact cell density was determined by Coulter particle counting using a ZB model or Multisizer II model (Coulter Electronics) with a 200 µm orifice tube and 3.3% NaCl as electrolyte and calibrated with 14.6 µm standard latex particles (Coulter, Electronics). The results were counterchecked microscopically by heamocytometry.

Daily growth rates (ml) deduced from population dynamics

During the exponential growth phase, the daily growth rate could be calculated using the following equation:

$$\mu_1 = 1/t\ln(N_j/N_{j-1}), \qquad (1)$$

where N_J = cell density at day j, and t = time (day).

Cell cycle of Alexandrium catenella:
Cell division

A dinoflagellate culture at the mid-exponential growth phase (density of 5000 cells ml^{-1}) was divided into five 250 ml sub-cultures. Culture conditions were the same as the stock culture. The cell density was monitored every two hours for 48 h by Coulter particle counting.

Phases of the cell cycle

Synchronization A dinoflagellate culture reaching early to mid-exponential growth phase (density of 1000 to 4000 cells ml^{-1}) was maintained in continuous darkness for 36 h to synchronized the cells. After exposion to two 14L/10D cycles, it was divided into 24 sterilized 250 ml sub-cultures for further analyses.

Sample preparation Samples were collected at 1 h intervals for 24 h. Cells in 30 ml aliquots of the culture medium were washed twice with phosphate buffer saline (PBS, pH 7.40) and concentrated to 1 ml final volume by centrifugation at $200 \times \boldsymbol{g}$ for 10 min at 4 °C. They were slowly injected into 2 ml cold (4 °C) absolute ethanol for overnight fixation and to wash free of the photosynthetic pigments.

The fixed samples were washed twice with PBS and spun down to 1 ml at $320 \times \boldsymbol{g}$ for 10 min to remove the ethanol and the soluble pigments. They were then incubated at 37 °C for 10 min. before adding RNAase (Type XII-A, Sigma) and propidium iodide (PI, excitation: 488 or 514 nm/emission: 550–640 nm, Sigma) to give final concentrations of 200 Kunitz units ml^{-1} and 40 μg ml^{-1} in PBS, respectively. A further 30 min. incubation was allowed and the tube was then transferred to 4 °C in the dark, and allowed at least 1 hour to reach equilibrium before analysis by flow cytometry. The entire procedure was completed within 4 hr.

Data acqusition and analysis Samples were passed through the Elite Flow Cytometer (Coulter Electronics) equiped with a 100 μm quartz flow cell. PI was excited with a 488 nm Argon laser (15 mW) and the red fluorescence (>625 nm) was collected by the appropriate photomultiplier tube. Cell numbers were plotted against the fluorescence intensity (DNA content in arbitrary units) to give a histogram. It was then analyzed to give the percentage of G_0/G_1, S, and G_2/M phase respectively using the 'Mulicycle' computer software (Phoenix Flow System) which was based on Gaussian distribution, nonlinear least square curve fitting and a model

of synchronous S phase (Marquardt, 1963; see Bevington, 1969; Dean & Jett, 1974; Fox, 1980).

Calculations Instant rate of change (r) of S and G_2/M fractions (%) were calculated according to the following equation:

$$r = \ln|[N_t - N_{t-1}]|/t \qquad (2)$$

where N_t= cell fraction in specific phase /100 at time t, and t = 1 hour.

The start and the end of S and G_2/M phase were determined by $r \geq r_{max}/2$, where r_{max} = maximum rate of change within the L/D cycle.

The daily mean specific growth rate (μ_2) was determined using the following equation (Carpenter & Chang, 1988):

$$\mu_2 = 1/(T_S + T_{G_2/M})n \sum_{j=1}^{n} \ln[1 + f_s(t_j) + f_{G_2/M}(t_j)], \qquad (3)$$

where
T_S = duration of the S phase;
$T_{G_2/M}$ = duration of the G_2/M phase;
$f_S(t_j)$ = fraction of S phase /100 at time t_j;
$f_{G_2/M}(t_j)$ = fraction of G_2/M phase /100 at time t_j;
n = number of smaples obtained in a 24 h cycle;
t_j = time of obtaining the jth sample.

The doubling time (D.T.), in days, were calculated using the following equation:

$$D.T. = \ln2/\mu \qquad (4)$$

where $\mu = \mu_1$ or μ_2.

Toxin production and determination

Dinoflagellate extracts

The dinoflagellate population density in the cultures was first determined by particle counting and a known volume was then filtered through a bacterial filter (0.22μm membrane filter, Millipore) to trap the cells. The filter membrane was extracted 3 times with acidified 95% ethanol (pH 4.0) and the extract concentrated by rotatory evaporation under reduced pressure. The residue was dissolved with ammonium acetate buffer (100 mM, pH 4.50), sub-divided in polypropylene tubes, lyophilized and stored at -80 °C. On the day of toxin quantification, the sample was prepared freshly with 1 ml ammonium acetate buffer (5 mM, pH 4.5).

The solution was filtered through a molecular-sieve membrane with a 10 000 MW cutoff (Centricon-10, Amicon) to remove proteins. A sample size of 100 μl was injected into the HPLC for fractionation.

Reference PSP standards

To enable us to compare toxicity levels of dinoflagellate samples collected over the period of study, a reference paralytic shellfish poison (PSP)-contaminated green mussel (*Perna viridis*) was set aside at the start. These shellfish samples were collected from the same *A. catenella* bloom of March–April 1989, and frozen at -80 °C. Extracts from this source was purified for reference until a calibrated PSP Reference Standard was obtained from the U.S. Food Toxin Laboratory in 1995.

The contaminated mussel samples were homogenized with a commercial meat blender. The homogenate was extracted three times with 0.1 N HCl and boiled for 5 min each time. After centrifugation ($4000 \times g$ or 30 min), supernatants were pooled and pH was adjusted to 4.5 with NaOH. After further centrifugation at $15\,000 \times g$ or 30 min and at $100\,000 \times g$ for 1 h at 4 °C, the extract was concentrated by lyophilization and redissolved in deionized water. Purification involved an ion-exchange chromatography step using Bio-Rex 70 ion exchange resin (-400 mesh, Na+ form) and eluted with 0.1N HCl. The PSP peak was identified by taking aliquots for fluorescent derivatization following the procedure of Sullivan & Iwaoka (1983). The toxin-containing fractions were pooled, lyophilized, re-dissolved in ammonium acetate buffer (5 mM, pH 4.5) and split into two portions. One portion was used for the determination of toxicity by the standard mouse bioassay (AOAC, 1984). The other was analyzed by HPLC using the post-column fluorescent derivatization system of Sullivan (1987) and Sullivan & Wekell (1987).

HPLC quantitation

Two reverse phase HPLC systems were used. One method was based on a μ-Bondapak C-18 column (dimension 300 mm × 3.9 mm, particle size 10 μm, pore size 12.5 μm) and isocratic elution with 10% methanol at a flow rate of 0.8 ml min^{-1}. The PSP-group toxins were eluted together and the peak area could be calibrated against the mouse toxicity results and later used to quantitate sub-lethal concentrations of the neurotoxin. The second method fractionated the PSP-group into individual components using a Hamilton PRP-1 (150 mm × 4.1 mm, particle size: 5μm) polystyrene divinylbenzene column under ion-pair conditions (1.5 mM of hexanesulfonic acid/heptanesulfonic acid sodium salts added in both eluants). Gradient elution was performed with ammonium phosphate at 1.5 mM (pH 6.70 ± 0.02) in eluant A, and 75% ammonium phosphate at 6.25 mM (pH 7.00 ± 0.02) – 25% acetonitrile in eluant B (Sullivan, 1987). With a flow rate of 1.3 ml/min, the gradient profile (A:B) was 100–0% at 0–4 min, 70–30% at 7–11 min, 10–90% at 11–17 min, and 0–100% at 17–17.5 min. The column was heated to 35 °C.

Effects of pH and salinity on Alexandrium catenella

Inocula in the mid-exponential growth phase (density $\approx$ 4000 cells ml^{-1}) from the same culture stock were applied to the sub-cultures (100 ml each) so that the initial cell density of each bottle was 400 ± 20 cells ml^{-1} as determined by Coulter counting and counterchecked by haemocytometry.

The range of pH and salinities tested is given in Table 1. On alternate days for 14 days, aliquots of the cultures were taken for cell density counts by Coulter counting.

Effects of temperature

The initial *A. catenella* cell density of 50 ± 5 cells ml^{-1} in the culture medium was achieved with suitable dilution. Batches were made up in 250 ml of 'K medium', and kept in illuminated environmental chambers with temperature controlled to ± 0.2 °C of the set temperature (Table 1). The cell density and toxin content were monitored every third day for 37 d.

Effect of photoperiodicity

After synchronizing the cells by exposure to 36 h continuous darkness as already described, *A. catenella* cultures in 'K medium' were exposed to specified photoperiod regimes as follows: 10L/14D, 12L/12D, 14L/IOD, 16L/8D, 18L/6D. Temperature was set at 25 °C. The same experimental procedures as described previously were used to determine cell density and the length of the G_0/G_1, S and G_2/M phase of the cell cycle. Aliquots were also taken for toxin analysis during different phases of the cell cycle.

Table 1. Physical conditions and nutrient concentrations tested for their effects on the population dynamics and toxin production in *A. catenella*.

Physical variates												
pH				7.5	8	8.38	8.5	9				
Photoperiod (Light/Dark cycle)				10/14	12/12	14/10	16/8	18/6				
Temperature (°C)				10	15	20	23	25	30			
Salinity (‰)				15	20	25	30	35	40	45		
Macronutrients												
NO^{3-} ($\times 10^{-4}$ M)	0.088	0.88	2.21	4.42	8.83	17.7	35.3	53	70.6	88.3	176	353
NH_4^+ ($\times 10^{-5}$ M)	0.05	0.5	1.25	2.5	5	10	20	30	40	50		
PO_4^{3-} ($\times 10^{-5}$ M)	0.01	0.1	0.25	0.5	1	2	4	6	8	10	20	
SiO_3^{2-} ($\times 10^{-5}$ M)	0.054	0.54	1.35	2.7	5.4	10.8	21.6	32.4	43.2	54	108	216
Vitamins												
B_{12} ($\times 10^{-10}$ M)	0.037	0.37	0.93	1.85	3.7	7.4	14.8	22.2	29.6	37	74	148
Biotin ($\times 10^{-9}$ M)	0.021	0.21	0.53	1.05	2.1	4.2	8.4	12.6	16.8	21	42	84
Thiamin ($\times 10^{-7}$ M)	0.03	0.3	0.75	1.5	3	6	12	18	24	30	60	120
Micronutrients												
Co^{2+} ($\times 10^{-8}$ M)	0.05	0.5	1.25	2.5	5	10	20	30	40	50		
Cu^{2+} ($\times 10^{-8}$ M)	0.01	0.1	0.25	0.5	1	2	4	6	8	10		
Fe^{2+} ($\times 10^{-5}$ M)	0.012	0.12	0.29	0.59	1.17	2.34	4.68	7.02	9.36	11.7		
Mn^{2+} ($\times 10^{-7}$ M)	0.09	0.9	2.25	4.5	9	18	36	54	72	90	180	360
MoO_4^{2-} ($\times 10^{-8}$ M)	0.03	0.3	0.75	1.5	3	6	12	18	24	30	60	120
Se^{2+} ($\times 10^{-3}$ M)	0.01	0.1	0.25	0.5	1	2	4	6	8	10	20	40
Zn^{2+} (10^{-3} M)	0.08	0.8	2	4	8	16	32	48	64	80	160	320
Chelator												
EDTA ($\times 10^{-4}$M)	0.01	0.1	0.25	0.5	1	2	4	6	8	10		

Effect of nutrients

Culture conditions were similar to that for testing the effect of salinity or pH. Individual macronutrient or micronutrient and the chelator EDTA was varied as given in Table 1, while keeping all other parameters the same as 'K medium'. Aliquots were taken for cell density on alternate days for 14 days.

To test the effect of N and P enrichment of natural seawater on the population growth and toxin production, 5 triplicate batch cultures were set up. The control contained the full nutrient content of 'K medium', the remaining received no additional or 10% of the amount of nitrogen or phosphate. Aliquots were taken for cell density and toxin content determination as in other groups.

Results

Laboratory cultures of A. catenlla in 'K medium'

Population dynamics
The population dynamics of laboratory cultures of *A. catenella* in 'K medium' showed typical sigmoidal curves. Starting with a low concentration of about 50 cells ml^{-1}, the lag phase lasted 14 days, when the density (mean ± standard deviation) reached 138 ± 19 cells ml^{-1} ($n = 9$) (Figure 1). The exponential growth phase then set in and continued for another 14 days, when the density reached 8303 ± 356 cells ml^{-1}. The population growth then slowed down and the shift marked the onset of the stationary phase. The maximal carrying capacity was approximately 13 000 to 15 000 cell ml^{-1}. The mean population growth rate (μ_1) and the mean doubling time (DT_1) were 0.28 ± 0.005 d^{-1} and 2.47 ± 0.04 d (mean ± standard deviation), respectively.

Cell divisions
A preliminary study on unsynchronized batch cultures was conducted at 2-hourly intervals under the 14L/10D

122

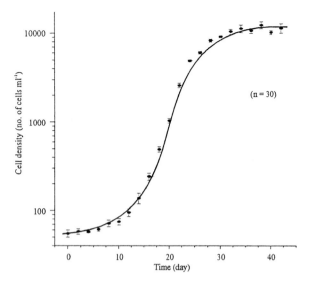

Figure 1. Population growth of *Alexandrium catenella* in 'K medium' at 25 °C, with a starting culture of 50 cells ml^{-1}. Note the lag phase lasting 14 days, the exponential growth phase lasting also 14 days, and the final stationary phase.

photoperiod cycle and the results showed that the cell density increased rapidly during the dark period (Figure 2). The increase continued until 3 h after the onset of the light phase. Thereafter, cell density remained almost level and even decreased at the end of the light phase. Thus cell divisions occurred predominantly in darkness, while cell death or encystment occurred mainly in the light period.

Phases of the cell cycle (G_0/G_1, S and G_2/M phases)
To study phases of the cell cycle using flow cytometry, cells were synchronized by exposing them to continuous darkness for 36 h. As the normal day/night cycle was turned back on, these cells were allowed to undergo through 2 cell cycles before study. Discrete S (DNA synthesis) and G_2/M (second growth phase and mitosis) phases were found in the synchronized culture (Figure 3). The S phase started at 4 hours after the onset of the dark phase and continued for 4 hours. The G_2/M phase started at 3 hours after the onset of the S phase and lasted for about 5 hours. For a 10-hour dark cycle, it ended shortly after the onset of the light phase. About 50% of the population entered the cell cycle and performed the cell division process. The specific growth rate (μ_2) and the doubling time (DT$_2$) calculated were 0.29 d^{-1} and 2.43 d, respectively (Table 2).

Physical factors affecting population dynamics

Effect of pH
All treatment groups showed an increase in cell density on day 4, except the group with pH of 6.50. In the latter group, there was a decrease starting from day 2 and reached 219 ± 11 on day 4 (Figure 4a). The population then regained the rapid growth rate and caught up with the group maintained at pH 7.00 on day 10.

On day 4, the cell yields of groups maintained at pH = 7.50, 8.00, 8.5 and 9.0 were in the range of 1000 to 1500 cells ml^{-1}. The group exposed to pH = 9.00 almost reached its full carrying capacity by day 4. As for the groups exposed to pH = 7.0, 7.5, 8.0 and 8.5, population growth continued after day 4. On day 10, the cell yields increased as pH shifted from 7.00 to 8.50 but dropped sharply at pH = 9.00. The optimal pH range was thus about pH = 8.50 and the growth of *A. catenella* was quite sensitive to changes in the pH of the sea water.

Effect of salinity
At the early stages of growth (day 2 to 6), salinity at 20‰ gave the best results with 35‰ being the second best (Figure 4b). However, from day 8 onwards, the latter group gave the best performance. Even cultures with salinity at 25 and 30‰ gave better population growth compared with the group at 20‰. On the other hand, high salinity (40 and 45‰) or low salinity (15‰) appeared to be unsuitable for *A. catenella*. Thus the optimal range was between 30 and 35‰.

Effect of temperature
Temperature appeared to have major effects on population growth in *A. catenella*. A water temperature of 20–25 °C was found to promote the highest growth rate (Figure 4c and 11). At the optimal temperature, the mean maximal carrying capactity reached a cell density of $16\,000 \pm 1250$ cells ml^{-1} ($n = 9$) with a mean population growth rate (equalled to 0.29 ± 0.04 d^{-1} during the exponential growth phase.

At 10 °C, the culture already reached its maximal carrying capacity (957 ± 98 cells ml^{-1}) on day 4. The mean population growth rate (μ_1) was 0.08 ± 0.04 d^{-1}, while mean doubling time was 13.12 ± 4.24 d.

High temperature was also unfavourable for population growth. At 30 °C, the maximal capacity was only 2600 ± 200 cells ml^{-1}, population growth rate (μ_1) was 0.13 ± 0.02 d^{-1}, and the mean doubling time was 8.57 ± 2.65 d.

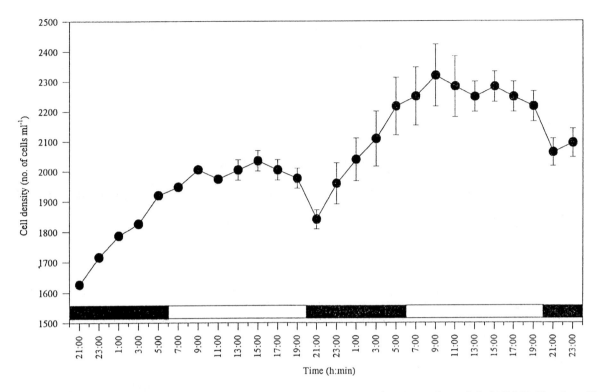

Figure 2. Diurnal changes in cell density of *Alexandrium cafenella* in 'K medium' at 25 °C, given a photoperiod of 14L/10D. Note the rapid increase in cell density only during the dark phase.

Table 2. Specific growth rate (μ_1 and μ_2), doubling time (DT_1 and DT_2) under different temperature and photoperiod regimes.

	Specific growth rate deduced from population growth μ_1 (d^{-1})	Doubling time DT_1 (d)	Specific growth rate deduced from cell cycle μ_2 (d^{-1})	Doubling time DT_2 (d)
Temperature (photoperiod: 14L/10D)				
10°C	0.08	13.12		
20°C	0.29	2.65		
30°C	0.13	8.57		
Photoperiod (temperature 25°C)				
18L/6D	0.27	2.53	0.26	2.69
16L/8D	0.29	2.39	0.29	2.38
14L/10D	0.28	2.47	0.29	2.43
12L/12D	0.27	2.60	0.27	2.56
10L/14D	0.24	2.85	0.21	3.25

Effect of photoperiod

Population growth Batch cultures under different light/dark cycles seemed to grow well and had similar cell yields at both day 4 (1100 to 1700 cells ml^{-1}) and day 10 (10 000 to 16 000 cells ml^{-1}) (Figure 4d). The optimal range was from 12L/12D to 16L/8D with the maximum shifting between 14L/10D and 16L/8D. Thus as far as population growth was concerned, *A. catenella* appeared to be relatively insensitive to photoperiodicity.

Cell cycle Synchronized batch cultures were exposed to five different light (L)/dark (D) cycles, i.e. 10L/14D, 12L/12D, 14L/10D, 16L/8D and 18L/6D. It was found

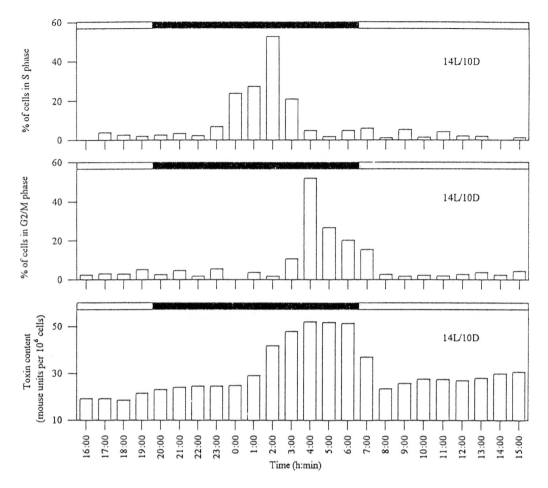

Figure 3. Duration of different components of the cell cycle: G1 = first growth phase, S = DNA synthesis phase, G₂/M = second growth phase and mitosis.

that in all photoperiod regimes, the G_1 phase was timed to end at approximately 3 hours after the onset of darkness and the S phase had clearly begun at 4 hours (Figure 5). In the 12L/12D cycle, the G_2/M phase was completed within the dark period. However, in 16L/8D photoperiod, the G_2/M phase actually extended into the light phase (Figure 5). The data clearly demonstrated that the onset of darkness was the main cue regulating phases of the cell cycle.

From the calculated specfic growth rate (μ_2) and percentage of cells entering the S and G_2/M phases, the optimal photoperiod regime was found to be 16L/8D and 14L/10D cycle (Table 3). There was a good correlation between the estimated growth rate and the doubling time ($r = 0.976$ and $r = 0.954$, respectively, $P<0.01$). The slowest growth rate and the longest doubling time were recorded for the 10L/14D cycle.

Nutritional factors affecting population growth

Macronutrients
Total ammonium (NH_4^+)   High concentrations of total ammonium (300 to 500 μM) greatly stimulated the population growth from day 0 to day 2, but slowed down immediately thereafter and almost reached their maximal carrying capacities (Figure 6a). Though low concentrations (0.5 to 25 μM) did not support good growth at the early stages (day 2 to day 6), the cell densities greatly increased by day 8 giving 2 to 3 times higher carrying capacities. The maximal density was observed in 100 μM from day 4 (3042 ± 398 cells ml^{-1}) to day 14 (25 368 ± 1100 cells ml^{-1}) and the yields dropped sharply as the concentration increased or decreased from that. The optimal range was therefore restricted between 25 to 200 μM.

125

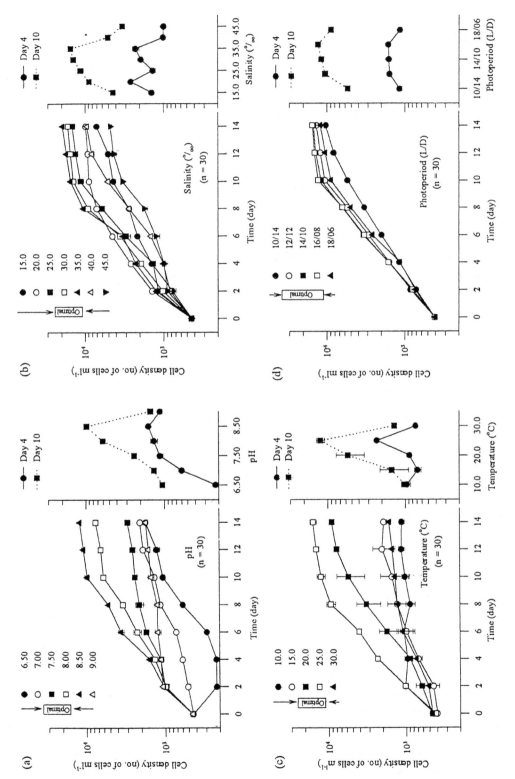

Figure 4. Population dynamics of *Alexandrium cafenella* under different (a) pH, (b) salinity, (c) temperature, and (d) photoperiod conditions.

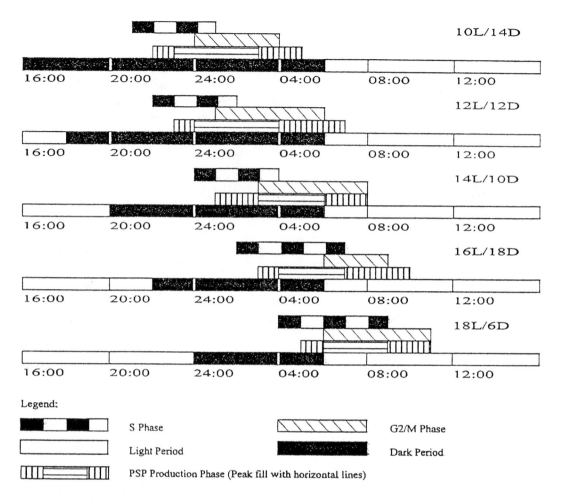

Figure 5. Cell cycle and toxin production in *Alexandrium catenella* exposed to varying photoperiod regimes: 10L/14D, 12L/12D, 14L/10D, 16L/8D, 18L/6D, PSP production.

Nitrate (NO_3^-) On day 4, the nitrate concentration became non-limiting when nitrate concentrations exceeded 0.221 mM: cell density reached 1800 to 3200 cells ml^{-1} in all groups tested (Figure 6b). At lower concentrations (0.088 to 0.0088 mM), significant reduction in the population growth (770 to 800 cells ml^{-1}) was observed on day 4. There was no obvious optimal concentration until day 10, at which a slightly higher yield (15 547 ± 509 cells ml^{-1}) was found in cultures containing 1.766 mM nitrate. Approximately 10-fold increase or decrease in nitrate concentrations gave a small decrease in cell density only. The optimal range was shown to be from 0.221 to 8.83 mM and the maximal cell density equalled to 18 006 ± 500 cells ml^{-1} on day 14 in 1.766 mM. Therefore, *A. catenella* could grow well in a wide range of nitrate concentration.

Glycerophopsphate At both day 4 and day 10, glycerophosphate concentrations exceeding 10 μM was found to be non-limiting for population growth (2300 to 2800 cells ml^{-1} and 10 000 to 13 000 cells ml^{-1}, respectively) (Figure 6c). On the other hand, concentrations lower than 5 μM resulted in half of the corresponding yields (1000 to 1200 cells ml^{-1} and 3900 to 7000 cells ml^{-1}). Maximal cell density was observed within the range of 40 to 60 μM. The maximal cell density was recorded as 16 669 ± 692 cells ml^{-1} on day 14 in 60 μM.

Silicate (SiO_3) From day 0 to day 8, there was no great difference in population growth within the concentration range of silicate studied (Figure 6d). When approaching day 10, higher cell yield was found between 54 to 540 μM and peaked at 216 μM with a

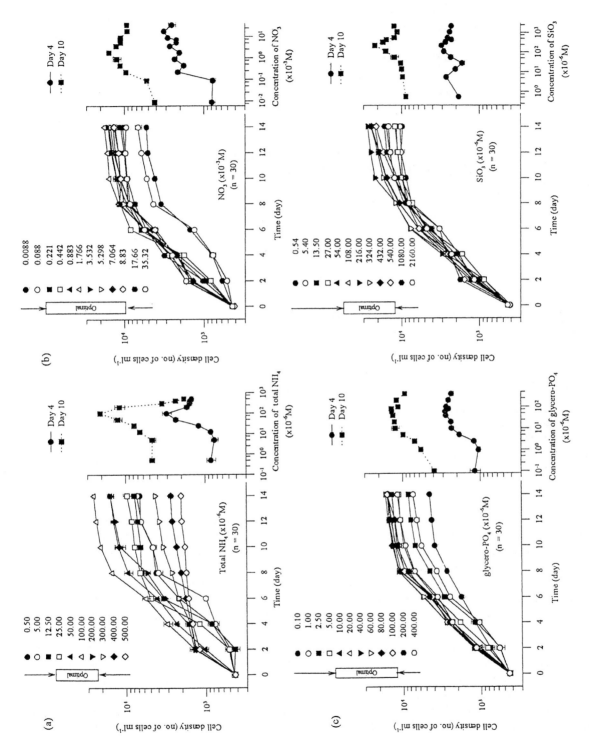

Figure 6. Macronutrient requirements for population growth in *Alexandrium catenella* maintained at 25 °C in a 14L/10D photoperiod under various concentrations of (a) ammonia, (b) nitrate, (c) glycerophosphate, and (d) silicate.

128

cell density reaching 21 681 ± 979 cells ml^{-1}, being 2-fold that of the 54 and 540 μM. Therefore, there seemed to be a long delay (8 days) before silicate enrichment showed its effect and silicate concentration could be rate-limiting during the late exponential phase only. The optimal range was demonstrated from 108 to 540 μM with a maximal cell density equalled to 26 642 ± 1023 cells ml^{-1} on day 14 in 216 μM.

Trace minerals
Cobalt (Co) An extended concentrations from 12.5 to 400 nM was found to be the optimal range (Figure 7a). The stimulatory effect was more prominent on day 4 than on day 10. Consequently, cobalt seemed to be more important during the exponential growth phase than the stationary phase. Concentration of 100 nM gave the best yield throughout the whole study period except day 6. The maximal cell density noted on day 14 was 17 705 ± 774 cells ml^{-1} ($n = 9$).

Copper (Cu) Cu requirement for growth showed a similar pattern to that for cobalt (Co), but the optimal concentration range was lower, starting from 2.5 to 40 nM (Figure 7b). Within this range, the percentage of increase in cell densities was greater on day 4 (about 40%) than day 10 (about 20%). Therefore, Cu also played an important role during the exponential phase. Statistical analysis showed that there was no siginificant difference ($P<0.05$) between the results of 5, 10 and 20 nM. The maximal cell density achieved on day 14 was 15 588 ± 1515 cells ml^{-1} at 10 nM ($n = 9$).

Iron (FeEDTA) It is important to note that chelated iron (FeEDTA) was used for the study. The optimal range determined from the resut was in the μM range, a 1000-fold higher than any other trace metals except manganese (Mn), ranging from 2.925 to 46.8 μM (Figure 7c). Maximal cell density was recorded in 23.4 μM on each sampling day, except day 2 and 6. In addition, the end of the exponential phase was extended from day 8 to day 10 while all the others were already in the stationary phase. The maximal cell density was noted as 24 513 ± 1168 cells ml^{-1} on day 14. At both day 4 and day 10, an increase in concentration from 0.117 to 23.4 μM caused an increase in the yields, but the extent of the increment was greater in the latter. Further increase in concentration caused a sharp decrease in cell densities. Actually, 70.2 to 117 μM were already in their full carrying capacities on day 4. On the whole, *A. catenella* required a relatively high concentration of iron (in the form of EDTA-chelate) than other trace metals and showed a high sensitivity towards its change.

Manganese (Mn) Same as for iron (Fe), *A. catenella* required a μM level of manganese to flourish (Figure 7d). The optimal range was found to be 0.225 to 7.2 μM. Concentrations at 1.8 μM gave the best yields on day 4 and day 10 to 14. Its exponential phase also extended from day 8 to day 10 while the others were already in the stationary phase. On day 14, the maximal cell denisty was found to be 21 848 ± 969 cells ml^{-1}. However, changes in concentration outside the optimal range merely caused a relatively moderate decrease in the yields.

Molybdenum (Mo) The optimal range was shown to be 30 to 600 nM. On day 4, the maximal cell density was found in 60 nM (Figure 7e). However, on day 6, it shifted to 180 nM and continued to day 14, at which the cell density equalled to 23 344 ± 621 cells ml^{-1}. On both days, a great decrease in cell density occurred as the concentration decreased from the maxima, while an increase in concentration only gave a relatively moderate decrease in growth. Thus, concentrations in or below the nM range would seriously slow down the population growth of *A. catenella*.

Selenium (Se) The optimal range (20 to 100 nM) for *A. catenella* was relatively narrow when compared to other trace metals (Figure 7f). The maximal cell denisty was noted at 40 nM in the time course except day 4, and the reading obtained on day 14 was 23 559 ± 1269 cells ml^{-1}. Concentration lower than 1 nM caused a 3-fold decrease in cell density when compared to the maxima on day 4 and day 10, while concentration higher than 100 nM also inhibited the population growth by 2/3.

Zinc (Zn) Zn showed a wider optimal concentration range: 40 to 1600 nM (Figure 7g). Within this range, the resulting cell density was almost the same level. Only a slightly higher yield was found at 480 nM throughout the experiment except on day 8 and 12. The maximal cell density on day 14 was 18 735 ± 577 cells ml^{-1}. Statistical analysis showed that there was no siginificant difference ($P>0.05$) between the results of 320 and 480 nM. Therefore, it seemed that *A. catenella* could survive in a wide range of Zn concentration.

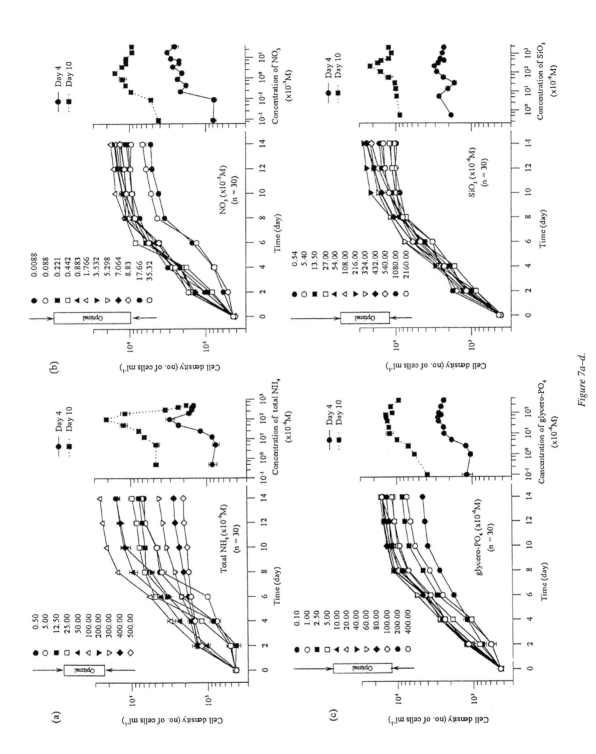

Figure 7a–d.

130

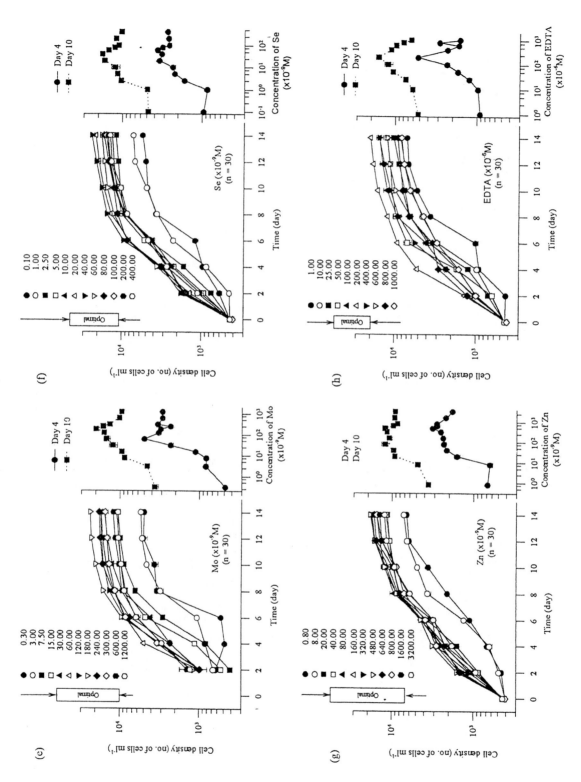

Figure 7e–h. Trace mineral requirements for population growth in *Alexandrium catenella* maintained at 25 °C in a 14L/10D photoperiod under various concentrations of (a) Co, (b) Cu, (c) Fe(EDTA), (d) Mn, (e) Mo, (f) Se, (g) Zn, and (h) the metal chelator EDTA.

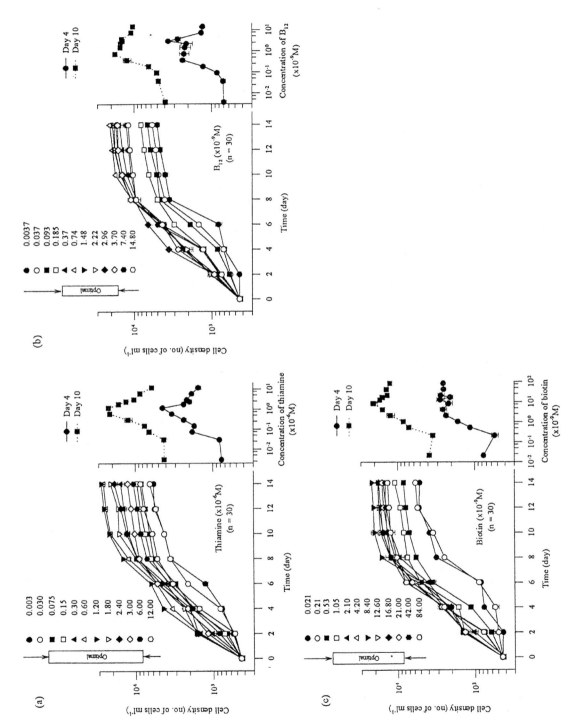

Figure 8. Vitamin requirements for population growth in *Alexandrium catenella* maintained at 25 °C in a 14L/10D photoperiod under various concentrations of (a) thiamine, (b) vitamin B_{12}, (c) biotin.

Vitamins

Thiamin An obvious optimal concentration for thiamin ($1.2 \mu M$) was noted on day 4 and continued until the end of the experiment (Figure 8a). Either increase or decrease in concentration from the optimum caused a rapid decline in the cell densities at both day 4 and day 10. The optimal range was determined as 0.075 to $6 \mu M$ and the maximal cell density reached on day 14 at $1.2 \mu M$ was $27\,952 \pm 1000$ cells ml^{-1}.

Vitamin B_{12} Low concentration of vitamin B_{12} (0.0037 to 0.185 nM) was not effective in stimulating the population growth, with the cell denisties approximately one-third that of the optimal (Figure 8b). On day 4, the maximal cell density was obtained at 2.96 nM while on day 10, it shifted towards a lower concentration of 0.74 nM. It seemed that higher concentration was necessary for stimulation at the early stage (day 4 and day 6), while lower one might be more suitable for the stationary phase (day 10 to day 14). The optimal range was found between 0.37 to 3.7 nM and within this range similar performance was observed. Although increases in concentration above optimal range (7.4 to 14.8 nM) showed some inhibitory effect, a fair yield was still obtained ($10\,957 \pm 468$ and $10\,387 \pm 580$ cells ml^{-1}) on day 10. The maximal cell denisty was $20\,945 \pm 964$ cells ml^{-1} in 0.74 nM on day 14. Therefore, *A. catenella* required the concentration of vitamin B_{12} in the nM level to flourish.

Biotin When exposed to low concentrations (<1.05 nM), the population seemed to reenter the lag phase from day 0 to day 6, while cells maintained at higher concentrations continued their growth rate in the mid-exponential phase (Figure 8c). On day 4, high cell density (2000 to 2800 cells ml^{-1}) was reached in the concentration ranging from 2.1 to 84 nM. On reaching day 8, an optimal concentration was found in 8.4 nM and persisted till the end of the experiment. The results showed an optimal range of 1.05 to 21 nM with a maximal cell density = $21\,859 \pm 571$ cells ml^{-1} achieved in 8.4 nM on day 14. Same as for vitamin B_{12}, *A. catenella* required a nM level of biotin for proper population growth.

Metal chelator

EDTA *A. catenella* was very sensitive to the concentration of the metal chelator EDTA. The effect was already very marked on day 4, at which, increase in concentration from 1 to $200 \mu M$ gave a sharply increase

in cell densities (Figure 7h). However, further increases from 200 to $1000 \mu M$ resulted in a totally opposite pattern. This trend persisted until the end of the experiment. The optimal range was between 5 to $40 \mu M$ and the maximal density was found always in the concentration of $20 \mu M$. The corresponding cell density on day 14 was $20\,084 \pm 683$ cells ml^{-1}.

Toxin production and accumulation

Standardization of HPLC with purified PSP

Crude PSP was partially purified from green mussels, *Perna viridis*. The starting materials (total of 400 mouse-units) were extracted three times with 0.1 N HCl and boiled for five minutes. After solvent partition with chloroform, the defatted extract gave a total activity of about 360 mouse-units (90% recovery). After ultracentrifugation, large amount of contaminants were removed including proteins and peptides. The toxins were further purified through a Bio-Rex 70 column. The total activity recovered was approximately 270 mouse units (67.5% recovery). The specific activity improved 1000-fold. Statistical analysis was done on the correlation between results from standard mouse bioassay method and μ-Bondapak C18 HPLC system. Extracts of food samples, toxic red tide seawater samples and partially purified PSP were used for the test. A good linear correlation ($R = 0.944$, $P < 0.001$) was found (Figure 9a). The calibrated HPLC values were used later to determine sub-lethal quantities of PSP-toxins isolated from small volumes ($100 \mu l$) of culture media.

Toxin production

The PSP-group profile of *A. catenella* is given in Figure 10. The principle components isolated, in descending order of abundamce, were GTX-4 > GTX-3 > GTX-1 > B2 (=GTX-6) > neoSTX = GTX2. Their relative toxicities as reported by Genenah & Shimmizu (1981) and Oshima et al. (1989) were: 1673: 2234: 1638: 180: 1038: 793 mouse-unit $\mu mole^{-1}$ respectively. The C1-4 derivatives were lumped together in the chromatogram, but they were relatively less toxic (for C2: toxicity was 430 mouse-unit $\mu mole^{-1}$).

PSP accumulation in A. catenella
Effect of temperature

For all temperature treatment groups, the toxin content $cell^{-1}$ (mouse-units. million $cells^{-1}$) increased as the cells went from the lag phase to the exponential phase

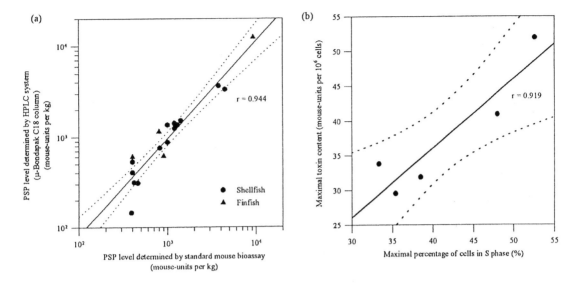

Figure 9. Correlation between (a) the toxin quantified by 11-Bondapak-Cl8-HPLC (eluted with methanol) vs the mouse bioassay results, and (b) the toxin content vs the % of cells entering the S-phase in the cell cycle.

and peaked in early exponential phase (Figure 11). It then slowly dropped as the cell population approached the stationary phase. At 20 and 30 °C, the toxin levels in cells in the stationary phase declined below those of the lag phase. The peak levels were 31.21 ± 2.87 and 25.93 ± 2.67 mouse-units million cells^{-1} at 10 and 20 °C, respectively. At 30 °C, there was a significantly higher value of 34.20 ± 2.93 mouse-units. million cells^{-1} ($P<0.05$).

The picture became quite different when toxin concentrations were expressed in mouse-units l^{-1} water, since temperature had different and dramatic effects on cell density. At 20 °C, the rapid rise in cell density during the mid-exponential phase resulted in a marked rise in toxin concentration which reached its maximum (107.6 ± 7.76 mouse-units l^{-1}) at the beginning of the stationary phase. In contrast, only about 10.86 ± 0.76 to 14.99 ± 1.29 mouse-units l^{-1} could be reached at 10 and 30 °C, respectively.

Effect of photoperiod
PSP production was found to occur at a precise time in each L/D cycle tested. Although the S-phase started at 4 hours after onset of darkness under most photoperiod regimes, it could be shifted to begin at 5 hours in short-light/long-dark periods, e.g. 10L/14D (Figure 5). The onset and duration of G$_2$/M phase also varied somewhat with the photoperiod. PSP production, however, started one hour after the appearance of the S phase

in all groups. The level remained high and dropped either at the beginning of or one hour after the end of the G$_2$/M phase. The PSP production phase shifted with the change of S and G$_2$/M phases in different photoperiods. There was thus a tight coupling between the level of PSP produced and the percentage of cells entering the S phase ($r = 0.919$) (Figure 9b).

At the optimal temperature, salinity, and pH, the highest level of PSP accumulated per cell was found near mid-night in a 10L/14D photoperiod cycle.

Effect of changing nitrogen and phosphate concentrations
Omitting N from the 'K medium' resulted in a drastic reduction in the population growth as well as PSP accumulation in *A. catenella* cultures (Figure 12). Reducing N to 10% of standard 'K medium' had little effect on population dynamics and toxin accumulation in the cells. Omitting PO$_4$ from the medium, in contrast, resulted in reduction in population growth only but toxin accumulation was still significant. Reducing the PO$_4$ to 10% of 'K medium' resulted in some reduction on population growth but the accumulation of toxin in the cells was drastically enhanced.

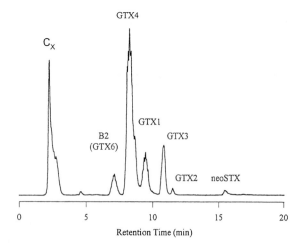

Figure 10. Profile of PSP-group toxins in *Alexandrium catenella* after separation by reverse phase ion-pair HPLC on a Hamilton PRP-1 polystyrene divinylbenzene column and eluted with a gradient formed by mixing ammonium phosphate buffer (pH = 6.70) with 75% ammonium phosphate (pH = 7.00) −25% acetonitrile, and containing the ion-pair agent hexanesulfonate-heptanesulfonate (15 mM).

Discussion

Seasonal variations in PSP in shellfish

The toxin levels in PSP-contaminated shellfish tend to vary with the season. Along the Pacific coast of the United States and Canada, all cases of PSP occurred between May and October (Meyer, 1953). The European and South African outbreaks took place during May through October (Meyer et al., 1928; Sapeika, 1958). The highest toxicity levels in shellfish in the New Brunswick and Nova Scotia areas occurred between mid-July and the latter part of September, with an overall maximun in lateAugust (Gibbard & Naubert, 1948). While in Chile, October and November (spring) would be the dangerous period. In these temperate regions, PSP tend to be high when the water temperature is around 15–20 °C. However, in some areas in Alaska, Chambers & Magnusson (1950) even found that *Saxidomus giganteus* maintained dangerous toxic levels throughout the year.

In sub-tropical and tropical areas of the Pacific, the outbreaks of PSP occurred during December through February in Papua New Guinea-New Britian (Maclean, 1977), during March in Sabah and Brunei (Beales, 1976; MacLean, 1979), and during April in the Philippines (Gacutan et al., 1985). The toxic dinoflagellate involved in these cases were different variaties of *Pyro-*

dinium bahamense. The apparent northward spread appeared to follow the sun's northward movement as it came out of winter solstice.

In Hong Kong and the South China coast, the main peak of PSP is in the months of March and April when the seawater temperature warms up to 20–23 °C. On several occasions, the causative agent had been identified as *A. catenella* or *A. tamaranse. A. catenella* has been reported in different parts of the Pacific Ocean (Hallegraeff et al., 1991), while *A. tamarense* had been reported in the North Atlantic Ocean and Japan (Okumura et al., 1994). The variation in PSP-contamination of shellfish during the year could result from (1) seasonal incidence of toxic dinoflagellate blooms; (2) variations in toxin production in the resident dinoflagellate; (3) or a combination of both.

Toxic red tide blooms

Dinoflagellates form durable cysts which over-winter in the sediments. It has been generally assumed that the primary factor which signals excystment is the temperature. As the temperature warms up in the spring, the number of dinoflagellates in the water increased. As a matter of fact, from our record of regular screening for PSP in shellfish covering a span of 18 years (Lam et al., 1989; Chan & Liu, 1991), the highest incidence of toxic red tide both in terms of number of episodes and severity of PSP-contamination was in the spring of 1989. The results of cell counts of dinoflagellates in the water has shown a steady decline since (Hong Kong Government Environmental Protection Department Annual Reports, 1988–1994), although most of the species were non-toxic. The level of PSP-contamination of shellfish had declined as well. However, on a number of ocassions, *A. catenella* cell count in the water was high but no PSP-contamination of shellfish was detectable. Cell count of toxic dinoflagellate species in the seawater sample alone, therefore, might not serve as a completely reliable index and early warning for PSP in seafood.

Population dynamics

Population growth in *A. catenella* was found to be quite sensitive to pH, salinity, and temperature regimes. The present findings provided an explanation for the distribution of PSP-contamination of local shellfish, which have been collected from 17 sites along the coast and screened by the mouse toxicity test on a monthly basis for over 5 years (1986–1991) (Chan & Liu, 1991). The

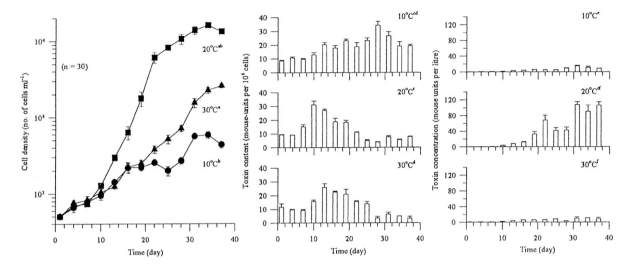

Figure 11. Effect of temperature on *Alexandrium catenella* population growth and toxin accumulation ([a,b,c,d,e,f] indicate significant difference, $p<0.05$).

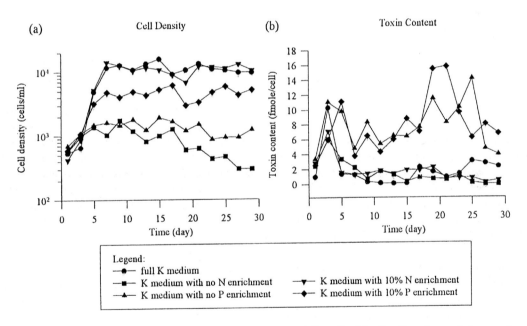

Figure 12. Effect of omitting altogether or reducing the level to 10% of that of 'K medium' for N or P on the population growth and toxin accumulation in *Alexandrium catenella*.

annual peak of PSP contamination concentrated in the eastern and southern side of Hong Kong's territorial waters in March–April, when sea surface temperatures are in the range of 18–23 °C, the salinity was around 34 ‰ and the pH of 8. During this period, the salinity and pH in the Western part of Hong Kong's territorial waters was low (e.g. salinity of 20‰ and pH = 7.55 at Tsim Bei Tsui located at the far Western part of

the New Territories). Water samples taken from that locality did not support the growth of *A. catenella* in the laboratory (<0.01% of full capacity) (Chan & Siu, 1995). Within the Victoria Harbour itself, where urban discharge was high (e.g. North Point), and the water had a high salinity (32‰) but a low pH of 7.6, shellfish taken from that locality was below the baseline level for PSP and water samples taken from this site only partially supported *A. catenella* growth (about 60% of normal capacity). During the autumn and the dry winter, the PSP-contamination of shellfish spreaded to the western side of Hong Kong's territorial waters when the flow of freshwater down the Pearl River retreated and the salinity and pH increased. Thus at Sham Tseng, Ma Wan, and Sandy Bay situated at the Western side of the Harbour facing the Pearl River, during the months of September–December the salinity rose to 32‰ and pH to 7.6–7.7. Some PSP-contamination of shellfish (About 50–100% above baseline) was recorded, although the level was still lower than that found in the eastern/southern end the Territory. Water samples taken from these sites during this time of the year were capable of supporting *A. catenella* growth.

Although *A. catenella* grow optimally at 25 °C, high cell densities need not lead to higher toxin production and PSP-contamination of local shellfish became reduced in May and remained low throughout the summer months at all sites along the coast of Hong Kong.

Nutrient levels also greatly affected dinoflagellate population dynamics. For *A. catenella*, our results showed that clean oceanic water did not carry sufficient nutrients to sustain the rapid exponential growth phase. This was true for both macro and micronutrients. However, Hong Kong harbour surface water is quite polluted, and the N, P levels (see Hong Kong Government EPD annual reports) were high enough to sustain the rapid growth phase. Si could be limiting, but with the rains in spring and early summer, large amounts of Si were washed down from land sites during this period. As for the trace minerals, the sediments in the seabed contained high levels of Fe, Zn, Cu, Cr, Mn, and other micro-nutrients. Since dinoflagellates can undergo diel vertical migration, they could gain access to these nutrients during the night, especially during the spring time when a thermocline generally developed, and other phytoplanktons which did not show the vertical migration would not be able to gain sufficient minerals for growth, thus paving the way for dinoflagellates to multiple rapidly and become the dominant species to produce a red tide.

Toxin profile in dinoflagellates

The total toxin content in the dinoflagellate itself has been found to vary with physical factors such as temperature (Hall, 1982; Ogata et al., 1987), light intensity (Hall, 1982; Ogata et al., 1987) and salinity (White, 1978). However, although differences between species have been reported, the toxin profile of individual species has been reported to be invariant by most workers (Hall et al., 1990; Boyer et al., 1987; Cembella et al., 1987; Ogata et al., 1987; Oshima et al., 1993; Franco et al., 1994). Toxins of *A. tamarense* (*Gonyaulax tamarensis* var. *excavata*) were found to contain more than seven components, including saxitoxin (Shimizu et al., 1975; Oshima et al., 1979; Ghazarossian et al., 1974), neosaxitoxin (Shimizu et al., 1978), sulfonic acid ester derivative of saxitoxin (11-hydroxysaxitoxin sulfate) (Boyer et al., 1978; Okumura et al., 1994). Cutures of *A. catenella* have been shown to contained mainly saxitoxin and neosaxitoxin (Schantz et al., 1966), but in small amounts only. The distinctive toxin chemistry of *A. minutum* was described as gonyautoxins GTX-1, 2, 3 and 4 (Oshima et al., 1989; Franco et al., 1994). The tropical dinoflagellate, *Pyridinium bahamense* var. *compressa*, was found to accumulate STX, neoSTX9, GTX-5, GTX-6. Our own strain of *A. catenella* did not contain detectable levels of saxitoxin, and neosaxitoxin occurred in trace amounts. The major components were GTXs, particularly GTX-4, which has high intrinsic toxicity.

Other workers have reported variations in toxin profile (Boyer et al., 1986) and toxin content (Alam et al., 1979; Shimizu, 1979; Hall, 1982; Maranda et al., 1985; Boyer et al., 1986; Cembella et al., 1987; Anderson et al., 1990) even among different strains or isolates of the same species (Kim et al., 1993). Thus among three isolates of *A. tamarense* examined, the first one contained a large proportion of the carbamyl-N-sulfo compounds or $N(21)$ sulfo toxins (B1, B2, C1, C2); the second one contained primarily GTX-1 through 4 (Boyer et al., 1986); while the last one was totally non-toxic (Boyer et al., 1986). Sako et al. (1992) examined toxin compositions and mating type of F1 progenies from crosses between *A. catenella* strains having different toxin profiles. In all strains used, the mole percentage of their toxin composition did not significantly change in any growth phase. One parental strain produced GTX-4 and C-4, while the other produced neosaxitoxin (neoSTX) and saxitoxin (STX) during all growth phases. F1 progenies showed one parental toxin composition and segregated independently with the mating type. These data strongly suggested that Mendelian inheritance of toxin profiles occurred in heterothallic individuals (Sako et al., 1992).

For each single isolate, the toxin content can vary dramatically under different growth conditions. Physiological studies have demonstrated that toxicity within a dinoflagellate would vary with the age of the cell culture (Prakash, 1967; Proctor et al., 1975; White &

Maranda, 1978; Oshima & Yasumoto, 1979; Schmidt & Loeblich, 1979). In general, the toxin content was low in the lag phase, peaked in mid-exponential growth phase, and decreased as cultures reached the stationary phase (Boyer et al., 1986; Boyer et al., 1986, 1987; Cembella et al., 1987; Boczar et al., 1988; Anderson et al., 1990). This change in toxin content per cell was due to a concomittant change in all toxin components and not due to metabolic interconversion of the toxin components or selective increase/decrease in one specific toxin analogue (Alam et al., 1979; Oshima et al., 1 982; Hall, 1 982; Boyer et al., 1986, 1987; Cembella et al., 1987; Ogata et al., 1987; Oshima et al., 1993). The factors responsible for the decline in total toxin content as the culture matures is still not clear, although it had been attributed to a shortage of CO_2 for photosynthesis in dense cultures or a shift in the pH as a result of accumulation of acidic metabolites (Anderson et al., 1990). Another mechanism proposed attributed the decline to an increased leakiness of the cell membranes, resulting in diffusional loss of toxin to the surrounding medium (Boyer et al., 1986, 1987). In any case, the constancy of the toxin profile in a given dinoflagellate culture was quite remarkable and probably resulted from a fixed genetic trait which could vary from one geographic region to another (Hall, 1982), but would not respond to a transient change in environmental conditions (Boyer et al., 1987). Thus the toxin profiles could be regard as potential taxonomic markers (Cembella et al., 1987; Oshima et al., 1993).

Toxin biosynthesis and nutrient supply

In the present study, the amount of toxin accumulated in the cell not only varied with the cell cycle and diurnal activity, being highest at around mid-night, but also varied drastically with changes in the physical and nutritional factors, in particular temperature and N/P ratios. The highest level of toxin that could be produced in a population of A. catenella in culture was obtained when the temperature was 20 °C, the salinity was 35‰, the pH was 8.38, and the photoperiod was 10L/14D. This level of population density (about 15–16000 cell ml^{-1}) and toxin level (50–80 mouse-unit ml^{-1}) actually matched the highest levels ever recorded in the field in Hong Kong, i.e. the toxic bloom of late March–early April of 1989 (Chan & Liu, 1991), from which our present stock of A. catenella was obtained.

Despite earlier reports to the contrary (Proctor et al. 1975; White, 1978; Hall, 1982; and Ogata et al., 1987), Anderson et al. (1990) recently found that,

under most growth conditions, toxins were produced at rates that varied in direct proportion to the growth rate. He therefore suggested that cells produced toxins at rates approximating those needed to replace 'losses' to daughter cells during cell division. In this model, the PSP toxins were assumed to have some physiological role to play in the dinoflagellate, and mechanisms have been developed to preserve their levels in the cell. Our present results supported this idea. In nutrient-replete media, the accumulation of PSP in A. catenella cells coincided with new DNA synthesis, and the level returned to the baseline level after mitosis.

Several speculative biosynthetic pathways have been considered for the biosynthesis of the tricyclic perhydropurine derivatives which form the backbone of the saxitoxin-group of neurotoxin. After numerous feeding experiments, the normal purine metabolic pathway has been ruled out as the origin of the purine moiety. Instead the involvement of arginine or its precursor, α-ketoglutarate, was first confirmed in the formation of the toxin skeleton (Shimizu et al., 1984). Subsequently it was established that ornithine, the direct precursor of arginine, could be incorporated into the toxin with a loss of the carboxyl group (Shimizu et al., 1984). Acetate (Shimizu et al., 1984) and methionine in the form of S-adenosylmethionine (SAM), also played a role in contributing the necessary carbons and nitrogens in the toxin skeleton (Shimizu, 1986). Although investigators have proposed biosynthetic pathways, but they have not isolated chemical intermediates or enzymes used only in toxin production (Anderson, 1994). Since toxin of the saxitoxin group is synthesized from arginine, acetate and the methyl group of methionine, toxin production must compete with other biochemical reactions which metabolize these precursors.

In the present study, the results clearly showed that the level of toxin could wax and wane with the cell cycle. Our data have now pinpointed the timing of PSP biosynthesis alongside the S-phase of the cell cycle, and shifted with this phase as the timing of the cell cycle was affected following exposure to different photoperiod regimes. This suggested that PSP biosynthesis occurred in parallel with new DNA synthesis. This link was quite tight. In our photoperiod experiments, as we increased the daylength, the onset of the S-phase shifted with the delayed in the timing of nightfall. The accumulation of PSP also shifted by the same period of time until in short nights, it proceeded well in the light phase of the following day. If the purine-based PSP-toxins competed with the nucleic acids for the same precursor N-compounds, and nucleic acids additionally require phosphorus for their synthesis, then changing the N/P supply to the cell would drastically alter the two events. In other studies, cellular toxicity has been found to vary with the availability of phosphorous and nitrogen as well (Hall, 1982; Boyer et al., 1987).

Phosphorus limitation caused an increase in total toxicity in experimental cultures (Hall, 1982; Anderson et al., 1990), reaching levels 3 to 4 times that observed in the control (Boyer et al., 1987). High toxin content possibly was resulted from de novo synthesis of arginine, the precusor for PSP biosynthesis, at levels in excess of cell division which was nearly suspended (Anderson et al., 1990).

Nitrogen limitation showed no effect, however, on the total toxicity per cell and remained constant throughout the treatment (Hall, 1982). These results are in contrast to those of Boyer et al. (1987) and Anderson et al. (1990), who found that N limitation resulted in low toxin production rates and low toxin contents. This decrease can be explained if the toxins were acting as nitrogen storage compounds (saxitoxin is 33% N on a molecular weight basis). It was due to the competition between the saxitoxin production and other important metabolic pathways for scarce nitrogen atoms (Anderson et al., 1990).

The level of toxicity per cell was observed to increase when a sub-optimal temperature was used for growth (Proctor et al., 1975; Ogata et al., 1980, 1987; Hall, 1982; Boyer et al., 1985; Anderson et al., 1990). The effect of low temperature could result from a reduction in the protein synthesis, resulting in a surpus of arginine (Arg) within the cell that could be used for toxin synthesis. It follows that the enzymatic reactions necessary for both Arg and toxin biosynthesis would be less affected by low temperature than those involved in general protein synthesis. An alternate explanation could simply be that low temperature slows cell division more than toxin synthesis, such that the longer duration of each cell generation allows more time for the toxin to accumulate (Anderson et al., 1990). In the present study, A. catenella also showed a high toxin accumulation at 20 °C, which was suboptimal to population growth (25 °C).

Along with a decrease in temperature, a decrease of light intensity caused an increase in the amount of toxin produced accompanied by a decrease in the growth rate (MacIsaac et al., 1979). This indicated that photosynthesis played an important role in the production of PSP toxins.

An early finding showed an elevation in toxin production with increasing salinity up to 37‰ (White, 1978). Recent work demonstrated that there was no significant change in toxin content, however, with either acclimated growth at elevated salinity, or with short term increases or decreases of salinity (Anderson et al., 1990).

Toxin production and population growth

Since cell divisions occur rapidly during the exponential population growth phase, one would expect that toxin production would also be maximal during this period. Our axenic cultures confirmed this. At 20 °C, the total level of toxin in the cells suspended in the medium reached 10 mouse-units l^{-1} when the cell density was only 600–700 cells ml^{-1} (Figure 11). The toxin level reached 70–80 mouse-units l^{-1} by the time the cell population reached 3000 cells ml^{-1}. These levels of toxin in the water have been obtained in moderate redtide blooms (from the point of view of cell density) associated with severe shellfish PSP contamination (Chan & Liu, 1991). As cells reached a density of 10 000 cells ml^{-1}, toxin production per million cells was reduced. These findings have important implications for food safety and public health. It has been generally assumed that dinoflagellate numbers provide a reliable index of PSP toxicity, and a general alarm may not be called for when dinoflagellates occur in moderate numbers below 1–2000 cells ml^{-1}. Our results showed that higher PSP per million cells occurred during the earlier stages of a bloom.

Conclusions

The strain of *A. catenella* isolated from Hong Kong's waters in 1989 produced in order of abundance, GTX-4, GTX-3, GTX-1, B2 (= GTX-6), neoSTX, GTX2 and also miscellaneous C-compounds, and hence is highly toxic.

The synthesis of PSP-group of toxins by *A. catenella* in culture, occurred during the S-phase of the cell cycle, which was timed to begin at approximately 4 h after the onset of darkness and lasted 4–5 h. Thus the best time to assay the presence of toxin in seawater associated with a bloom of the this species is between mid-night and 4 a.m.

Since DNA-synthesis competes with PSP biosynthesis for N source, while P is not required for PSP biosynthesis, N supply is rate-limiting for both cell division and toxin production in the cell, while toxin production and accumulation is stimulated when P concentration is limiting while N is non-limiting. Thus the N:P ratio is expected to have a marked influence on the production of toxin during an *A. catenella* bloom.

Clean sea water does not provide sufficient nutrients for *A. catenella* growth. The surface inshore waters of Hong Kong provide the macronutrients (N, P) especially after heavy rains (Si), and the bottom sediments contained more than adequate supplies of trace minerals. When nutrients are non-limiting, the main limiting factor for *A. catenella* population growth and toxin production is temperature (optimal range is 20–25 °C), salinity (30–35‰) and pH (8.0–8.5). These optimal conditions prevail in the western and southern parts of Hong Kong during the spring (March–April) and spreads to the western parts during the autumn (September).

Under optimal conditions, the maximal carrying capacity was approximately 13 000 to 15 000 cell ml^{-1}. The mean population growth rate (μ_1) and the mean doubling time (D.T.$_1$) were 0.28 d^{-1} and 2.47 d, respectively. However, toxin biosynthesis was highest during the early exponential population growth phase, and PSP contamination of shellfish could occur when the cell density of *A. catenella* was only 3000 cells ml^{-1}.

Acknowledgements

This study was support by the grant from the Hong Kong Research Grants Council.

References

Alam, M. I., C. P. Hsu & Y. Shimizu, 1979. Comparison of toxins in three isolates of *Gonyaulax tamarensis* (Dinophyceae). J. Phycol. 15: 106–110.

Anderson, D. M. & T. P. O. Cheng, 1988. Intracellular localization of saxitoxins in the dinoflagellate *Gonyaulax tamarensis*. J. Phycol. 24: 17–22.

Anderson, D. M., D. M. Kulis, J. J. Sullivan & S. Hall, 1990. Toxin composition variations in one isolate of the dinoflagellate *Alexandrium fundyense*. Toxicon 28: 885–893.

Anderson, D. M., 1994. Red tides: Many experts believe these blooms of toxic algae have recently become more prevalent, posing a greater threat to human and marine health. Scient. Am., August 1994: 52–58.

Beales, R. W., 1976. A red tide in Brunei's coastal waters. Brunei Mus. J. 3: 167–182.

Bevington, P. R., 1969. Data Reduction and Error Analysis for the Physical Sciences. McGraw-Hill, New York.

Boczar, B. A., M. K. Beitler, J. Liston, J. J. Sullivan & R. A. Cattolico, 1988. Paralytic shellfish toxins in *Protogonyaulax tamarensis*

and *Protogonyaulax catenella* in axenic culture. Pl. Physiol. 88: 1285–1290.

Boyer, G. L., E. J. Schantz & H. K. Schnoes, 1978. Characterization of 11-hydroxysaxitoxin sulfate, a major toxin in scallops exposed to blooms of the poisonous dinoflagellate *Gonyaulax tamarensis*. J. chem. Soc. (London) chem. Comm: 889–890.

Boyer, G. L., J. J. Sullivan, R. J. Anderson, F. J. R. Taylor, P. J. Harrison & A. D. Cembella, 1986. Use of high performance liquid chromatography to investigate the production of paralytic shellfish toxins by *Protogonyaulax* spp. in culture. Mar. Biol. 93: 361–369.

Boyer, G. L., J. J. Sullivan, R. J. Anderson, P. J. Harrison & F. J. R. Taylor, 1987. Effects of nutrient limitation on toxic production and composition in the marine dinoflagellate *Protogonyaulax tamarensis*. Mar. Biol. 96: 123–128.

Carpenter, E. J. & J. Chang, 1988. Species-specific phytoplankton growth rates via diel DNA synthesis cycles. I. Concept of the method. Mar. Ecol. Prog. Ser. 43: 105–111.

Chambers, J. S. & H. W. Magnusson, 1950. Season variations in toxicity of butter clams from selected Alaska beaches. U. S. Fish Wildlife Serv., Spec. sci. Rep., Fish. No. 53: 19.

Chan, D. K. O. & S. J. Liu, 1991. Effect of paralytic shellfish poison on the cardioventilatory function of the eel. In N. De Pauw & J. Joyce (eds), Aquaculture and the Environment, Europ. Aquacult. Soc. Spec. Pub. 14: 64–65.

Chan, D. K. O. & K. Y. Siu, 1995. Paralytic shellfish poison contamination in finfish mariculture. Europ. Aquacult. Soc. Spec. Pub. 23: 245–246.

Cembella, A. D., J. J. Sullivan, G. L. Boyer, F. J. R. Taylor & R. J. Anderson, 1987. Variations in paralytic shellfish toxin composition within the *Protogonyaulax tamarensis/catenella* species complex; red tide dinoflagellates. Biochem. syst. Ecol. 15: 171–186.

Dean, P. & J. Jett, 1974. Mathematical analysis of DNA distributions derived from flow micro-fluorometry. Cell Biol. 60: 523.

Fox, M. H., 1980. A model for the computer analysis of synchronous DNA distributions by flow cytometry. Cytometry 1: 71–77.

Franco, J. M., P. Fernandez & B. Reguera, 1994. Toxin profiles of natural populations and cultures of *Alexandrium minutum* Halim from Galician (Spain) coastal waters. J. appl. Phycol. 6: 275–279.

Gacutan, R. Q., M. Y. Tabbu, E. J. Aujero & F. Icatlo, Jr., 1985. Paralytic shellfish poisoning due to *Pyrodinium bahamense* var. *compressa*. in Mati, Davao Oriental, Philippines. Mar. Biol. 87: 223–227.

Genenah, A. & Y. Shimizu, 1981. Specific toxicity of paralytic shellfish poisons. J. agric. Food Chem. 29: 1289–1291.

Gibbard, J. & J. Naubert, 1948. Paralytic shellfish poisoning on the Canadian Atlantic coast. Am. J. publ. Health 38: 550–553.

Hall, S., G. Strichartz, E. Moczydlowski, A. Ravindran & P. B. Reichardt, 1990. The saxitoxins: sources, chemistry, and pharmacology. In S. Hall & G. Strichartz (eds), Marine Toxins. Am. chem. Soc. Symp. Ser. 418: 29–65.

Hallegraeff, G. M., C. J. Bolch, S. I Blackburn. & Y. Oshima, 1991. Species of the toxigenic dinoflagellate genus *Alexandrium* in Southern Australian waters. Bot. Mar. 34: 575–587.

Hong Kong Government Environmental Protection Department, 1989–1994. Marine water quality in Hong Kong for 1989, 1990, 1991, 1992, 1993, 1994. Government Printer, Hong Kong.

Keller, M. D. & R. R. L Guillard, 1985. Factors significant to marine dinoflagellate culture. In D. M. Anderson, A. W. White, & D. G. Baden (eds), Toxic Dinoflagellates. Elsevier, New York: 113–116.

Keller, M. D., R. C. Selvin, W. Claus & R. R. L. Guillard, 1987. Media for the culture of oceanic ultraphytoplankton. J. Phycol. 23: 633–638.

Kim, C. H., Y. Sako & Y. Ishida, 1993. Variation of toxin production and composition in axenic cultures of *Alexandrium catenella* and *A. tamarense*. Nipp. Suis. Gakk. 59: 633–639.

Lam, C. W. Y., M. Kodama, D. K. O. Chan, T. Ogata, S. Sato & K. C. Ho, 1989. Paralytic shellfish toxicity in shellfish in Hong Kong. In T. Okaichi, D. M. Anderson & T. Nemoto (eds), Red Tides: Biology, Environmental Science and Toxicology. Proc. First Int. Symp. Red Tides, Japan, 1987, Elsevier, New York: 455–458.

MacIsaac, J. J., G. S. Grunseich, H. E. Glover & C. M. Yentsch, 1979. Light and nutrient limitation in *Gonyaulax excavata*: nitrogen and carbon trace results. In D. L. Taylor & H. H. Seliger (eds), Toxic Dinoflagellate Blooms, Elsevier North Holland, New York: 107–110.

MacLean, J. L., 1977. Observations on *Pyrodinium bahamense*, a toxic dinoflagellate in Papua New Guinea. Limnol. Oceanogr. 22: 234–254.

MacLean, J. L., 1979. Indo Pacific red tides. In Taylor & H. H. Seliger (eds), Toxic Dinoflagellate Blooms. D. L. Elsevier North Holland, New York: 173–178.

Marquardt, D. W., 1963. An algorithm for least-squares estimation of nonlinear parameters. Soc. ind. appl. Math. 11, 431–441.

Meyer, K. F., H. Sommer & P. Schoenholz, 1928. Mussel poisoning. J. prev. Med. 365–394.

Meyer, K. F., 1953. Medical Progress: food poisoning. New Engl. J. Med. 249: 765–773, 804–812, 843–852.

Ogata, T., T. Ishimaru & M. Kodama, 1987a. Effect of water temperature and light intensity on growth rate and toxicity changes in *Protogonyaulax tamarensis*. Mar. Biol. 95: 217–220.

Ogata, T., M. Kodama & T. Ishimaru, 1987b. Toxin production in the dinoflagellate *Protogonyaulax tamarensis*. Toxicon 25: 923–928.

Okumura, M., S. Yamada, Y. Oshima & N. Ishikawa, 1994. Characteristics of paralytic shellfish poisoning toxins derived from short-necked clams (*Tapes japonica*) in Mikawa Bay. Nat. Toxins 2: 141–143.

Oshima, Y. & T. Yasumoto, 1979. Analysis of toxins in cultured *Gonyaulax excavata* cells originating in Ofunato Bay, Japan. In D. L. Taylor & H. H. Seliger (eds), Toxic Dinoflagellate Blooms, Elsevier, New York: 377–380.

Oshima, Y., M. Hirota, T. Yasumoto, G. Hallegraeff, S. Blackburn & D. Steffensen, 1989. Production of paralytic shellfish toxins by the dinoflagellate *Alexandrium minutum* Halim from Australia. Nipp. Suis. Gakk. 55: 925.

Oshima, Y, S. I. Blackburn & G. M. Hallegraeff, 1993. Comparative study on paralytic shellfish toxins profiles of dinoflagellate *Gymnodinium catenatum* from three different countries. Mar. Biol. 16: 471–476.

Prakash, A., 1967. Growth and toxicity of a marine dinoflagellate, *Gonyaulax tamarensis*. J. Fish. Res. Bd Can. 24: 1589–1606.

Proctor, N. H., S. L. Chan & A. J. Trevor, 1975. Production of saxitoxin by cultures of *Gonyaulax catenella*. Toxicon 13: 1–9.

Sako, Y., C. H. Kim & Y. Ishida, 1992. Mendelian inheritance of paralytic shellfish poisoning toxin in the marine dinoflagellate *Alexandrium catenella*. Biosci. Biotechnol. Biochem. 56: 692–694.

Sapeika, N., 1958. Mussel poisoning: a recent outbreak. S. african med. J. 32: 527.

Schantz, E. J., J. M. Lynch, G. Vayvada, K. Matsumoto & H. Rapoport, 1966. The purification and characterization of the poison produced by *Gonyaulax catenella* in axenic culture. Biochemistry 5: 1191–1194.

Schmidt, R, J, & A. R. Loeblich III, 1979. A discussion of the systematics of toxic *Gonyaulax* species containing paralytic shellfish poison. In D. L. Taylor & H. H. Seliger (eds), Toxic Dinoflagellate Blooms. Elsevier, New York: 83–88.

Shimizu, Y., M. Alam, Y. Oshima & W. E. Fallon, 1975. Presence of four toxins in red tide infested clams and cultured *Gonyaulax tamarensis* cells. Biochem. Biophys. Res. Commun. 66: 731–737.

Shimizu, Y., C. Hsu, W. E. Fallon, Y. Oshima, I. Miura & I. Nakanisha, 1978. Structure of neosaxitoxin. J. am. chem. Soc. 100: 6791–6793.

Steidinger, K. A., 1975. Basic factors influencing red tides. In V. R. LoCicero (ed.) Proc. first int. Conf. on toxic Dinoflagellates 1974. The Massachusetts Science and Technology Foundation, Wakefield, Mass.: 153–159.

Sullivan, J. J., 1987. The determination of PSP toxins by HPLC and Autoanalyzer. In Am. chem. Soc. Symp. Ser. 486: 66–77.

Sullivan, J. J. & W. T. Iwaoka, 1983. High pressure liquid chromatographic determination of the toxins associated with paralytic shellfish poisoning. J. Assoc. off. anal. Chem. 66: 297.

Sullivan, J. J. & M. M. Wekell, 1987. The application of high performance liquid chromatography in a paralytic shellfish poisoning monitoring program. In D. E. Kramer & J. Liston (eds), Seafood Quality Determination, Elsevier North Holland, New York: 357–371.

White, A. W., 1978. Salinity effects on growth and toxin content of *Gonyaulax excavata*, a marine dinoflagellate causing paralytic shellfish poisoning. J. Phycol. 14: 475–479.

White, A. W. & L. Maranda, 1978. Paralytic shellfish toxins in the dinoflagellate *Gonyaulax excavata* and in shellfish. J. Fish. Res. Bd Can. 35: 397–402.

Hydrobiologia **352**: 141–147, 1997.
Y.-S. Wong & N. F.-Y. Tam (eds), Asia–Pacific Conference on Science and Management of Coastal Environment.
©1997 *Kluwer Academic Publishers. Printed in Belgium.*

Are changes in N:P ratios in coastal waters the key to increased red tide blooms?

I. J. Hodgkiss[1] & K. C. Ho[2]
[1]*Department of Ecology and Biodiversity, The University of Hong Kong, Hong Kong*
[2]*School of Science and Technology, The Open Learning Institute of Hong Kong, Hong Kong*

Key words: N:P ratios, coastal waters, red tides

Abstract

There is mounting evidence of a global increase in nutrient levels of coastal waters through riverine and sewage inputs, and in both the numbers and frequency (as well as the species composition) of red tides. However, it is still not possible to conclude the extent to which the increase in red tides in coastal waters can be attributed to the increase in nutrient levels, since so many other factors are involved.

Undoubtedly, a relationship exists between red tides and the N and P load of coastal waters, and many nutrient enrichment experiments have shown that marine phytoplankton blooms are often nutrient limited. What is now becoming clear, however, is that although in classical Liebigian terms minimum amounts can be limiting, nutrient ratios (such as N:P and Si:P) are far more important regulators.

This paper reviews evidence collected by the authors from Tolo Harbour, Hong Kong together with data collected in Japanese and North European coastal waters by various authors, which indicates that both long term and relatively short term changes in the N:P ratio are accompanied by increased blooms of non-siliceous phytoplankton groups and, furthermore, that the growth of most red tide causative organisms in Hong Kong coastal water is optimized at a low N:P (atomic) ratio of between 6 and 15.

Introduction

So long as growth is not overprolific, algal blooms are of benefit to both natural fisheries and marine culture operations since the animals' food supply is plentiful. Sometimes, however, the algae occur in such large numbers that they colour the sea surface. The best known are the red tides which result, though others like brown tides can also occur depending on the algae involved. Such blooms may have a negative effect and cause economic losses to fisheries and aquaculture and, indeed, they may have impacts on human health. Sometimes, the fish kills which result from these harmful algal blooms are simply a result of the pure numbers of algae involved, when high algal respiration or high bacterial decomposition of the bloom lead to oxygen depletion. On other occasions, however, the algae have the ability to produce toxins which can kill fish or find their way through fish or shellfish to humans, resulting

in various kinds of poisoning. Another type of harmful algal bloom (often termed noxious algal blooms) involves the algae damaging the fishes gills, and this can lead to severe losses in cage fish.

Algal blooms are natural phenomena, which have occurred throughout recorded history. However, in the past few decades there has been mounting evidence from around the world to suggest that coastal marine phytoplankton blooms have increased in frequency, intensity and geographic distribution. For example, at Halskov Rev (Nielsen & Aertebjerg, 1984; Gargas et al., 1980); in the Gulf of Finland (Niemi, 1974); in the Baltic Sea (Nehring et al., 1984; Renk et al., 1988); in Dutch Coastal waters (Cadée, 1986b); along South African coasts (Horstman, 1981); in Tolo Harbour, Hong Kong (Lam & Ho, 1989); and in Chinese coastal waters (Qi et al., 1995). Ho & Hodgkiss (1991) in a review of red tides in subtropical coastal waters from 1928 to 1989 showed that the number of recorded occurrences had escalated from 1 or 2 every 10 years at the beginning of this period to over 220 between 1980 and 1989; and that 19 countries in the subtropical region were now badly affected.

Reviewing the evidence available, Anderson (1989) concluded that there had been 'a major global expansion ...... of red tides throughout the world' and Smayda (1990) referred to a 'global epidemic of algal blooms'.

Smayda & White (1990) argued that there were not enough long term data at the global level to conclude with certainty that toxic and noxious algal blooms have been increasing in frequency and intensity over the past 20 years, and Smayda (1990) challenged the view that a global bloom epidemic is currently occurring on the grounds that it may be merely an artifact of increased monitoring, improved analytical techniques or increased awareness of toxic outbreaks accompanying increased mariculture of finfish and shellfish. However, amongst the data sets now available, there is clear evidence of a long term increase in phytoplankton blooms and red tides in many coastal waters. Even Smayda (1990) noted that there was considerable evidence to suggest that significant changes in phytoplankton species occurrences, biomass and productivity, novel species occurrences, unusual blooms, and shifts in predominance have occurred in regions of the world as far apart as the North Sea, South Africa and Hong Kong. He concluded that, at least, 'a potential for global expansion of algal bloom problems is now occurring'.

It is clear that we need to better understand such algal blooms so that we might be able to prevent their occurrence, reduce their damaging effects or, at least, be able to predict when and where they might occur.

Hallegraeff (1993) has explored four explanations for the apparent global increase in algal blooms, namely: increased scientific awareness; increased utilization of coastal waters for aquaculture; transport of dinoflagellate cysts in ships' ballast waters or associated with shellfish imports; and stimulation of plankton blooms by 'cultural eutrophication' and/or unusual climatological conditions. He concludes that it may not be possible, for some time, to have a conclusive answer regarding the relative importance of each of these, but that nutrient loading has an obvious relationship with increased blooms.

Since we have no further data to add to those which Hallegraeff (1993) reported regarding his first three 'explanations' and, since he himself highlights the obvious and important relationship between nutrient loading and increased blooms, we have turned our attention to the possible stimulation of plankton blooms by eutrophication. This review explores, therefore, the mounting evidence for an increased frequency, intensity and geographic distribution of such coastal phytoplankton blooms – the so called 'global spread'; looks at the evidence for nutrient loading being related to these increased coastal blooms; and investigates the possibility of a nutrient ratio linkage with algal blooms (rather than simply a Liebigian link between increased nutrient supply and increased blooms).

Nutrient loading, nutrient ratios and algal blooms in Hong Kong coastal waters

There is general agreement that nutrient loads have been increasing in many coastal waters, usually as a result of increased riverine or sewage inputs. For example in the USA, Smith et al. (1987); in the Dutch Wadden Sea (Fransj, 1986); in the Seto Sea, Japan (Okaichi, 1989); in the Baltic Sea (Fonselius, 1972; Nehring, 1984; Niemi & Aström, 1987; Kononen, 1988); and in Tolo Harbour (Hodgkiss & Chan, 1987; Lam & Ho, 1989).

Smayda (1989) points out that there is a general consensus that phytoplankton growth in the sea is often nutrient limited; that the uptake of nutrients such as N, P and Si follow the Redfield Ratio (that is, they exhibit a stoichiometric proportionality); and that numerous nutrient enrichment experiments have shown that nutrient addition tends to relax nutrient limitation.

In Tolo Harbour, Hong Kong, a 10 fold increase in mean dissolved phosphate levels and a 5 fold increase in mean dissolved nitrate levels between 1978 and 1985 coincided with a very large increase in standing crop of phytoplankton ($\times 8.5$), an increase in red tide blooms from 2 in 1977/78 to 17 in 1984 and an increased contribution of dinoflagellate abundance from 26 to 66% of the phytoplankton population in the outer harbour and from 11 to 26% in the inner harbour (Chan & Hodgkiss, 1987; Hodgkiss & Chan, 1983, 1986, 1987). Lam & Ho (1989) showed that an 8 fold increase in red tides in Tolo Harbour between 1976 and 1986 was related to a 6 fold increase in the catchment population, which had brought about a 2.5 fold increase in nutrient loading.

Chiu, Hodgkiss & Chan (1994) studied a relatively unspoilt area of Hong Kong (Tai Tam Bay) and found that, compared to their earlier study (Chan, Chiu & Hodgkiss, 1991), there had been an increase in algal biomass from 1987 to 1991; a small shift in community dominance (diatoms fell from 98.7 to 96.5% of the

Table 1. Optimal N:P (atomic) ratios for the growth of various red tide causative organisms.

Causative Species	Optimal N:P Ratio for Growth
Alexandrium catenella (Whedon & Kofoid) Balech	15–30:1
Ceratium furca (Ehrenberg) Claparède & Lachmann	12–22:1
Gonyaulax polygramma Stein	4–8:1
Gymnodinium nagasakiense Takyama & Adachi	11–16:1
Noctiluca scintillans (Macartney) Kofoid & Swezy	8–14:1
Prorocentrum dentatum Stein	6–13:1
Prorocentrum minimum (Pavillard) Schiller	4–13:1
Prorocentrum sigmoides Bohm	4–15:1
Prorocentrum triestinum Schiller	8–15:1
Scrippsiella trochoidea (Stein) Balech	6–13:1
Skeletonema costatum (Greville) Cleve	15–30:1

population and dinoflagellates rose from 1.2 to 3.4%); and also a small shift in nutrient loading.

The classical approach to explain such nutrient regulation of marine phytoplankton growth has been based upon Liebig's Law of the Minimum. However, it is now becoming clear that nutrient ratios can be far more important regulators, particularly in terms of species selection. Numerous investigators have superficially looked at the N:P ratio in their attempts to answer the question as to whether N or P was the more limiting nutrient. This use of the Redfield Ratio has been very successful. However, it is now clear that a more important question is whether altered nutrient ratios accompanying eutrophification of coastal waters are an important factor in the increased blooming of some species.

In Hong Kong, data from the Environmental Protection Department for 1984–1990 (Figure 1) indicate that when dissolved N levels were greater than 0.1 mg l^{-1} and dissolved P levels greater than 0.02 mg l^{-1}, red tide occurrences were highly probable.

Hodgkiss & Chan (1987) and Chan & Hodgkiss (1987) reported a decline in N:P ratio in Tolo Harbour, Hong Kong and, at the same time, a change in the phytoplankton community, with the red tide causative organisms (mainly dinoflagellates) taking over dominance from the diatoms. Thus, in inner Tolo Harbour, changes in the species composition of the phytoplankton from 1981 to 1990 (Table 2) reflected this change in the harbour's phytoplankton community as the N:P ratio declined from an annual mean of around 20:1 to 11:1 during the period 1982 to 1989 (Figure 2).

Environmental Protection Department data also show that with this gradual decrease in the N:P ratio there is a significant increase in red tide occur-

rences during this time period (Figure 2). Regression analysis of this data, which yielded an F value of 0.35 (P = 0.583), showed the significant relationship between the two. Interestingly, from this same figure, it can also be seen that whenever a small drop in the N:P ratio occurs, the red tide occurrences increased correspondingly.

In the 1980s, this harbour was seriously affected by red tides and the major species were *Prorocentrum micans* Ehrenberg, *P. sigmoides* Bohm and *P. triestinum* Schiller (Ho & Hodgkiss, 1995). Bottle test biossay of limiting factors indicated that favourable N:P atomic ratios for these 3 species were 5–10; 4–15 and 8–15:1 respectively. The N:P ratio in Tolo Harbour fell from a mean of 20.3:1 in 1983 to 11.05:1 in 1989 (Figure 2 applies) as these species increased in abundance.

Ho & Hodgkiss (1993) had previously demonstrated that the growth of most Hong Kong red tide causative organisms was optimized at a low N:P (atomic) ratio of between 4 and 16 (Table 1).

Discussion

Evidence is accumulating from all over the world (not just Hong Kong) that 'accelerated eutrophication' (i.e. the speeding up of this natural process by man's input of domestic, agricultural and industrial wastes) can stimulate algal growth in coastal waters. Thus, in the Gulf of Finland, Niemi (1974) was able to correlate increasing concentrations of phosphorus in the surface layer with increasing annual primary production rates; and Lassig et al. (1978) showed a similar strong positive correlation between these factors. Cadée (1984,

144

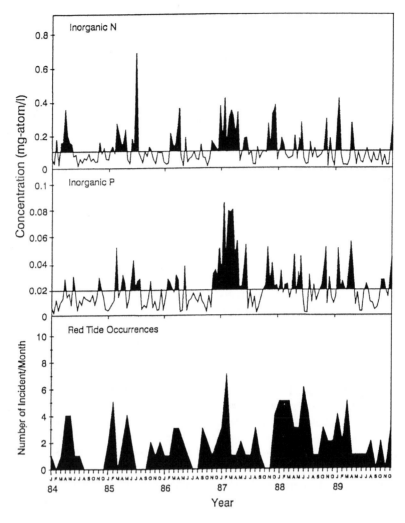

Figure 1. Variations in surface inorganic N and inorganic P concentrations in relation to red tide occurrences in inner Tolo Harbour for the years 1984–1989. (Original data courtesy of the Director, Environmental Protection Department, Government of Hong Kong).

1986a) showed that in the Dutch Wadden Sea (part of the southern North Sea) primary production and phosphate concentrations had both exhibited long term increases. Qi et al. (1995) linked dramatic red tide increases in Chinese coastal waters between 1980 and 1991 with increased nitrate and phosphate levels. In the Black Sea, long term (1957–1984) coincident increases in mean phosphate concentration and maximal abundance of red tide have been noted along the Roumanian Coast by various authors (Bodeanu & Usurelu, 1979; Mihnea, 1979). In the Seto Inland Sea, Japan, a 7 fold increase in red tide outbreaks between 1965 and 1976 was associated with increased levels of coastal nutrient enrichment (Prakash, 1987; Yanagi, 1988).

Thus, a regionally consistent and persistent pattern has emerged of a long term increased frequency of blooms associated with coastal nutrient enrichment (Smayda, 1990). This considerable evidence of significant changes in phytoplankton species occurrences, biomass and productivity, as well as shifts in predominance, occurring in regions as far apart as the North Sea and Hong Kong support the hypothesis that phytoplankton blooms are increasing in coastal waters on a global scale and that they are linked to long term increases in coastal nutrient levels. Anderson (1989) summed this up when he pointed out that the literature undeniably documents a global increase in the frequency, magnitude and geographic extent of coastal algal blooms and red tides over the past 2 decades and that

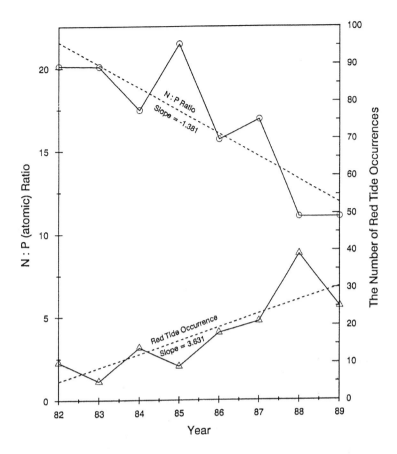

Figure 2. Variations in the N:P ratio versus red tide occurrences in inner Tolo Harbour during the years 1982–1989. (Original data courtesy of the Director, Environmental Protection Department, Government of Hong Kong).

there is very strong correlation between the number of red tides and the degree of coastal pollution.

Furthermore, the hypothesis that nutrient ratios are involved in increased blooms and species changes is, in our view, a viable one. Obviously, as Smayda (1989) put it, other factors will be involved in regulating bloom events, but nutrient ratios can be considered to be of major importance. At present unfortunately there are only limited data available in the literature to address this question for the marine phytoplankton, whereas much more is available for freshwater species (Sommer, 1987). Smayda (1989, 1990) has indicated how long term declines in the Si:N and Si:P ratios in the Baltic, North and Black Seas in response to nitrification, are accompanied by increased blooms of non siliceous phytoplanktonic groups. He presents evidence of such decreased Si:P ratios from the German Bight (Berg & Radach, 1985); Baltic Sea (Niemi & Aström, 1987); Gulf of Finland and Gulf of Bothnia (Pitkänen, 1978; Pitkänen & Malin, 1980); and the Belt

Sea (von Bodungen, 1986). He notes that significantly, not only have blooms of non-silica requiring groups increased during this same period, but they have even replaced the diatoms as the dominant biomass group in some areas.

Smayda (1990) has rehearsed the arguments for evaluating N and P ratios with Si, since this simplifies the evaluation into an issue of diatom blooms versus non diatom blooms versus ionic ratios. However, there is also evidence to link changes in the N:P ratio both with increases in blooms and with species changes. For example, Philips and Tanabe (1989) suggested that a shift in the N:P ratio of the loads entering marine coastal waters contributed to an alteration in species dominance in the phytoplankton population, with diatoms being gradually replaced by dinoflagellates and our data from Hong Kong indicate similar changes in species dominance and a consequent increase in red tides in relation to shifting N:P ratios.

Table 2. Changes in phytoplankton species composition in inner Tolo Harbour from 1978 to 1990

Year	Dominant Phytoplankton Species	
1981	Diatoms:	*Chaetoceros affine*
		C. costatum
		C. curvisetum
		Leptocylindrus danicus
		L. minumus
		Nitzschia delicatissima
		Skeletonema costatum
1983	Diatoms:	*Chaetoceros costatum*
		Leptocylindrus danicus
		L. minumus
		Nitzschia delicatissima
		N. longissima
		N. seriata
		Skeletonema costatum
1985	Diatoms:	*Leptocylindrus danicus*
		L. minimus
		Nitzschia delicatissima
		N. longissima
		N. seriata
		Skeletonema costatum
	Dinoflagellates:	*Prorocentrum sigmoides*
		P. triestinum
1988	Diatoms:	*Leptocylindrus danicus*
		L. minimus
		Nitzschia delicatissima
		N. seriata
		Skeletonema costatum
	Diaoflagellates:	*Gonyaulax polygramma*
		Gymnodinium nagasakiese
		Noctiluca scintillans
		Prorocentrum sigmoides
		P. triestinum
1990	Diatoms:	*Leptocylindrus danicus*
		L. minimus
		Nitzschia seriata
		Skeletonema costatum
	Dinoflagellates:	*Gonyaulax polygramma*
		Gymnodinium nagasakiense
		Noctiluca scintillans
		Prorocentrum sigmoides
		P. triestinum

Indeed, the results of our studies have given us an effective strategy for the management of red tides in Hong Kong. The control of P input **in addition** to the control of N input into Tolo Harbour was obviously required since, as a result of N control, the N:P ratio had fallen and so led to excessive red tides. Such falling N:P could possibly be used in other locations as a marker for predicting increased red tide occurrences and so for directing management actions.

Whether or not this 'local' strategy for red tide management in Hong Kong coastal waters can be applied globally (or indeed to oceanic as well as coastal waters) remains to be seen. What is clear, however, is that our results from Hong Kong clearly indicate that there has been an expansion of algal bloom problems; that these can be related to human activities and, in particular to the N:P ratio; and that increased international research efforts are needed to evaluate man's involvement in the possible global expansion of algal bloom problems and the relationship between nutrient ratios (particular N:P) and the appearance of blooms of many novel and harmful species.

References

Anderson, D. M., 1989. Toxic algal blooms and red tides: A global perspective. In Okaichi, T., D. M. Anderson & T. Nemoto (eds), Red Tides: Biology, Environmental Science, and Toxicology. Elsevier, New York, Amsterdam, London: 11–16.

Berg, J. & G. Radach, 1989. Trends in nutrient and phytoplankton concentrations at Helgoland Reede (German Bight) since 1962. ICES C.M. 1985/L:2.

Bodeanu, N. & M. Usurelu, 1979. Dinoflagellate blooms in Romanian Black Sea coastal waters. In Taylor, D. L. & H. H. Seliger (eds), Toxic Dinoflagellate Blooms. Elsevier/North Holland, Inc., New York: 151–154.

von Bodungen, B., 1986. Annual cycles of nutrients in a shallow inshore area, Kiel Bight – Variability and trends. Ophelia 26: 91–107.

Cadée, G. C., 1984. Has input of organic matter to the western Wadden Sea increased during the last decades? In Laane, R. W. P. M. & W. J. Wolff (eds), The Role of Organic Matter in the Wadden Sea. Neth. Inst. Sea Res. Publ. Ser. 10: 71–82.

Cadée, G. C., 1986a. Increased phytoplankton primary production in the Marsdiep area (western Dutch Wadden Sea). Neth. J. Sea Res. 20: 285–290.

Cadée, G. C., 1986b. Recurrent and changing seasonal patterns in phytoplankton of the westernmost inlet of the Dutch Wadden Sea from 1969–1980. Mar. Biol. 93: 281–289.

Chan, B. S. S., M. C. Chiu & I. J. Hodgkiss, 1991. Plankton dynamics and primary productivity of Tai Tam Bay, Hong Kong. Asian Mar. Biol. 8: 169–192.

Chan, B. S. S. & I. J. Hodgkiss, 1987. Phytoplankton productivity in Tolo Harbour. Asian Mar. Biol. 4: 79–90.

Chiu, M. C., I. J. Hodgkiss & B. S. S. Chan, 1994. Ecological studies of phytoplankton in Tai Tam Bay, Hong Kong. Hydrobiologia 273: 81–94.

Fonselius, S. H., 1972. On biogenic elements and organic matter in the Baltic. Ambio. Spec. Rept. 1: 29–36.

Fransj, H. G., 1986. Effects of freshwater inflow on the distribution, composition and production of plankton in the Dutch coastal waters of the North Sea. In Skreslet, S. (ed.), The Role of Freshwater Outflow in Coastal Marine Ecosystems. Springer-Verlag, Berlin: 241–249.

Gargas, E., S. Mortensen & G. Aertebjerg Nielsen, 1980. Production and photosynthetic efficiency of phytoplankton in the open Danish waters 1975–1977. Ophelia. Supp. 1: 123–144.

Hallegraeff, G. M., 1993. A review of harmful algal blooms and their apparent global increase. Phycologia 32: 79–99.

Ho, K. C. & I. J. Hodgkiss, 1991. Red tides in subtropical waters: An overview of their occurrence. Asian Mar. Biol. 8: 5–23.

Ho, K. C. & I. J. Hodgkiss, 1993. Assessing the limiting factors of red tide by bottle bioassay. Asian Mar. Biol. 10: 77–94.

Ho, K. C. & I. J. Hodgkiss, 1995. A study of red tides caused by *Prorocentrum micans* Ehrenberg, *P. sigmoides* Bohm and *P. triestinum* Schiller in Hong Kong. In Morton, B., G. Xu, R. Zou, J. Pan & G. Cai (eds), The Marine Biology of the South China Sea II. World Publishing Corporation, Beijing, PRC: 111–118.

Hodgkiss, I. J. & B. S. S. Chan, 1983. Pollution studies on Tolo Harbour, Hong Kong. Mar. envir. Res. 10: 1–44.

Hodgkiss, I. J. & B. S. S. Chan, 1986. Studies on four streams entering Tolo Harbour, Hong Kong in relation to their impact on marine water quality. Arch. Hydrobiol. 108: 185–212.

Hodgkiss, I. J. & B. S. S. Chan, 1987. Phytoplankton dynamics in Tolo Harbour. Asian Mar. Biol. 4: 103–112.

Horstman, D. A., 1981. Reported red water outbreaks and their effects on fauna of the west and south coasts of South Africa, 1959-1980. Fish. Bull. S. Afr. 15: 71–88.

Kononen, K., 1988. Phytoplankton summer assemblages in relation to environmental factors at the entrance to the Gulf of Finland during 1972–1985. Kieler Meeresforsch., Sonderh. 6: 281–294.

Lam, C. W. Y. & K. C. Ho, 1989. Red tides in Tolo Harbour, Hong Kong. In Okaichi, T., D. M. Anderson & T. Nemoto (eds), Red Tides: Biology, Environmental Science and Toxicology. Elsevier, New York, Amsterdam, London: 49–52.

Lassig, J., J-M. Leppänen, A. Niemi & G. Tamelander, 1978. Phytoplankton primary production in the Gulf of Bothnia in 1972–1975 as compared with other parts of the Baltic Sea. Finn. Mar. Res. 244: 101–115.

Mihnea, P. E., 1979. Some specific features of dinoflagellate *Exuviaella cordata* Ostf. blooming in the Black Sea. In Taylor, D. L. & H. H. Seliger (eds), Toxic Dinoflagellate Blooms. Elsevier/North Holland Inc., New York: 77–82.

Nehring, D., 1984. The further development of the nutrient situation in the Baltic Proper. Ophelia, Suppl. 3: 167–179.

Nehring, D., S. Schulz & W. Kaiser, 1984. Long-term phosphate and nitrate trends in the Baltic Proper and some biological consequences: a contribution to the discussion concerning eutrophication of these waters. Rapp. P-v. Reún. Cons. Int. Explor. Mer 183: 193–203.

Nielsen, A. & G. Aertebjerg, 1984. Plankton blooms in Danish waters. Ophelia, Suppl. 3: 181–188.

Niemi, A., 1974. Primärproduktionen som kriterium vid uppskattningen av recipienters föroreningsgrad. Nordforsk Miljövardssekretariatet Publ. 4: 173–188.

Niemi, A. & A. M. Astrom, 1987. Ecology of phytoplankton in the Tvärminne area, SW coast of Finland. IV. Environmental conditions, chlorophyll *a* and phytoplankton in winter and spring 1984 at Tvärminne Stärfjrd. Ann. Bot. Fenn. 24: 333–352.

Okaichi, T., 1989. Red tide problems in the Seto Inland Sea, Japan. In Okaichi, T., D. M. Anderson & T. Nemoto (eds), Red Tides: Biology, Environmental Science and Toxicology. Elsevier, New York, Amsterdam, London: 137–142.

Phillips, D. J. H. & S. Tanabe, 1989. Aquatic pollution in the Far East. Mar. Pollut. Bull. 20: 297–303.

Pitkänen, H., 1978. The wintertime trends in some physical and chemical parameters in the Gulf of Bothnia 1966–1967. Finn. Mar. Res. 244: 76–83.

Pitkänen, H. & V. Malin, 1980. The mean values and trends of some water quality variables in winter in the Gulf of Finland 1966–1978. Finn. Mar. Res. 247: 51–60.

Prakash, A., 1987. Coastal organic pollution as a contributing factor to red-tide development. Rapp. P.-v. Réun. Cons. Int. Explov. Mer 187: 61–65.

Qi, Y., Y. Hong, S. Lu & H. Qian, 1995. An overview of harmful algal bloom (red tide) occurrences along the coast of china and research upon them. In Morton, B., G. Xu, R. Zou, J. Pan & G. Cai (eds), The Marine Biology of the South China Sea II. World Publishing Corporation, Beijing, PRC: 107–110.

Renk, H., J. Nakonieczny & S. Ochocki, 1988. Primary production in the Southern Baltic in 1985 and 1986 compared with long-term mean seasonal variation. Kieler Meeresforsch., Sonderh. 6: 203–209.

Smayda, T. J., 1989. Primary production and the global epidemic of phytoplankton blooms in the sea: A linkage? In Cosper, E. M., V. M. Bricelj & E. J. Carpenter (eds), Novel Phytoplankton Blooms. Springer Verlag, Berlin, Heidelberg, New York: 449–483.

Smayda, T. J., 1990. Novel and nuisance phytoplankton blooms in the sea: evidence for a global epidemic. In Granéli, E., B. Sundström, L. Edler & D. M. Anderson (eds), Toxic Marine Phytoplankton. Elsevier, New York, Amsterdam, London: 29–40.

Smayda, T. J. & A. W. White, 1990. Has there been a global expansion of algal blooms. If so, is there a connection with human activities? In Granéli, E., B. Sundström, L. Edler & D. M. Anderson (eds), Toxic Marine Phytoplankton. Elsevier, New York, Amsterdam, London: 516–517.

Smith, R. A., R. B. Alexander & M. G. Wolman, 1987. Water-quality trends in the nation's rivers. Science 235: 1607–1615.

Sommer, U., 1987. Factors controlling the seasonal variation in phytoplankton species composition – A case study for a deep nutrient-rich lake. Prog. Phycol. Res. 5: 124–178.

Yanagi, T., 1988. Preserving the inland sea. Mar. Pollut. Bull. 19: 51–53.

Hydrobiologia **352**: 149–158, 1997.
Y.-S. Wong & N. F.-Y. Tam (eds), Asia–Pacific Conference on Science and Management of Coastal Environment.
©1997 *Kluwer Academic Publishers. Printed in Belgium.*

A comparison of marine planktonic and sediment core diatoms in Hong Kong with emphasis on *Pseudo-nitzschia*

Mike Dickman & Tom Glenwright
Department of Ecology & Biodiversity, University of Hong Kong, Pokfulam Road, Hong Kong

Key words: Thanatocoenoses, Hong Kong, sediment cores, diatoms, phytoplankton, dissolution

Abstract

Potentially toxic diatoms belonging to the genus *Pseudo-nitzschia* were observed for the first time in plankton samples from Hong Kong collected in 1996. To determine whether potentially toxic diatoms had become more common during the last six decades, three gravity cores were taken from the anaerobic sediments of Kowloon Bay in Victoria Harbour. Anaerobic sediments are thought to be ideal for palaeoecological reconstructions because their vertical stratigraphy is undisturbed by bioturbation. Analysis of the Kowloon Bay sediment cores indicated that very few individual diatoms belonged to the genus *Pseudo-nitzschia*, even though *Pseudo-nitzschia* was found in abundance in many of the plankton samples taken from a nearby site. The relative absence of *Pseudo-nitzschia* frustules was interpreted as indicating that these thin walled, poorly silicified, planktonic diatoms failed to preserve in the saline (32–34‰), slightly alkaline (pH 7.6–7.8), anaerobic sediments of Kowloon Bay.

Dissolution of thinly silicified diatoms rather than predation was believed to be the reason for their virtual absence in the core. The anaerobic conditions near the bottom of Kowloon Bay and the shallowness of the Bay, 12 m, makes predation an unlikely explanation.

Diatom abundance declined in the sediment cores below a depth of 15 cm (ca 1955). This was attributed to the decrease in nutrient loading to Victoria Harbour prior to 1955 rather than enhanced diatom dissolution in the deeper sediments. Benthic diatoms became proportionately more abundant below the 15 cm core depth.

Introduction

The stimuli for red tide formation are complex but it is generally agreed that nutrients are one of the key factors in fueling cell reproduction. The sea is enriched with nutrients from a variety of sources including direct sources such as sewage and nutrient-rich industrial waste to indirect sources such as diffuse agricultural runoff. These nutrients stimulate red tide formation. What is not known is whether nutrients 'preferentially' stimulate the proliferation of toxic species. Since it is known that nutrient loading to Victoria Harbour is greater today than it was sixty years ago, it was decided to test the hypothesis that the number and kinds of toxic diatoms in a sediment core from Victoria Harbour were greater nearer the top of the core than at its base.

We also attempted to determine if the frequency of diatoms, as indicated from their sediment accumula-

tion rates in a core from Victoria Harbour, had changed over the last sixty years.

The genus *Pseudo-nitzschia* contains nearly 20 species of diatoms (Hasle, 1993). At present, only three of these have been classified as producing a toxin (domoic acid). The three toxic species are *Pseudo-nitzschia multiseries* Hasle (formerly *Pseudo-nitzschia pungens* (Grun.) Hasle f. *multiseries* (Hasle) Hasle), *Pseudo-nitzschia australis* Frenguelli and *Pseudo-nitzschia pseudodelicatissima* (Hasle) Hasle (Martin et al., 1990).

Diatoms represent the most widely distributed and ecologically diverse of the algae. Diatom identification is based on cell size, shape and sculpturing of their cell walls or frustules (one frustule is composed of two interlocking halves or valves) (Round et al., 1992). Because of their siliceous nature, diatoms are often preserved in sedimentary deposits unless their valves are very thinly silicified (Round et al., 1992). Although *Pseudo-nitzschia* valves are thinly silicified, we believed that a look at their preservation in Kowloon Bay sediments was justified.

Methods

Three sediment cores were taken in October, 1995 from a site approximately 20 m west of the mooring buoy (Buoy #38) which is located adjacent to the Kai Tak airport runway (denoted by the 'X' in Figure 1). A Kajak gravity coring device with a core liner having an internal diameter of 7.5 cm was used to take the cores. The water depth at the site was 12 m. Sediments were anaerobic as indicated by the presence of iron sulphide and there was a smell of hydrogen sulphide over the entire length of the cores. The first core was 20 cm long whilst the second was 43 cm long. The third core (33 cm) has yet to be analysed. No invertebrate life was observed in any of the core sections. The second core was used for lead dating while diatoms were analysed in both cores. Diatom data for core one are reported here.

Core sectioning and slide preparation

In the laboratory, the sea water above the core's mud-water interface was removed using a U-shaped siphon. In this manner, almost all the sea water was siphoned away without disturbing the flocculant surface layer. The sediments were extruded from the core liner by placing the core on a stand in such a way that the top of the perspex core liner tube fitted snugly into a metal sectioning plate. The sediment inside the core was then pushed up until the sediment's surface layer reached the same height as the top of the metal sectioning plate. A 1 cm high circular perspex collar, which had the same diameter as the corer, was placed on top of the metal sectioning plate so that the sediment could be extruded into it. A flat stainless steel blade was used to cut and transfer each 1 cm thick section to labeled sterile Whirl-Pak® bag. The metal sectioning plate, perspex collar and sectioning spatula were washed to avoid cross-contamination of the sediment sections. This procedure was repeated at 1 cm intervals until all the sediment within the perspex core liner had been extruded.

Diatoms were separated from the sediment matrix by treatment with strong oxidising acids which digested the associated organic matter. The oxidation step was followed by several rinses in distilled water to remove the acid and any salts which might crystallise on drying. Suspensions of the acid cleaned diatoms were then dispersed onto standard coverslips using special sedimentation plates (Battarbee, 1973, 1987). The cover slips were then air dried and mounted in Canada Balsam mounting medium. The resulting permanent slide preparations (slide mounts) were scanned under an Olympus BX 50 research microscope equipped with Nomarski interference optics. Diatoms were counted along line transects until a minimum of 200 individuals had been counted for each of the four replicate slides at each of the five sediment depths. On one core (#2), 50% hydrogen peroxide was substituted for the strong acids to determine if *Pseudo-nitzschia* frustule preservation was improved.

^{210}Lead analysis

^{210}Pb activity was determined through the extraction and counting of the daughter isotope, Po^{210}. Extraction of ^{210}Po was based on methods developed by Flynn (1968). The ^{210}Po and a ^{208}Po internal trace was plated onto silver planchets and counted on an alpha spectrometer. Supported ^{210}Po and was then calculated from ^{226}Ra measured by a gamma spectrometer. The sediment age was calculated using the c.r.s. model of Appleby & Oldfield (1978) according to the pattern of ^{210}Pb activity-depth curve (Han, 1986; Xiang & Han, 1992).

Calculating absolute abundance of diatoms

Each 1 cm thick core section had a volume of 44 178 cm^3 and a dry weight of 30 768 g. At a sedimentation rate of 3.15 mm yr^{-1} the amount of sediment deposited on the bottom of Kowloon Bay over an area of 1 cm^2 would be 0.696 g dry weight.

The number of diatoms accumulating per gram dry weight is equal to $(C \times N)/(V \times S \times T)$ where V is the volume of sediment added to the sedimentation tray, T is the number of ocular fields counted, S is the area of one ocular field, N is the total number of diatoms counted per coverslip and C is the area of the sedimentation tray (Meng, 1994).

The accumulation rate of diatoms was calculated by multiplying the sedimentation rate, presented as a sediment flux rate (g cm^{-2} yr^{-1}) by the organic matter content. The result was then multiplied by the concentration of the diatoms estimated for each 1 cm depth interval.

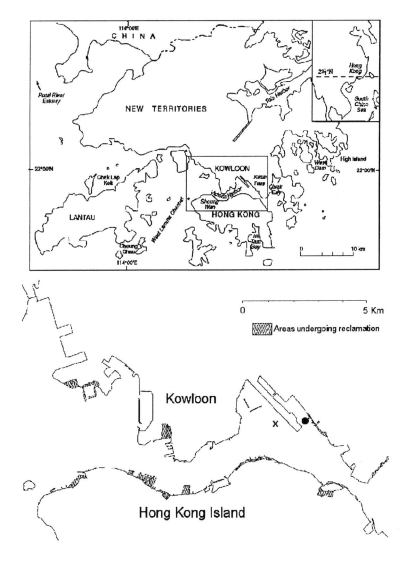

Figure 1. The general situation figure has the Victoria Harbour area magnified below. 'X' and '•' represent the sediment core retrieval and plankton sampling sites respectively. The point of land between sampling locations in the Kowloon Bay area of Victoria Harbour is the Kai Tak airport runway.

Plankton samples

Plankton samples were collected at weekly intervals from the surface waters at Kwun Tong which is located only about 1.3 km from the Kowloon Bay sample site. Plankton samples were collected by passing 100 l of surface water through a ten micron mesh net. Water chemistry was analysed using a Horiba model U-10 water quality analyser. Plankton samples were preserved with Lugol's iodine solution and observed under an Olympus research microscope equipped with Nomarski interference optics. Samples containing *Pseudo-nitzschia* were dried on coverslips which were then mounted on aluminium stubs and coated with Au/Pd. The stubs were examined in a Cambridge S360 analytical scanning electron microscope (SEM) at the Electron Microscope Unit of the University of Hong Kong.

Results

The 25 most common phytoplankton species from Kwun Tong were compared with the 25 most common

Table 1. Comparison of the 25 most common diatom species observed in kwun Tong plankton samples and the surface sediment core section from Kowloon Bay.

Kwun Tong phytoplankton species	Kowloon Bay sediment core sample (0–1 cm) species
Achnanthes affinis (-)	
Achnanthes javanica var. *subconstricta* (-)	
	Actinoptychus pericavatus (-)
Bacillaria paxillifera (+)	
Chaetoceros concavicorne (-)	
Chaetoceros costatum (-)	
Chaetoceros curvisetum (-)	
Chaetoceros debile (-)	
Chaetoceros decipiens (-)	
	Coscinodiscus blandus (+)
	Coscinodiscus excentricus (+)
	Coscinodiscus inclusus (+)
	Coscinodiscus lineatus (+)
	Coscinodiscus minor (+)
	Coscinodiscus radiatus (-)
	Coscinodiscus rothii (+)
	Coscinodiscus subconcavous var. *tenuior* (-)
	Coscinodiscus temperii (-)
	Coscinodiscus wittianus (+)
Cyclotella comta	*Cyclotella comta*
	Cyclotella straita (+)
Cyclotella stylorum	*Cyclotella stylorum*
Cylindrotheca closterium (-)	
Diploneis bombus	*Diploneis bombus*
Eucampia zoodiacus	
	Gomphonema sphaerophorum (-)
Lauderia borealis (-)	
Licomophora ehrenbergii (-)	
Melosira nummuliodes (+)	
Navicula sp.	
	Neodelphineis pelagica (-)
Nitzschia sp.	
Nitzschia cocconeiformis	*Nitzschia cocconeiformis*
Paralia sulcala	*Paralia sulcata*
Plagiotropsis lepidoptera (-)	
Pleurosigma elongatum (-)	
Pseudonitzshia spp. (+)	
	Stephanopyxis weyprechtii (+)
	Rhaphoneis miocenica (-)
Skeletonema costatum	*Skeletonema costatum*
Thalassonema nitzschioides	*Thalassionema nitzschioides*
Thalassiothrix frauenfeldii (+)	
	Trachyneis antillarum (+)

(+) denotes the same diatom species was found in the opposite sample but in very low numbers.
(-) denotes this diatom species was absent from the opposite sample.

Table 2. Accumulation rates of diatom material expressed as numbers cm² yr⁻¹.

Accumulation rate	Sediment core depth (cm)				
	0–1	4–5	9–10	14–15	19–20
Mean diatom	196.073	336.350	215.230	125.131	128.775
	± 83.948	± 92.969	± 49.877	± 28.248	± 16.188
Mean centric fragments	216.457	532.044	343.957	191.500	155.872
	± 81.404	± 138.915	± 36.177	± 83.504	± 38.786
Mean pennate fragments	47.753	84.174	118.376	57.948	51.097
	± 27.226	± 24.612	± 26.554	± 22.863	± 15.457
Chaetoceros spines	53.505	73.178	183.765	39.785	56.625
	± 38.192	± 16.852	± 74.315	± 21.943	± 14.336
Chaetoceros spores	13.706	44.936	19.447	38.164	7255
	± 7.223	± 22.198	± 13.115	± 50.625	± 4692
Diatom: fragment ratio	1348	1832	2148	1993	1607

diatom species from a sediment core taken from the nearby Kowloon Bay (Table 1). The '−' in Table 1 indicates species found in the plankton samples that were not in the sediment samples and *vice versa*. The '+' indicates species found in abundance in the plankton samples that were quite rare in the sediment samples and *vice versa*. Ten to sixteen transects were counted for each of the 4 replicates and the 25 most common diatoms listed (Table 1).

Downcore diatom accumulation rates in surface and subsurface sediments

In order to determine if the number of diatom valve fragments increased downcore, their absolute accumulation rates were calculated for five sediment depths (Table 2). The number of pennate and centric diatom fragments were also enumerated to determine whether predation or dissolution were the major reason for the production of the fragments.

The number of *Chaetoceros* auxospores and spines were also counted as whole *Chaetoceros* diatoms were rarely found (Table 2).

²¹⁰ *Lead estimates of sedimentation rates in Kowloon Bay*

Based on the ²¹⁰Pb results the estimated sedimentation rate in Kowloon Bay, Victoria Harbour, Hong Kong is 3.15 mm yr⁻¹ (Figure 2).

Based on this estimate, the age of the 0–1 cm section would be about 3.2 years, i.e. the sediments at a depth of 0.5 cm were likely to have been deposited in

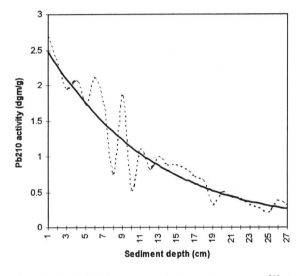

Figure 2. The dashed line represents the observed unsupported ²¹⁰Pb activity for sediment from 1 cm depth to the top of 27 cm of core #2. The solid black line represents the 'smoothed' line curve used to estimate the sediment accumulation rate.

1993. The 4–5 cm section would have an approximate age of 14.5 years and the 9–10 cm section would have an isotope estimated age of about 30 years (ca 1965) (Figure 3).

Pseudo-nitzschia characteristics

The valves of *Pseudo-nitzschia* are long and narrow (Figures 4–6). Valves of *Pseudo-nitzschia pungens* and *Pseudo-nitzschia multiseries* are of similar size, (ca 4 μm wide and 75 μm long), and could only be distinguished from one another with SEM analysis.

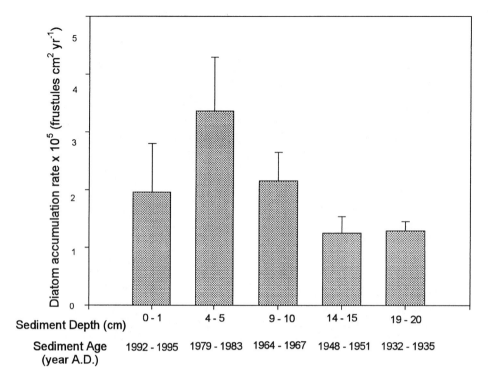

Figure 3. Changes in diatom accumulation rates for core #1.

The ratio of toxic to non-toxic *Pseudo-nitzschia* was inferred from SEM analyses of plankton samples taken from Victoria Harbour on 16 January, 1996 when both species were present in a ratio of 1 potentially toxic individual to 89 non-toxic individuals. This study constitutes the first time that *Pseudo-nitzschia multiseries* has been identified from Hong Kong's Kowloon Bay.

Discussion

Estimates of sedimentation rates in Kowloon Bay

According to Chalmers (1984), inputs of sediments into Victoria Harbour are primarily derived from sewage solids, land reclamation and dredging projects. Very little sediment from the Pearl River accumulates in Victoria Harbour (Yim, 1983). During the past 8000 years the mean sedimentation rate has been estimated at between 1.25 to 2.5 mm yr^{-1} (Yim, 1983). Sedimentation rates have increased to 3.15 mm yr^{-1} today. A similar pattern was reported by Wong & Li (1990) for Deep Bay, Hong Kong. From 1898 to 1949 the Deep Bay sediment accumulated at a rate of 8 mm

yr^{-1}. After 1949 the sedimentation rates in Deep Bay doubled. 15 mm yr^{-1} are currently deposited in the sediments of Deep Bay (Wong & Li, 1990). Our estimated sedimentation rate for Kowloon Bay of 3.15 mm yr^{-1} falls between the estimates made by Yim (1983) and Chalmers (1984). Our reported isotope mean sedimentation rate was based on the estimates from ^{210}Pb of 3.15 mm yr^{-1} (Figure 2).

Diatom accumulation rates

Reductions in the downcore diatom accumulation rate was interpreted as indicating that significantly fewer diatoms were present in the water column 60 years ago than 15 years ago. To determine whether this decline in the number of diatom frustules was a result of increased dissolution, the number of diatom fragments were counted at each depth (Table 2). If intact frustules were breaking up as a result of valve dissolution, it was reasoned that the fragment counts would have increased. The fragment counts did not increase with depth (Table 2). The number of diatom fragments expressed per gram of sediment dry weight were rough-

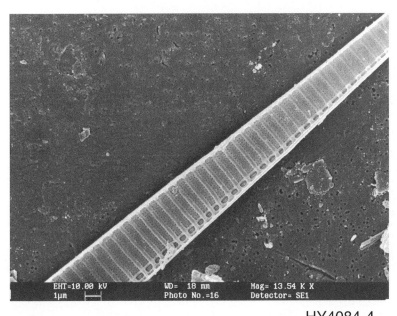

HY4084-4

Figure 4. SEM photomicrograph of *Pseudo-nitzschia multiseries* magnified 13 540 times (82 μm long and 4 μm wide).

ly proportional to the number of diatoms counted at each depth (Table 2).

In total, 13 of the 25 common planktonic diatoms failed to show up in the surface sediments. The relative abundance and numbers of benthic species was higher in the deeper sediments and this was interpreted as indicating that Victoria Harbour waters had lower nutrients and turbidity 50–60 years ago. We inferred from this that light penetration was probably greater in Victoria Harbour than it is today. As a result, benthic diatom species such as *Achnanthes, Diploneis, Navicula, Pleurosigma* and *Cocconeis* were able to coexist with planktonic species in a lower nutrient and higher transparency conditions which likely existed 50–60 years ago.

Diatom dissolution

The absence of *Pseudo-nitzschia* frustules from the sediment core (Table 2) indicates that these diatoms fail to preserve in the marine, alkaline, anaerobic sediments of Kowloon Bay. The mechanism of diatom dissolution involves Si-O bonds being broken by hydroxyl ions which arise from the hydrolysis of metal carbon-

ates (Stumm & Morgan, 1970). The role of anoxic sediments is poorly understood although humic acids produced by decomposing organic matter are assumed to reduce the dissolved silica saturation within pore water and therefore anaerobic conditions speed up diatom dissolution rates (Flower, 1993).

Twelve other common planktonic diatom taxa displaying poorly silicified valves failed to show up in the sediments of Kowloon Bay. Most notable amoung these were 5 species of *Chaetoceros* and *Skeletonema costatum* that were common in the plankton but poorly represented in the sediments of the core. Few, if any, *Eucampia* and *Pseudo-nitzschia* frustules were present in the sediment cores although most taxa were common in the plankton samples taken at Kwun Tong, a nearby site in Victoria Harbour.

Dissolution of the more delicate frustules rather than predation was the most likely explanation for their presence in the plankton and absence in the sediment core. The anaerobic conditions in the bottom of Kowloon Bay and the shallowness of the bay, 12 m, makes predation an unlikely explanation. *Pseudo-nitzschia* was also absent in well aerated surface sedi-

156

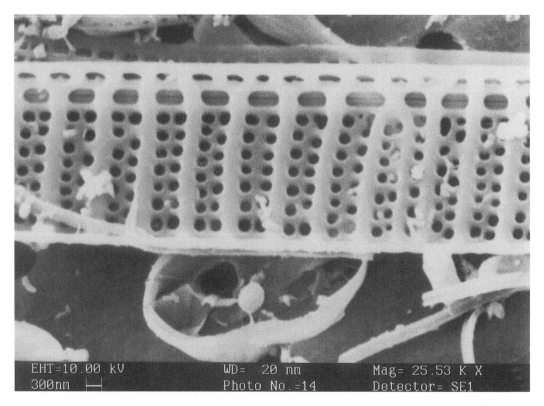

Figure 5. SEM photomicrograph of *Pseudo-nitzschia pungens* magnified 25 530 times. Note the two (and sometimes partial third row) of intercostal poriods. There about 8–10 poriods across the width of the valve.

ment samples taken by Glenwright & Dickman (1997) at several sites in Hong Kong.

To determine if acid cleaning of the diatoms was contributing to their dissolution, sediment samples for core #2 were cleaned with hydrogen peroxide instead of acid. The results indicated that the change from strong acid to hydrogen peroxide in the cleaning operation marginally improved the number of *Pseudo-nitzschia* observed, although they still remained virtually absent.

According to Round (1981) 'Very many marine planktonic species are too delicate to survive the journey to the ocean floor'. The proportion of preserved species is only 30–40% in most marine sediments (Round, 1981; Boltovskoy, 1994). The rapid recycling of silica means that diatoms reaching the sediment-water interface, only account for an estimated 0.06–6% (Guillard & Kulkham, 1977) of the diatoms deposited from biocoenoses.

Dissolution and fragmentation hinder and complicate the interpretation of diatom assemblages collected from bottom sediments (Johnson, 1974; Mikkelsen, 1977; Shemesh et al., 1989; Pichon et al., 1992;

Boltovskoy, 1994). Furthermore, palaeoecological investigation of specific species of concern, such as toxic *Pseudo-nitzchia*, is also severely restricted by the lack of preservation.

Conclusions

1. The first published observation of a toxic species of *Pseudo-nitzschia* (*Pseudo-nitzschia multiseries*) was based on our SEM photographs of plankton samples collected from Lamma Island and Victoria Harbour in Hong Kong in the winter of 1995/96.

2. The ratio of non-toxic to toxic *Pseudo-nitzschia* was inferred from the SEM analyses of the plankton samples taken from Victoria Harbour on 16 January, 1996 when both species were present in a ratio of 89 to 1.

3. The rarity of *Pseudo-nitzschia* frustules in the sediments of Victoria Harbour (0–20 cm) indicates that these delicate, thin walled planktonic diatoms fail to preserve well in marine sediments as do many other thinly silicified planktonic diatoms.

157

Figure 6. SEM photomicrograph of *Pseudo-nitzschia multiseries* magnified 37 510 times. Note the three to four rows of intercostal poriods. There are about 20 poroids across the width of the valve (9 in 1 μm). The valve is 80 μm long and 4 μm wide.

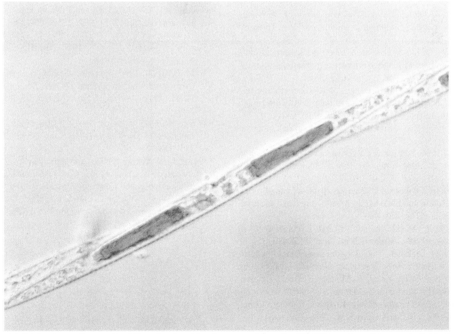

Figure 7. Light photomicrograph of *Pseudo-nitzschia* cf. *pungens*. Note the 20–25% overlap of the two individuals.

158

4. The high numbers of benthic species and their increased relative abundance in the deeper sediments of Kowloon Bay was interpreted as indicating that Victoria Harbour waters had lower nutrients and turbidity 50–60 years ago than today. As a result, light penetration was greater than today and benthic species were able to coexist with planktonic species in Victoria Harbour. Today, Secchi transparency of Victoria Harbour rarely exceeds 1.5 m and benthic diatom taxa are rarer in the harbour's surface sediments than its deeper (15–20 cm) sediments.

Acknowledgments

We are grateful to the Electron Microscope Unit of The University of Hong Kong for their assistance in preparing and taking the SEM photographs of *Pseudonitzschia* and to Lily Ng of The University of Hong Kong for her assistance in gathering and counting the plankton samples. We are also grateful to Dr Ji Shen of the Nanjing Institute of Geography and Limnology for ^{210}Pb analyses of the sediment core. This research was funded by the Hong Kong Committee on Research and Conference Grants.

References

Appleby, P. G. & F. Oldfield, 1983. The assessment of Pb^{210} data from sites with varying sediment accumulation rates. Hydrobiologia 13: 29–35.

Battarbee, R. W., 1973. A new method of estimating absolute microfossil numbers with special reference to diatoms. Limnol. Oceanogr. 18: 647–653.

Battarbee, R. W., 1987. Relative composition, concentration and calculated influx of diatoms from a sediment core from Lough Erne, Northern Ireland. Pol. Arch. Hydrobiol. 25: 9–16.

Boltovskoy, D., 1994. The sedimentary record of pelagic biogeography. Prog. Oceanogr. 34: 135–160.

Chalmers, M. L., 1984. Preliminary assessment of sedimentation in Victoria Harbour, Hong Kong. Geol. Soc. Hong Kong 1: 117–129.

Flower, R. J., 1993. Diatom preservation: experiments and observations on dissolution and breakage in modern and fossil material. Hydrobiologia 269/270: 473–484.

Flynn, W. W., 1968. The determination of low levels of Po^{210} in environmental materials. Analyt. Chem. Acta 43: 221–227.

Glenwright, T. & M. Dickman, 1997. Diatom assemblages in surficial sediments along a transect between Nine Pin Island Group and Kowloon Bay, Hong Kong. In Morton, B. (ed.), Proceedings of the Fourth International Marine Biological Workshop. The Marine Flora and Fauna of Hong Kong and Southern China. Hong Kong University Press, Hong Kong, 7 pp. (in press).

Guillard, R. R. & P. Kulkham, 1977. Ecology of the marine diatoms. In Werner, D. (ed.), The biology of the diatoms. Bot. Monogr. 13: 402.

Han, X., 1986. Pb^{210} dating from Boyang Lake, China. Journal of Hohai University 15: 220–225 (in Chinese).

Hasle, G. R., 1964. *Nitzschia* and *Fragilariopsis* species in the light and electron microscopes II. The group *Pseudo-nitzschia*. Skrifter Utgitt av Det Norske Videnskaps-Akademi I Oslo I. Mat.-Naturv. Ny 18: 5–43.

Hasle, G. R., 1993. Nomenclature notes on marine planktonic diatoms. The family Bacillariaceae. Nova Hedwigia Beih. 106: 315–321.

Hasle, G. R. & Syvertsen, 1996 Marine diatoms. In Tomas, C. R. (ed.), Identifying Marine Diatoms and Dinoflagellates. Academic Press Inc., New York, 598 pp.

Health and Safety Laboratory, 1977. Final tabulation of monthly Sr^{90} fallout data: 1954–1976. HASL-329. U.S. Energy Research and Development Administration. Chalk River, Tennessee, USA.

Martin, J. L., K. Haya, L. E. Burridge & D. J. Wildish, 1990. Nitzschia pseudodelicatissima: a source of domoic acid in the Bay of Fundy, eastern Canada. Mar. Ecol. Prog. Ser. 67: 177–182.

Meng, L., 1994. How accurate is the random settling method for quantitative diatom analysis? A test using *Lycopodium* spore tablets. Micropaleontology 40: 261–266.

Mikkelsen, N., 1977. Silica dissolution and overgrowth of fossil diatoms. Micropaleontology 23: 223–226.

Pichon, J. J., G. Bareille, M. Labracherie, L. D. Labeyrie, A. Baudrimont & J. L. Turon. 1992. Quantification of the biogenic silica dissolution in the Southern Ocean sediments. Quat. Res. 37: 361–378.

Round, F. E., 1981. The ecology of the algae. Cambridge University Press, 499 pp.

Shemesh, A., L. H. Burkle & P. N. Froelich. 1989. Dissolution and preservation of antarctic diatoms and the effect on sediment thanatocoenoses. Quat. Res. 31: 288–308.

Smith, J. C., 1993. Toxicity and *Pseudonitzschia pungens* in Prince Edward Island, 1987–1992. ISSN 0020-7918. Harmful Algae News 6: 1–3.

Stumm, W. & J. J. Morgan, 1970. Aquatic chemistry. Wiley and Sons, New York: 583 pp.

Todd, E. C. D., 1990. Amnesic shellfish poisoning: a new seafood toxin syndrome. In Graneli, E., B. Sunderstrom, L. Edler & D. M. Anderson (eds), Toxic Marine Phytoplankton. Elsevier, New York: 504–508.

Wong, S. H. & Y. S. Li, 1990. Hydrographic surveys and sedimentaion in Deep Bay, Hong Kong. Envir. Geol. Wat. Sci. 15: 111–118.

Xiang, L. & X. Han, 1992. Comparison of ^{210}Pb dating models: a case study of Chaohu Lake, China. Memoirs of Nanjing Institute of Geography and Limnology, Academa Sinica. (in Chinese).

Yim, W. W. S., 1984. Geochemical mapping of bottom sediments as an aid to marine waste disposal in Hong Kong. Conservation and Recycling 7: 309–320.

Hydrobiologia **352**: 159–166, 1997.
Y.-S. Wong & N. F.-Y. Tam (eds), Asia–Pacific Conference on Science and Management of Coastal Environment.

The essential elements of science and management in coastal environmental managements

Chua Thia-Eng
International Maritime Organization Department of Environment and Natural Resources Visayas Avenue, Diliman, Quezon City Philippines

Key words: integrated coastal management, science and management

Abstract

Coastal environmental management requires timely and appropriate policy, management and technological interventions to address a host of interrelated environmental problems arising from unplanned and unregulated coastal developments. These interventions can only be effective if they are based on sound scientific (including socioeconomic) information. Thus, science plays a significant role in the environmental management of coastal and marine areas.

This paper highlights the essential components of environmental coastal management that require scientific interventions, particularly in providing the scientific basis for policy interventions, and preventive and/or mitigating measures. The integration and packaging of scientific information for management actions require an interdisciplinary effort to address identified management problems. Scientific research should respond to management needs and make contributions to the better understanding of ecosystems and their responses to human interventions. Scientific information urgently required for management includes an understanding of interactions in coastal resource systems, identification, evaluation and prioritization of management issues, management strategies and actions, and development of methodologies and tools for environmental management. Obstacles to management-oriented research have also been identified to include defective perception among scientists, defective communication between scientists and decision makers, intellectual and cultural arrogance, and inadequate technical and management capability at the local level.

Introduction

Coastal ecosystems in many parts of the world are being degraded as a result of rapid economic development. The coastal area, which is defined as the interface between land and sea and where many important fragile ecosystems are located, has long been overlooked in national land-use planning and natural resource management. In Southeast Asia, large areas of mangroves, seagrass beds and coral reefs have either been removed or damaged (Paw & Chua, 1991). Similarly, 1/3 of Panama's mangrove resources have been lost while, in Ecuador, more than 30% of the valuable mangrove swamps have been converted into shrimp ponds (Goldberg, 1994). These activities threaten the existence of these tropical ecosystems, which are amongst the

world's highest in terms of marine biodiversity (Dubinsky, 1990).

The main purpose of coastal environmental management is to protect the functional integrity of these ecosystems and to maintain a sustainable flow of goods and services generated by them. In order to achieve this, human activities which affect the well being of the ecosystem, must be regulated and adequately managed.

The degradation of the coastal environment is attributed to a combination of causes including inadequate planning and management on the use of the coastal resources, lack of coordination amongst resource users, governance and management agencies, lack of policy and functional integration pertaining to the development of coastal and marine resources in national economic planning, ad hoc management interventions, inadequate legislation, poor law enforcement, inadequate political commitment, and lack of public awareness on the importance and the status of the resource systems (Chua & Scura, 1991; Chua & Garces, 1994).

Recognizing the high human population density and multisectoral economic activities in most coastal areas, a holistic and integrated management approach is essential. All stakeholders must be involved in the planning and management of these common resources (Scura et al., 1992). Integrated management is the recommended approach for coastal and marine management in achieving sustainable development goals (Chapter 17, Agenda 21) but surely it is also a difficult one to implement.

A fundamental management requirement is the availability of a reliable and scientific database that could be used by the coastal managers to develop appropriate coastal policies and corresponding management interventions for mitigating any adverse environmental change. Thus, there exists a close and mutually supportive relationship between research and management. Such a relationship is best manifested through the application of an interdisciplinary approach to resolve many of the teething management problems in coastal areas.

Elements of coastal management

Environmental management can be visualized in a three dimensional cube denoting processes, issues and actions. That is, proper management follows a logical process through planning, implementation, monitoring and evaluation in addressing single or a host of environmental management problems through a series of combined actions. Management actions could be categorized into three distinct areas, namely, those related to institutional and organizational arrangements, those related to incentives and/or regulation to change human behavior, and those that require direct public involvement/investment (Scura et al., 1992).

The lack of a comprehensive, systematic, procedural and integrated approach often results in plans not executed. Or, if they were implemented, most plans fail because of poor interagency coordination, insufficient policy and functional integration or, inadequate feedback due to the lack of monitoring and evaluation.

Planning is a fundamental component of management. The planning process enables adequate identification and prioritization of management problems, the determination of management objectives and targets, the development of policy and management strategies, and the establishment of monitoring and evaluation protocols.

While a proactive approach should be the essence of environmental management, reactive measures are often more necessary to arrest the already worsening environmental condition. Thus, in most instances, both preventive and mitigative measures are to be undertaken simultaneously. Prevention is one important strategy in environmental management. Preventive measures can be instituted through area (or resource) allocation and regulation for their optimal use. The establishment of a functional zonation scheme in the coastal area is one such example. Mitigating measures can be manifested through regulation and/or economic incentives or disincentives. The public sector plays a pivotal role in environmental management in terms of institutionalizing management measures. Considerable public involvement or investment is needed in terms of providing technical assistance at the local level, promoting research to fill critical information gaps for management interventions, promoting public awareness at all levels, and creating employment opportunities to remove environmental stress. Irrespective of the target area of management, whether it is overfishing, use conflicts or pollution, similar categories of management interventions can be applied (see Table 1).

Integration and coordination are two major unifying elements of environmental management (Chua, 1993). All the key management requirements, including planning, institutional arrangements, legislation, finance, law enforcement, monitoring, and evaluation, need to be closely integrated in the process of plan formulation and implementation. Policy systems and functional integration are essential in achieving best management results. Even research should be closely integrated to serve management goals (Figure 1).

A coordination mechanism must be in place to bring about a closer dialogue between line agencies and major stakeholders on proposed management actions. The views of the stakeholders should be considered at the various stages of planning and implementation. Activities of line agencies must be coordinated, including the conduct of public awareness campaigns, in order not to send conflicting messages to the general public.

Table 1. Comparison of elements of planning and management interventions through the application of an integrated coastal management (ICM) framework.

Issues	Process Planning	Implementation	Monitoring and Evaluation	Actions Institutional and Organizational Arrangements	Incentives/Regulations to Change Behavior	Direct Public Involvement/ Investment
Pollution 1	Proper identification and prioritization of management issues Local participation Formulation of management strategies and policy options Formulation of action plan for implementation; monitoring and evaluation protocols Evaluation, decision and adoption	Integration of essential issues into development plans Execution of activities Coordination of relevant actors Local participation	Conduct periodically during both planning and implementation Follow progress, assess performance toward meeting objectives Identify unanticipated, social, economic and environmental consequences of management actions Recommend adjustments in actions and/or objectives	Clarification of legislation and responsibilities, e.g., implementation of MARPOL Convention Clarification of organizational jurisdiction and responsibilities, e.g., Department of Environment, Marine Department, Harbour Department Monitoring and enforcement	Effluent charges Tradeable permits Deposit-refund system Subsidies/taxes on inputs/outputs Subsidies for use of substitutes and abatement inputs Performance bond Regulation of emissions Emission standards Regulation of equipment, processes, inputs/outputs Standards for inputs or processes Bans and quotas on inputs or outputs Land/water-use zoning	Waste collection/ treatment/disposal Public awareness and education Technical assistance Research and development Capacity building
Aquaculture Use Conflicts[2]	Identification and prioritization of management issues Environmental profiling Formulation of policies and management strategies Formulation of action plans Developing monitoring and evaluation protocols Program evaluation, public consultation, decision and adoption	Establishment of coordination, management and financial mechanisms Implementation of legislation Execution of action plans	Periodic undertakings during planning and implementation Follow progress, assess performance towards meeting objectives Identify unanticipated, social, economic and environmental consequences of management actions Recommended adjustments in actions and/or objectives	Clarification of legal rights and responsibilities, e.g., traditional or customary use rights Clarification of organizational jurisdiction and responsibilities Interagency coordination, policy and functional integration Enforcing and monitoring	Functional zonation, permit system Tax rebates, subsidies, other economic incentives User fees, fines, taxes Regulation on use of sites, culture systems, seasons, waste treatment Business opportunities	Research and development Technical assistance Capacity building Education and public awareness Alternative employment options Protecting water quality of culture site
Overfishing[3]	Proper identification and prioritization of management issues Local participation Formulation of management strategies and policy options Formulation of action plan for implementation; monitoring and evaluation protocols Evaluation, decision and adoption	Integration of essential issues into development plans Execution of activities Coordination of relevant actors Local participation	Conduct periodically during both planning and implementation Follow progress, assess performance toward meeting objectives Identify unanticipated, social, economic and environmental consequences of management actions Recommend adjustments in actions and/or objectives	Clarification of legal rights and responsibilities, e.g., traditional or customary use rights (TURF) Clarification of organizational jurisdiction and responsibilities Monitoring and enforcement	Tradeable fishing permits or quotas Taxes, marketing board margins, license fees Non-tradeable fishing permits or quotas Fishing bans Regulation of gear, vessels, season	Restocking Fisheries enhancement Research and development Technical assistance Education and public awareness Alternative livelihood programs Marine protected areas

Sources: [1] Chua (1995); [2] Chua (in press); [3] Scura et al. (1992).

Elements of science in coastal environmental management

Scientific research is obviously important to provide critical information needed for the formulation of policy and management interventions. Its role can be seen from Figure 2 which describes the key elements generally needed for the preparation of a comprehensive and integrated environmental management action program. Baseline information is critical in environmental profiling while specific information is needed to address specific management issues identified. The amount of database required is enormous and must continue to build up as the program proceeds.

As illustrated in Figures 1 and 2, information needs cover a wide area of disciplines and demand expertise in legislation, social sciences, economics, ecology, public administration, coastal engineering, planning, and management. It is only through the close and coordinated efforts of these experts, that the needed management information can be developed.

Information relevant to management needs

Based on the required information for coastal management, the ASEAN/US Coastal Resources Management Project (Chua & Scura, 1992) summarized the key areas for research (Table 2). A review of the

Table 2. Key research areas for CAM (after Scura et al., 1992).

The ASEAN-US CRMP experience suggests that research should be directed at the following areas:

Understanding Interactions in Coastal Resource Systems
- Understanding the ecological function of critical aquatic resources, habitats or ecosystems (e.g., coral reefs, mangroves, marshes/wetlands).
- Identification of trends in supply and demand for goods and services derived from coastal resources and habitats, and social and economic factors influencing these.
- Interpretation of implications for management of carrying capacity or assimilative capacity of aquatic habitats and systems. Identification and documentation of critical threshold levels and indices for management.

Identification and Prioritization of Management Issues
- Identification, physical quantification and evaluation of tradeoffs and trends of impacts in coastal areas.
- Identification of a general typology relating occurrence of specific management issues with biophysical, socioeconomic and institutional and organizational factors.
- Valuation of social and environmental benefits and costs of sectoral activities.
- Identification of management priorities through evaluation of the sustainable level of output, adverse impacts, and associated net benefits and costs.

Identification and Evaluation of Management Strategies and Actions
- Identification and evaluation of appropriate policies and management strategies to mitigate negative impacts and maximize human welfare benefits.
- Identification of general guiding principles with respect to appropriate management action elements including (1) market-based incentives, (2) regulations, (3) direct public involvement or investment and (4) institutional and organizational arrangements.

Development of Methodologies and Tools
- Evaluation, integration and packaging of appropriate, cost-effective methodologies and techniques to facilitate the inventory of resource distribution, utilization and impacts within the coastal resource system.
- Evaluation, integration and packaging of appropriate, cost-effective methods and techniques for the evaluation of benefits and costs of coastal activities and management interventions.

---

specific research requirements indicated in the table, reveal that most of the critical information needed by environmental managers is in fact not available or not well developed. For example, there is little reliable information pertaining to the optimal area of coral reefs or mangrove swamps that should be preserved in order to maintain the functional integrity of these ecosystems. Similarly, information on carrying capacity is limited and the methods for estimation are still not well developed (GESAMP, 1986). Despite general public recognition of the socioeconomic contributions of these ecosystems, there are few reliable data on their direct and indirect values. Although socioeconomic information is now being gathered, justification of environmental protection and preservation of biodiversity have not been convincing enough to galvanize political commitment. While biological information is voluminous, most is not of direct relevance to management needs. Management decisions are difficult to make due to the complexity of coastal resources in terms of multisectoral competition for the limited common resources, the complexity of multisectoral and interagency coordination, and the difficulties in poli-

cy and system integration. To help decision making, a wide range of scientific information and management tools are needed to consider trade-offs, options, and consequences.

Opportunities for interdisciplinary research

The research areas described above open up enormous opportunities in terms of scope and areas for interdisciplinary research. Environmental research at the conceptual level to test hypotheses is equally as important as applied research at the local level to complete information gaps in any of the key areas indicated in Figure 2. Both basic and applied research must focus on generating the basic information needed for scientific management of the coastal environment. Many basic research topics can be generated from the outline of key research areas listed in Table 2. These topics are generally included in environmental management programs at the national or even local level. When adopted by the government, it opens up opportunities for financial resources in undertaking the needed research. In the present situation of scarce financial

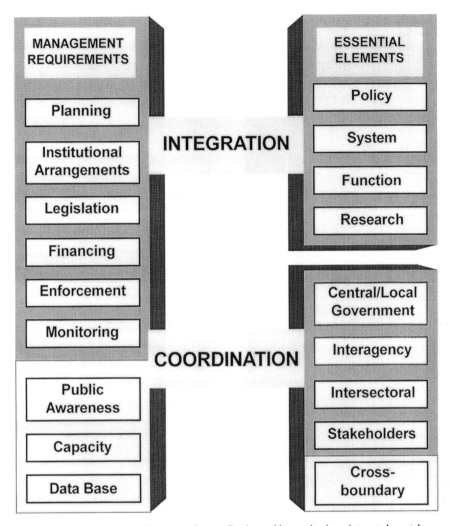

Figure 1. Management requirements and essential elements for coordination and integration in an integrated coastal management program.

resources for research in developing countries and even in some developed countries, researchers should orient their research focus to answer the pressing management needs of the nation and the world at large.

Obstacles to management-oriented research

There have been obvious obstacles to management-oriented research in the past. and hopefully these obstacles will be subdued in the future. The main obstacles identified include defective perception, defective communication, intellectual and cultural arrogance, lack of integration of various branches of science, and insufficient capacity at the local level.

Defective perception amongst scientists

The lack of a clear perception on roles in management have led many scientists to focus their research on generating information for the sake of knowledge. They have no or little concern that their research results be used for management decisions as they consider management's sole function is that of making social and political decisions. Moreover, a great number of scientists consider that interdisciplinary work in environmental management will not generate the type of scientific papers normally acceptable in prime literature. In an age of 'publish or perish', such a perception indeed becomes a mental obstacle to the involvement of both natural and social scientists to collectively answer some of the most challenging and scientifical-

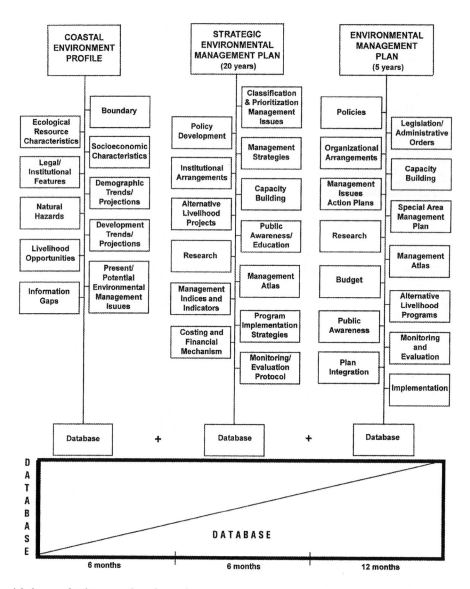

Figure 2. Essential elements for the preparation of coastal environmental profiles and environmental management plans with an indicative timeframe.

ly relevant management problems such as the absorptive capacity of the coastal environment to economic development, quantification of carbon flow in the land-sea interaction, environmental accounting, or even the development of reliable and appropriate multivariate decision-tools for managers. It is ironic that in the world of modern technology, management decisions are still confined to the realm of personal experiences and visions.

Defective Communication Between Scientists and Managers/Decision-makers Clearly there are serious communication gaps between scientists and those mak-

ing management decisions. Scientists are supposed to generate precise information and databases which are to be used for management decisions. The fact is, many of the environmental disasters are the results of economic activities which have no or inadequate scientific inputs.

Defective communication between scientists and decision-makers was highlighted in the 1994 Royal Colloquium on Tropical Coastal Zones chaired by His Majesty, King Carl XVI in Stockholm (Royal Colloquia, 1994): '...scientists are accused of obscurantism and decision-makers of intellectual superficiality and dependence on public emotion. Scientists dismiss politicians as time-serving, and politicians dismiss scientists as incapable of applying their knowledge in the real world.'

While the gap still persists, there are trends of change in both the developed and developing nations where, increasingly, development and management decisions are based on sound scientific and ecological considerations.

Intellectual and cultural arrogance

There exists a wealth of traditional and indigenous knowledge pertaining to the sustainable use of natural resources and their protection and management. This is true for coastal inhabitants whose livelihoods traditionally depend on the goods and services provided by the marine ecosystem. Unfortunately, most of this valuable indigenous knowledge has not been adequately recorded or presented in the scientific literature. Such knowledge is often ignored in the formulation of economic development programs especially in developing nations which receive external aid. This obstacle was considered a major detriment to sustainable development in the 1994 Royal Colloquium, to quote 'Little respect has been paid to the existing knowledge systems in recipient areas for Western assistance. Similarly, little consideration has been given to indigenous traditions that are likely to be better adapted to a specific region than are ideas from outside. ....Sustainable development is further aggravated by the gulf between national leaders in developing countries (who have absorbed Western educational and cultural standards) and their own rural poor.'

Defective interdisciplinary integration between various branches of science

Management-oriented research requires a close functional integration of research efforts among various disciplines. However, interdisciplinary research has not been easy. Researchers still maintain their conventional disciplinary approach despite being a member of a multidisciplinary team. In some serious cases, teams of researchers embarking on scientific investigation in the same geographical area are often in ignorance of what the others have been doing. The lack of understanding of the requirements of environmental management is an obstacle to promotion of better interdisciplinary cooperation and collaboration. Very often, scientists fail to see the broad picture and management needs,

they also fail to identify a niche where they can play. There are also conceptual differences between natural scientists and social scientists with respect to how the coastal environment should be managed. The natural scientists are concerned with the protection of the biological and physical environments while the social scientists place their focus on the people or human interest. Integrating the various views should be the function of environmental managers in developing policy and management options. Management-oriented research on coastal and marine environments is a relatively new endeavor. It requires a change of perception and attitude amongst the disciplinary scientists in forging a collective effort towards solving a common problem. This may take as long as one or two years to realize as experienced by the author in his efforts to promote interdisciplinary, management-oriented research in Southeast Asia and recently in other East Asian countries.

Lack of technical and management capacities at the local level

The real action on environmental management is at the local level. The lower the level of government unit, the higher the chance of achieving sustainable development as there is less complexity in terms of resource competition, and less interagency conflicts in resource governance and management within the confines of the administrative boundary. However, technical and management capabilities are usually the weakest at this level of government. In many developing countries, technical and management capabilities may exist in the central government or, at most, at the provincial government levels. This has added constraints on the local government with respect to implementing national environmental and resource management regulations as well as undertaking local environmental planning and management. Importation of expertise from the central or provincial level or from outside the country is only a temporary measure. The long term solution is the immediate upgrading of the local capacity to plan and manage their own resources and the environment.

Conclusion

Science can and should definitely play an important role in the successful management of the coastal environment. Interdisciplinary research is essential to address many complex management problems that

166

require the efforts of natural and social scientists. Sound management of the coastal area requires the deployment of an integrated planning and management approach to develop both proactive and reactive management measures which can be undertaken at the local level.

Management-oriented research provides a host of research and financial opportunities which can and should be explored to promote quality research aimed at resolving some of the most challenging scientific and managerial questions. Such research includes ecosystems research, development of practical and applicable models to forecast environmental changes, and to assist in management decisions. Obstacles to management-oriented research have been identified. Unless a concerted effort to remove these obstacles is established, the use of scientific results for management will be limited. To achieve this, both scientists and those making social and political decisions must make a combined effort to change their perception and attitude.

References

Chua, T. E., 1993. Elements of integrated coastal zone management. J. Ocean. Coast. Mgmt 21: 81–108.

Chua, T. E., 1995. Marine pollution: Developments since UNCLOS III and prospects for regional cooperation. In SEAPOL Singapore Conference on Sustainable Development of Coastal and Ocean Areas in Southeast Asia: Post-Rio Perspective. K. L. Koh, R. C. Beckman & L. S. Chia (eds), National University of Singapore: 144–176.

Chua, T. E., in press. Sustainable aquaculture and integrated coastal management. In Bardach, J. E. (ed.), Sustainable Aquaculture, John Wiley & Sons, Inc.

Chua, T. E. & L. F. Scura (eds), 1991. Managing ASEAN's coastal resources for sustainable development: Roles of policymakers, scientists, donors, media and communities. Department of Science and Technology; Department of Agriculture, Department of Environment and Natural Resources, Department of Tourism, Philippines, and International Center for Living Aquatic Resources Management, Philippines. ICLARM Conference Proceedings 30, 125 pp.

Chua T. E. & L. F. Scura (eds), 1992. Integrative framework and methods for coastal area management. ICLARM Conf. Proc. 37, 169 pp.

Chua T. E. & L. R. Garces, 1994. Marine living resources management in the ASEAN region: Lessons learned and the integrated management approach. Hydrobiologia. 285: 257–270.

Dubinsky, Z., 1990. Evolution and zoogeography of coral reefs. In Achituv, Y. & Z. Dubinsky (eds), Ecosystems of the World 25 – Coral Reefs. Elsevier Press, Amsterdam: 1–9.

Group of Experts on the Scientific Aspects of Marine Pollution (GESAMP), 1986. Environmental capacity – An approach to marine pollution prevention. Rep. and Studies No. 30, 49 pp.

Goldberg, E. D., 1994. Coastal zone space-prelude to conflicts? United Nations Educational, Scientific and Cultural Organization (UNESCO), Paris, 138 pp.

Paw, J. N. & T. E. Chua, 1991. An assessment of the ecological and socioeconomic impact of mangrove conversion in Southeast Asia. In Chou, L. M., T. E. Chua, H. W. Khoo, P. E. Lim, J. N. Paw, G. T. Silvestre, M. J. Valencia, A. T. White & P. K. Wong (eds), Towards an Integrated Management of Tropical Coastal Resources. National University of Singapore, Singapore National Science and Technology Board, Singapore and International Center for Living Aquatic Resources Management, Philippines. ICLARM Conf. Proc. 22: 201–212.

Royal Colloquia, 1994. The 1994 Royal colloquium on tropical coastal zones: Political approaches to sustainable development – Going from knowing to doing. Stockholm, 24 pp.

Scura, L. F., T. E. Chua, M. D. Pido & J. N. Paw, 1992. Lessons for integrated coastal zone management: The ASEAN experience. In Chua, T. E. & L. F. Scura (eds), Integrative Framework and Methods for Coastal Area Management. ICLARM Conf. Proc. 37: 1–70.

Hydrobiologia **352**: 167–180, 1997.
Y.-S. Wong & N. F.-Y. Tam (eds), Asia–Pacific Conference on Science and Management of Coastal Environment.
©1997 *Kluwer Academic Publishers. Printed in Belgium.*

Status, problems and prospects of stock enhancement in Taiwan

I Chiu Liao

Taiwan Fisheries Research Institute, 199 Hou-Ih Road, Keelung, Taiwan 202

Key words: Taiwan, Japan, stock enhancement, sea-farming

Abstract

Stock enhancement started in Taiwan with the building and casting of artificial reefs in 1973. It was only in late 1987, however, that an integrated program on the operation and establishment of a stock enhancement system was developed.

In addition to building artificial reefs and establishing resource protection zones to create fishing grounds, the current stock enhancement program in Taiwan aims at restocking broodstock and fry or seeds. So far, seven species of finfish, four species of mollusks, and six species of crustaceans have been restocked. A total of 5.8 million finfish fry, 5 million molluskan seeds, and 30 million crustacean larvae with some eel and prawn broodstocks have been released up to 1996.

Although there are plans to establish sea-farming centers in Taiwan, the organization of the system has not been established completely. Furthermore, regulations on resource augmentation, protection, management of coastal fisheries and stock enhancement are absent or lacking. The success of stock enhancement depends on the pollution control that is in place in coastal areas and the fishermen's awareness of the importance of resource conservation and their involvement. The paper discusses the status, problems and prospects of stock enhancement in Taiwan, with some viewpoints in population genetics. It also refers to some experience in Japanese sea-farming.

Introduction

Stock enhancement, in its earliest form, commenced in Japan during the Meiji Period (1867–1912) and much earlier in the West. Oshima (1984) defined stock enhancement as the improvement of the productivity of fisheries by technological refinement, the cultivation of aquatic resources and, in the process, the reform of the fisheries structure. Oshima (1984) presented four major approaches to enhance aquatic resources or stocks. These are

- To increase the quantity by man-made recruitment.
- To produce the fingerlings under controlled conditions.
- To construct habitats and artificial substrates.
- To preserve and improve the natural environment.

In the first approach, the actual procedures include

1. Ranching and transplantation of finfish and shellfish fingerlings, artificial or collected from natural waters.

2. Sowing of the stone-bed with seaweed spores, transplantation of mother seaweed, plantation of subterranean stems of eelgrass, among others.
3. Protection of spawners and eggs laid.
4. Promotion of the settling of shellfish spats and seaweed spores on the bottom or substratum.

The second approach achieves its aims by promoting and protecting the survival and growth of finfish and shellfish juveniles. This includes the following:

1. Providing nursery for finfish juveniles.
2. Conserving eelgrass zone and seaweed beds for making a marine forest.
3. Constructing an artificial intertidal habitat for shellfish juveniles.

To construct habitats and artificial substrates for fishery resources, the following may be involved:

1. Artificial reefs.
2. Artificial stone and concrete-block beds for the fishery resources.
3. Blasting of useless reefs to improve the habitat.

4. Reconstruction of the reef-laver zone by means of cement spreading.

Finally, in the preservation and improvement of the natural environment approach, the following may be done:

1. Construction of finfish-paths and finfish-ladders in rivers.
2. Changing of the water-routes.
3. Breakwater works, such as submerged dikes and floating wave suppressors.
4. Improvement of bottom conditions to make a more suitable condition for bottom-dwelling living organisms.
5. Nutritional supply for invertebrates and finfish juveniles.
6. Biological control, e.g., extermination of predators and introduction of natural enemies against destructive organisms.

The first two approaches fall within sea-farming. Matsuoka (1989a) defined sea-farming as the artificial efforts to lend a hand to the growth of larvae, to tide them over the critical period, and to increase the yield of useful marine organisms by restocking them to the natural sea. Sea-farming thus aims at establishing a new genre of production, consisting of a four-stage cycle, namely fingerling production, fingerling restocking, growth under natural environments and recovery by fishing accompanied with stock management. Stock enhancement, as defined and used in this paper, goes beyond this by providing the necessary infrastructure to support the cycle, such as the construction of the habitat for fishery resources and the improvement of environmental conditions.

Status

Development of stock enhancement in Taiwan

Promotion of stock enhancement

Since 1973, artificial reefs have been cast each year at selected sites in Taiwan's coastal waters to establish new fishing grounds. The cost of casting artificial reefs from 1974 to 1995 is shown in Table 1. The budget came from the central government (through the Council of Agriculture, COA), provincial government (Taiwan Fisheries Bureau) and prefectural governments (respective Agriculture or Fishery Sections). The expenditures increased from merely US$219 000 in 1974 to US$9 322 000 in 1995. The total expenditure

Table 1. Expenditures (in US$1,000) of casting artificial reef in Taiwan (1974–1995)

Year	Central government	Provincial government	Prefectural governments	Total
1974	219			219
1975	145			145
1976	221			221
1977	268			268
1978	257		72	329
1979	367		186	553
1980	110		229	339
1981	38			38
1982	67		123	190
1983	74		158	232
1984	74	294	107	475
1985	184	294	165	643
1986	147	294	162	603
1987	151	331	151	633
1988	239	331	151	721
1989	356	331	165	852
1990	294	850	188	1332
1991	1103	1199	247	2549
1992	1103	1382	333	2818
1993	3676	1750	344	5770
1994	7353	2118	355	9826
1995	6147	2853	322	9322
Total	22 593	12 027	3458	38 078

Exchange rate: US$1 = NT$ 27.2; (NT) New Taiwan.

during the 22-year period amounted to US$38 078 000. From this, the various levels of government, that is, from the central government to the prefectural governments, are increasingly putting more emphasis on the casting of artificial reefs as a method to construct habitat or substrate for fishery resources. Figure 1 shows the casting sites. To determine the status of these cast artificial reefs, the suitability of the casting sites and the effectiveness of casting artificial reefs, several programs on the 'Investigation of the Artificial Reefs' have been conducted since 1987 by various agencies, such as the Taiwan Fisheries Research Institute (TFRI), National Taiwan Ocean University and National Kaohsiung Institute of Marine Technology (Chen et al., 1989).

So far, a total of 25 fisheries resource protection zones have been set up and declared. Figure 2 shows the distribution of these 25 fisheries resource protection zones and the kinds of fisheries resources protected. The fisheries resources protected include finfish (anchovy), crustaceans (lobster, kuruma prawn, red-

Figure 1. Distribution of artificial reefs in Taiwan.

tail shrimp, grass prawn), mollusks (small abalone, hard clam, *Tapes* spp., purple clam, blood cockle, top shell, Pinctata), echinoderm (sea urchin), and seaweeds (Porphyra, Gelidium, Meristotheca). To investigate the ecological environment and evaluate the effectiveness of these fisheries resource protection zones, a program called 'Protection and Enhancement of Coastal Fisheries Resources' was initiated in 1978 (Taiwan Fisheries Bureau, 1989). The expenditure under this program increased from US$29 000 in 1978 to US$826 000 in 1995 (Table 2).

Fingerling production and restocking
TFRI had been releasing fingerlings and adult marine aquatic species since 1976. Table 3 and Appendix 1 show the species and the number released. So far, TFRI had released 7 finfish species (red sea bream, 90 500; black sea bream, 160 840; goldlined sea bream, 335 000; thornfish, 1039; gray snapper, 50 000; Japanese eel, 25 859; marbled eel, 15 000), 6 prawn and crab species (kuruma prawn, 21 100 000; grass prawn, 5 800 000; sand shrimp, 100 000; bear prawn, 800 000; redtail prawn, 1 200 000; swim-

Table 2. Expenditures (in US$1,000) under the 'Protection and Enhancement of Coastal Fisheries Resources' program (1978–1995)

Year	Central government	Provincial government	Prefectural governments	Others	Total
1978	29				29
1979	83		11		94
1980	56		52		108
1981	46		68		114
1982	33		57	31	121
1983	65		33	44	142
1984	156	37	46	55	294
1985	159	55	61	74	349
1986	184	74	39	6	303
1987	149		21		170
1988	221		37		258
1989	401		42		443
1990	441		46		487
1991	551	88	57		696
1992	500	88	57		645
1993	551	88	72		711
1994	735	85	79		899
1995	561	74	189	2	826
Total	4921	589	967	212	6689

Exchange rate: US$1 = NT$27.2; (NT) New Taiwan.

Table 3. Fingerlings, subadults, adults of different species released by the Taiwan Fisheries Research Institute (TFRI) (1976–1995)

Category	No. of species	Quantity ($\times 10^3$)	Species (common name)
Finfish	7	693	Red sea bream, black sea bream, goldlined sea bream, thornfish, gray snapper, Japanese eel, marbled eel
Crustacean	6	29,050	Kuruma prawn, grass prawn, sand shrimp, bear prawn, redtail prawn, swimming crab
Mollusk	1	510	Small abalone

ming crab, 50000) and 1 molluskan species (small abalone, 510000). The released species were mostly fingerlings except for Japanese eel (adults) and grass prawn (included subadults also). Two- to three-year old Japanese eel have been released since 1976; a total of 41 326 have been released as of 1995 (Table 4). Before their release, the eels receive hormone injections to induce maturation (Liao et al., 1994). Table 5 shows the number of prawns released and recaptured from Taiwan's southern coast (Su et al., 1990).

Aside from TFRI, the prefectural governments have also been releasing fingerlings or adults every year since 1978. Table 6 shows the number of fingerlings or adults released.

Development of sea-farming in Japan

Promotion of sea-farming

To improve the sluggish aquatic animal production in the coastal waters around Japan in general and in the inland sea in particular, the Japanese government decided to implement an ambitious program. On the request of the government, the Seto Inland Sea Sea-Farming Association was established to manage sea-farming centers. The Association was composed of 12 prefectures around the Seto Inland Sea and 12 fishermen's cooperatives (Kitada, 1985).

The Seto Inland Sea Sea-Farming Association was eventually reorganized into the Japan Sea-Farming Association (JASFA) to establish national sea-farming

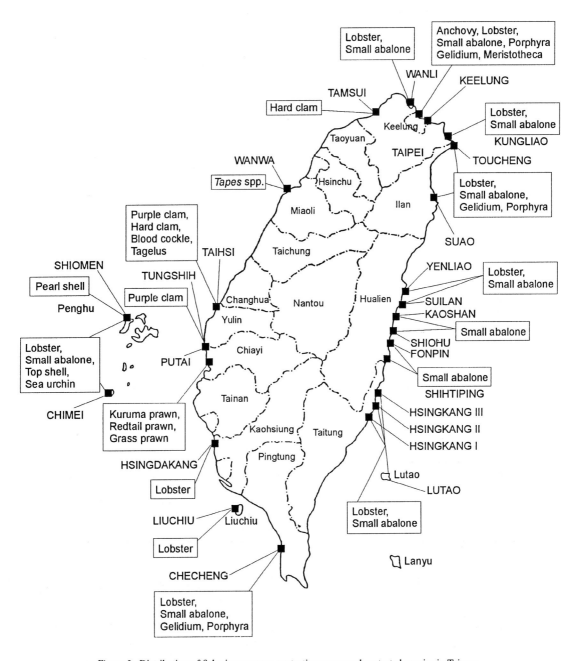

Figure 2. Distribution of fisheries resource protection zones and protected species in Taiwan.

centers. The number of national sea-farming centers increased from the initial two to the 16 now, distributed in 16 sea areas from Akkeshi, Hokkaido in the north to Ishigaki-jima, Okinawa Prefecture in the south (Figure 3). In these centers, JASFA is engaged in various technological developments by holding training sessions on a wide range of topics and by giving guidance and advice to various projects promoting sea-farming.

The primary role of the national sea-farming centers is technological development related to oceanic-migrating finfish and to those finfish that demonstrate regional migration. In addition, these centers are also engaged in the development of fingerling production

Table 4. Adult Japanese eel released by the Taiwan Fisheries Research Institute (TFRI) (1976–1995)

Year	Quantity
1976	3393
1978	3484
1979	1125
1981	4500
1982	675
1983	2290
1986	2250
1987	1440
1988	2050
1989	1474
1990	1840
1991	3500
1992	2250
1993	6350
1994	2200
1995	2505
Total	41326

■ National sea-farming centers

● Prefectural sea-farming centers

Figure 3. Distribution of sea-farming centers in Japan.

technology for local finfish species that have not yet been well studied. Prefectural sea-farming centers conduct mass production projects based on technologies developed at the national level. They are also engaged in the development of production technology and in mass production and restocking of fingerlings of local

species. Furthermore, some fishermen's cooperative associations, subsidized by the national, prefectural and municipal governments, are mass-producing and even restocking fingerlings of benthic animals.

Fingerling production and restocking
Fingerling production under JASFA in 1995 incorporated 20 finfish species, 5 crustacean species and 2 molluskan species (Appendix 2). Of these fingerlings, the Japanese flounder was the most abundantly produced. Table 7 and Appendix 3 show the fingerling production and fingerlings released in Japan in 1994. The production included 30 finfish, 13 crustacean, 23 molluskan, and 7 echinoderm species, while the fingerling restocked included 34 finfish, 12 crustacean, 25 molluskan and 8 echinoderm species.

In restocking volume, Japanese flounder accounted for the largest, followed by red sea bream. Among crustaceans, kuruma prawn led the group, followed by swimming crab. Among mollusks, short-necked clam was on the top, followed by scallops. In recent years the restocking of sea urchins has increased.

Problems and prospects

Enforcement of fisheries management and resource conservation

The major method of stock enhancement is the releasing of fingerlings. The fingerlings released must be protected until they grow to a suitable size (Matsuoka, 1989b). Furthermore, they must be regulated even if they grow to market size and are ready for exploitation. In Taiwan, the number of fishing vessels for use in coastal fisheries is currently not strictly regulated. Likewise, the mesh size, fishing ground and fishing season are also not controlled effectively. Illegal fishing methods, such as the use of electricity, poison and dynamite, have not yet been totally eliminated. The purpose of the fisheries resource zones is not attained. The system of fishery statistics is not established. Laws concerning stock enhancement have not yet been enacted.

Therefore, there is an urgent need for new laws on stock enhancement, increase of the fishery administration manpower, strict enforcement of regulations on fishery resources, improvement of the fishery statistics system of coastal fisheries, and enforcement of the

Table 5. Data of recaptured subadults of *P. monodon* released from Tungkang coast from August 1983 to June 1984

Date of release		No. of prawns released	No. of prawns recaptured	Recapture rate (%)	Sources of prawns for release
1983	Aug. 14	500	3	0.60	private farm
	Aug. 24	169	7	1.14	private farm
	Aug. 25	105	2	1.90	private farm
1984	Jan. 13	283	2	0.70	private farm
	Jan. 14	425	11	2.59	private farm
	Jan. 26	276	13	4.71	private farm
	Feb. 25	488	1	0.20	private farm
	Mar. 7	580	3	0.52	private farm
	Mar. 10	575	2	0.35	private farm
	Mar. 11	392	14	3.57	private farm
	Mar. 15	581	2	0.34	private farm
	Apr. 3	567	3	0.53	private farm
	Apr. 12	477	2	0.42	private farm
	May 3	604	1	0.17	private farm
	May 23	119	6	5.04	private farm
	June 6	199	31	15.57	TML
Total		6340	103		

Su et al. (1990)

Table 6. Fingerlings or adults released by prefectural governments in Taiwan (1978–1995)

Year	Small abalone ($\times 10^3$)	Short-necked (kg)	Purple clam (kg)	Tapes spp. (kg)	Prawns* ($\times 10^3$)	Grass prawn ($\times 10^3$)	Black sea bream ($\times 10^3$)	Other Sparids**
1978	22							
1979	28							
1980	56							
1981	25							
1982	30	172	1150					
1983	215	466	125					
1984	385	3135	19	98				
1985	380	35823		182	2000			
1986	100			166	2850			
1987	283			166				
1988	451			250	3070		17	
1989	480			285	2900	43	38	5
1990	333				2847	52	60	180
1991	336				3480	75	173	234
1992	361				3750	400	609	70
1993	839					5350	973	120
1994	382					3000	1354	210
1995	415					3000	1393	318
Total	5121	39596	1294	1147	17757	11920	4617	1137

* Kuruma prawn, bear prawn and sand shrimp.
** Include red sea bream and goldlined sea bream.

Table 7. Fingerling production and fingerlings released in Japan (1994)

Category	No. of species	Fingerling production ($\times 10^3$)	Fingerlings released ($\times 10^3$)	Species (common name)
Finfish	35	84 970	69 034	Herring, cod, blanquillo, striped jack, yellow amberjack, yellowtail, horse mackerel, rainbow fish, blue emperor, Japanese parrotfish, rock porgy, Japanese sea bass, chicken grunt, Japanese croaker, red spotted grouper, blue spotted grouper, black sea bream, red sea bream, sandfish, bluespotted mud hopper, fat cod, bar tailed flathead, black rock fish, jacopever, rock fish, devil stinger, Japanese flounder, roundnose flounder, dab, marbled sole, slime flounder, barfin flounder, spotted halibut, tiger puffer
Crustacean	14	597 037	342 047	Kuruma prawn, green tiger prawn, yellow sea prawn, speckled shrimp, northern prawn, pink shrimp, Japanese spiny lobster, Hanasaki king crab, horsehair crab, mangrove crab, green mangrove crab, swimming crab, hair crab
Mollusk	27	2 998 015	13 964 732	Ear shell, abalone, button shell, top shell, Japanese pearl oyster, marine snail, ark shell, native mussel, scallop, ezo giant scallop, freshwater clam, cockle, smooth giant clam, common hard clam, shield clam, short-necked clam, hen clam, northern clam, gaper, sand clam, razor shell
Others	8	74 253	75 466	Octopus, sea urchin, sea cucumber

whole planning and establishment of fisheries resource protection zones.

Control of pollution in coastal waters and reconstruction of fishing grounds

Estuarine areas are very important nursery grounds for many fishery resources. If these areas are polluted, the fingerlings released would not survive or grow. Therefore, it is very important to establish an automatic monitoring system and a network of water quality control centers in coastal areas, to effectively prevent the occurrence of water pollution (Hwang et al., 1991).

After years of operation by trawlers, the fishing grounds along the coastal areas have become virtual deserts. To provide a good habitat for fishery resources, the casting of artificial reefs is considered to be an effective way. Therefore, suitable sites for casting and the effectiveness of artificial reefs should be studied and evaluated.

Continuation of fundamental studies on stock enhancement

Many fundamental studies on stock enhancement have already been undertaken (Council of Agriculture, 1991). These studies dealt with such topics as the species and size of the fingerlings and locations that are suitable for releasing, optimal number of fingerlings for releasing, suitable season for releasing, most effective release methods, most efficient ways to enforce fishery regulations and exploit the resources, and evaluating the effectiveness of restocking. To solve these problems, long-term studies on environmental change, life history of the target species, and population dynamics are necessary.

Despite the yearly release of fingerlings, there is still a need to increase the diversity and quantity of the species released. The relationship, however, between stock enhancement and biodiversity must be clearly defined before mass restocking is conducted. The Convention on Biological Diversity launched during the United Nations Conference on Environment and Development ('Rio Summit') of 1992 in Rio de Janeiro, Brazil is meant to maintain and/or enhance the diversity of ecosystems, the diversity of species, the diversity of the gene pool, to use the constitutive factors of the biodiversity, and to distribute equally the profit from the gene resources. Some scientists have shown their concerns about the problems of stock enhancement and biodiversity in recent years. When promoting stock enhancement, these problems should be considered and avoided. The number of spawners needed to get a sufficient number of eggs and the appropriate number to be released to keep the biodiversity of the ecosystem should be determined.

The significance of restocking is most widely seen with chum salmon in Japan, where fingerlings have been restocked for more than a century (Matsuoka, 1989b). Almost all of the chum salmon taken in the adjoining seas of Japan and its rivers and streams are those artificially reared and restocked. The effects of restocking red sea bream, giant scallop, kuruma prawn, and some others, are also being seen in various localities in Japan.

In Taiwan, the species and quantity are not sizeable as compared with Japan. But still, more and more prefectural governments are releasing black sea bream each year. Despite the annual release by TFRI of adult Japanese eel, TFRI is still being asked to increase the number of Japanese eel released, to meet the rapid development and high demand of the eel culture industry. Because of the depleted grass prawn spawner stock from the waters surrounding Taiwan, pond-reared subadults have been released with some success (Su et al., 1990). There is still a need to study further the significance of restocking, particularly on the efficiency and effectiveness of the released species.

Education of fishermen

The direct involvement of fishermen is an important link in the development of stock enhancement. Many tasks, such as artificial reef casting, fingerling release, resource protection, and fishery regulation, depend on the fishermen. The government, therefore, should organize the fishermen, hold occasional training courses on the concept of stock enhancement and resource conservation and the actual procedures that need to be followed. The fishermen, especially the local ones, after understanding the importance of stock enhancement, can then actually engage in the tasks of releasing the fingerlings, abandoning illegal fishing, pollution control, fisheries resource protection, and fishery regulation (Liu & Su, 1993).

Establishment of a sea-farming association and centers

To promote stock enhancement effectively in Taiwan, a strong organization like JASFA is urgently needed. Although discussions on the establishment of such an association in Taiwan have been going on for a long time already, the association has yet to be founded. COA had entrusted the Taiwan Fisheries Consultants, Inc. (TFCI), a private think tank, to plan the establishment of sea-farming centers. Already, TFCI has identified three areas for the establishment of these centers but so far, preliminary preparations are only being done in one area. Related studies, such as on the life histories of resources, experimental restock-

ing, and evaluation of the effectiveness of restocking, have been conducted by TFRI, concerned universities and other research organizations. These studies can be turned over to form the association and sea-farming centers, once they are created. Therefore, with multi-sectoral cooperation from fishermen, fishery administrators and fishery researchers, stock enhancement can be promoted effectively and steadily.

Stock enhancement uses the seas to grow marine organisms, as agriculture uses the land to grow crops and livestocks. Stock enhancement is an extension of aquaculture and a new model of fishery resource management. The fingerling production and growth are undertaken on facilities inland and then in sea cages to nurture them in preparation for the natural sea environment. They are then released to certain locations where artificial reefs have been cast and seaweeds planted. The fingerlings inhabit there, grow and reproduce, transforming the area into a marine ranch. In other words, using stock enhancement management through marine ranching is to reach the final goal of sustainable aquaculture and fisheries.

Stock enhancement through population genetics

In addition to considering the biodiversity, the fishermen should also consider population genetics when undertaking stock enhancement. Jean (1995) proposed several suggestions on going about this. First, the prospective fingerlings for release should be produced from wild breeding stock. Using the progenies of completely cultivated stock in ponds should be avoided. Second, breeding stock with high genetic variation should be selected and using more breeders should be encouraged. Third, the distribution and population structure of any prospective species for release should be investigated before the release of fingerlings. Since each population is well adapted to its appropriate habitat, the release of fingerlings of one particular population into the habitats of different populations should be avoided. Fourth, to maintain ecological balance, the release of fingerlings of any species into the new habitat where this species has never inhabited should also be avoided.

There is an indication that recent R&D efforts in molecular biology will bring an upsurge in the use of mitochondrial DNA analysis, genetic divergent loci, microsatellite markers, among others, thus helping population geneticists to carry out advanced research. Current protocols are dependable, precise, and show potential and thus it is expected that they will be adopted to a greater degree.

Acknowledgments

This paper was prepared for the 'Asia-Pacific Conference on Science and Management of Coastal Environment', 25–28 June 1996, Hong Kong. I am grateful to Chin-Lau Kuo, Mei-Cheng Wu, and Jonathan VA A. Nuñez for their help in preparing and revising the manuscript.

References

Chen, C. C., S. C. Ou, C. C. Liu & C. J. Chang (eds), 1989. Collected Reprints on the Investigation of the Artificial Reefs. COA Fish. Ser. 22, 156 pp.

Council of Agriculture, 1991. Collected Reprints on Cultivated Fisheries (I). COA Fish. Ser. 26, 193 pp.

Hwang, D. F., Y. Tamao, C. C. Chen, M. Y. Wu & S. S. Jeng (eds), 1991. Environmental Production of Coastal Ecosystem. COA Fish. Ser. 23, Council of Agriculture, R.O.C., 416 pp.

JASFA, 1996. SAIBAI, No. 77, Japan Sea-Farming Association (JASFA), 54 pp. (In Japanese)

Jean, C. T., 1995. Studies on interspecific phylogenetic relationships and intraspecific population genetics of fishes of the Subfamily Sparinae (Perciformes:Sparidae) from the coastal waters of Taiwan. Ph.D. Thesis. Natl. Taiwan Ocean Univ., Keelung, Taiwan, 199 pp. (In Chinese, English abstract)

Kitada, S., 1985. Present state and prospects for seafarming. Technocrat 18: 10–17.

Liao, I. C., C. L. Kuo, T. C. Yu & W. N. Tzeng, 1994. Release and recovery of Japanese eel, *Anguilla japonica*, in Taiwan. J. Taiwan Fish. Res. 2: 1–6.

Liu, H. C. & M. S. Su, 1993. Research and development of stock enhancement in Taiwan. In Liao, I. C., J. H. Cheng, M. C. Wu & J. J. Guo (eds), Proceedings of the Symposium on Aquaculture Held in Beijing, 153–162. TFRI Conf. Proc. 3, Taiwan Fisheries Research Institute, Keelung, Taiwan, R.O.C., 280 pp. (In Chinese, English abstract)

Matsuoka, T., 1989a. Japan Sea-Farming Association (JASFA). Int. J. aquat. Fish. Technol. 1: 90–95.

Matsuoka, T., 1989b. Current state of affairs and problems facing sea-farming – with emphasis placed on technical problems of fingerling production. Int. J. aquat. Fish. Technol. 1: 324–332.

Oshima, Y., 1984. Status of 'fish farming' and related technological development in the cultivation of aquatic resources in Japan. In Liao, I. C. & R. Hirano (eds), Proceedings of ROC-Japan Symposium on Mariculture, 1–11. TML Conf. Proc. 1. Tungkang Marine Laboratory, Tungkang, Taiwan, R.O.C., 224 pp. (In English, Japanese abstract)

Su, M. S., I. C. Liao & R. Hirano, 1990. Restocking of subadult grass prawn, *Penaeus monodon*, in the coastal waters of southwest Taiwan. In Hirano, R. & I. Hanyu (eds), 99–102. The Second Asian Fisheries Forum. Asian Fisheries Society, Manila, Philippines: 991 pp.

Taiwan Fisheries Bureau, 1989. Investigations of Ecological Environment and Evaluation of Effectiveness in Fisheries Resources Reservation Waters of Taiwan. Taiwan Fish. Bur. Ser. 3. 250 pp. (In Chinese)

Appendix 1. Fingerlings or adults released by the Taiwan Fisheries Research Institute (TFRI) (1985–1995)

Year	TFRI unit	Common name	Scientific name	Quantity ($\times 10^3$)
1985	PB	Red sea bream	*Pagrus major*	0.5
		Kuruma prawn	*Penaeus japonicus*	1000
1986	LB	Japanese eel	*Anguilla japonica*	2.3
	TML	Grass prawn	*Penaeus monodon*	300
	PB	Kuruma prawn	*Penaeus japonicus*	2000
1987	LB	Japanese eel	*A. japonica*	1.4
	TML	Grass prawn	*P. monodon*	500
	PB	Kuruma prawn	*P. japonicus*	1500
1988	LB	Japanese eel	*A. japonica*	2
	TML	Grass prawn	*P. monodon*	600
	PB	Kuruma prawn	*P. japonicus*	2000
1989	DFB	Black sea bream	*Acanthopagrus schlegeli*	2
	LB	Japanese eel	*A. japonica*	1.5
	TML	Grass prawn	*P. monodon*	1000
	TB	Small abalone	*Haliotis diversicolor*	10
	PB	Kuruma prawn	*P. japonicus*	2000
		Goldlined sea bream	*Sparus sarba*	50
1990	DFB	Black sea bream	*A. schlegeli*	11
	LB	Japanese eel	*A. japonica*	1.8
	TML	Grass prawn	*P. monodon*	1000
	TB	Black sea bream	*A. schlegeli*	1.4
		Thornfish	*Terapon jarbua*	1
	PB	Kuruma prawn	*P. japonicus*	6000
		Goldlined sea bream	*S. sarba*	50
		Red sea bream	*P. major*	30
1991	DFB	Black sea bream	*A. schlegeli*	5
	LB	Japanese eel	*A. japonica*	3.5
	TML	Grass prawn	*P. monodon*	2000
	PB	Kuruma prawn	*P. japonicus*	2000
		Goldlined sea bream	*S. sarba*	50
		Red sea bream	*P. major*	30
1992	DFB	Black sea bream	*A. schlegeli*	12
	LB	Japanese eel	*A. japonica*	2.3
	TML.	Grass prawn	*P. monodon*	200
	PB	Kuruma prawn	*P. japonicus*	1000
1993	DFB	Black sea bream	*A. schlegeli*	18
	LB	Japanese eel	*A. japonica*	6.4
		Marbled eel	*A. marmorata*	15
	PB	Goldlined sea bream	*S. sarba*	120
		Kuruma prawn	*P. japonicus*	3400
		Small abalone	*H. diversicolor*	500
	TML	Kuruma prawn	*P. japonicus*	100
		Redtail prawn	*P. penicillatus*	1200
	TB	Black sea bream	*A. schlegeli*	60
		Goldlined sea bream	*S. sarba*	40
1994	DFB	Black sea bream	*A. schlegeli*	32
	LB	Japanese eel	*A. japonica*	2.2
	TML	Grass prawn	*P. monodon*	200
	PB	Kuruma prawn	*P. japonicus*	100

178

Appendix 1. Continued

Year	TFRI unit	Common name	Scientific name	Quantity ($\times 10^3$)
1995	DFB	Black sea bream	*A. schlegeli*	20
		Goldlined sea bream	*S. sarba*	5
	LB	Japanese eel	*A. japonica*	2.5
	TML	Sand shrimp	*Metapenaeus ensis*	100
		Bear prawn	*Penaeus semisulcatus*	800
		Grey snapper	*Lutjanus argentimaculatus*	50
	PB	Red sea bream	*P. major*	30
		Goldlined sea bream	*S. sarba*	20
		Swimming crab	*Portunus trituberculatus*	50

(PB) Penghu Branch; (LB) Lukang Branch; (TML) Tungkang Marine Laboratory; (DFB) Department of Fishery Biology; (TB) Taitung Branch.

Appendix 2. Fingerling production under the Japan Sea-Farming Association (JASFA) (1995)

Family name	Scientific name	English name	Japanese name	Fingerling production ($\times 10^3$)
Finfish				
Clupeidae	*Clupea pallasi*	Herring	Nishin	1888
Gadidae	*Gadus macrocephalus*	Cod	Madara	240
Branchiostegidae	*Branchiostegus japonicus*	Blanquillo	Akaamadai	1
Carangidae	*Caranx delicatissimus*	Striped jack	Shimaaji	719
	Seriola aureovittata	Yellow amberjack	Hiramasa	44
	Seriola dumerili	Amberjack	Kanpachi	8
	Seriola quinqueradiata	Yellowtail	Buri	399
Percichthyidae	*Lateolabrax japonicus*	Japanese sea bass	Suzuki	194
Seranidae	*Epinephelus akaara*	Red spotted grouper	Kijihata	19
	Epinephelus moara	Kelp grouper	Kue	1
	Plectropomus leopardus	Blue spotted grouper	Sujiara	4
Sparidae	*Pagrus major*	Red sea bream	Madai	1584
Trichodontidae	*Arctoscopus japonicus*	Sandfish	Hatahata	891
Scorpaenidaes	*Sebastes schlegeli*	Jacopever	Kurosoi	559
Synanceiidae	*Inimicus japonicus*	Devil stinger	Oniokoze	44
Paralichthyidae	*Paralichthys olivaceus*	Japanese flounder	Hirame	5044
Pleuronectidae	*Eopsetta grigorjewi*	Roundnose flounder	Mushigare	61
	Verasper moseri	Barfin flounder	Matsukawa	19
	Verasper variegatus	Spotted halibut	Hoshigarei	7
Tetraodontidae	*Takifugu rubripes*	Tiger puffer	Torafugu	614
Crustacean				
Pandalidae	*Pandalus hypsinotus*	Coonstrip prawn	Toyamaebi	328
Lithodidae	*Paralithodes brevipes*	Hanasaki king crab	Hanasakigani	519
Atelecyclidae	*Erimacrus isenbeckii*	Horsehair crab	Kegani	216
Portunidae	*Scylla tranquebarica*	Mangrove crab	Nokogirigazami	1030
	Scylla oceanica	Green mangrove crab	Amimenokogirigazami	75
Mollusca				
Sepiidae	*Sepia latimanus*	Giant cuttlefish	Kobushime	40
Ostopodidae	*Octopus vulgaris*	Octopus	Madako	36

Data from JASFA (1996)

Appendix 3. Fingerling production and fingerlings released in Japan (1994)

Family name	Scientific name	English name	Japanese name	Fingerling production ($\times 10^3$)	Fingerling released ($\times 10^3$)
Finfish					
Clupeidae	*Clupea pallasi*	Herring	Nishin	2178	2270
Gadidae	*Gadus macrocephalus*	Cod	Madara	143	84
Branchiostegidae	*Branchiostegus japonicus*	Blanquillo	Akaamadai	4	
Carangidae	*Caranx delicatissimus*	Striped jack	Shimaaji	333	250
	Seriola aureovittata	Yellow amberjack	Hiramasa	64	37
	Seriola quinqueradiata	Yellowtail	Buri	743	359
	Trachurus japonicus	Horse mackerel	Maaji		186
Labridae	*Halichoeres poecilepterus*	Rainbow fish	Kyusen		1963
Lethrinidae	*Lethrinus nebulosus*	Blue emperor	Hamafuefuki	153	109
Oplegnathidae	*Oplegnathus fasciatus*	Japanese parrotfish	Ishidai	17	19
	Oplegnathus punctatus	Rock porgy	Ishigakidai	11	15
Percichthyidae	*Lateolabrax japonicus*	Japanese sea bass	Suzuki	556	404
Pomadasyidae	*Parapristipoma trilineatum*	Chicken grunt	Isaki	644	606
Sciaenidae	*Nibea japonica*	Japanese croaker	Onibe	228	228
Serranidae	*Epinephelus akaara*	Red spotted grouper	Kijihata	58	14
	Plectropomus leopardus	Blue spotted grouper	Sujiara	92	
Sparidae	*Acanthopagrus schlegeli*	Black sea bream	Kurodai	9720	6521
	Acanthopagrus sivicolus	Black sea bream	Minamikurodai	360	177
	Pagrus major	Red sea bream	Madai	26650	21309
Trichodontidae	*Arctoscopus japonicus*	Sandfish	Hatahata	5208	5052
Gobiidae	*Boleophthalmus pectinirostris*	Bluespotted mud hopper	Mutsugoro		1
Hexagrammidae	*Hexagrammos otakii*	Fat cod	Ainame		773
Platycephalidae	*Platycephalus indicus*	Bar tailed flathead	Kochi	143	66
Scorpaenidae	*Sebastes inermis*	Black rock fish	Mebaru		536
	Sebastes vulpes	Jacopever	Kurosoi	2734	1969
	Sebastiscus marmoratus	Rock fish	Kasago	455	310
Synanceiidae	*Inimicus japonicus*	Devil stinger	Oniokoze	237	174
Paralichthyidae	*Paralichthys olivaceus*	Japanese flounder	Hirame	27489	21556
Pleuronectidae	*Eopsetta grigorjewi*	Roundnose flounder	Mushigarei	30	11
	Limanda herzensteini	Dab	Magarei	573	176
	Limanda yokohamae	Marbled sole	Makogarei	3818	2098
	Microstomus achne	Slime flounder	Babagarei	3	1
	Verasper moseri	Barfin flounder	Matsukawa	9	13
	Verasper variegatus	Spotted halibut	Hoshigarei	56	23
Tetraodontidae	*Takifugu rubripes*	Tiger puffer	Torafugu	2344	1722
Crustacean					
Penaeidae	*Penaeus japonicus*	Kuruma prawn	Kurumaebi	473 800	277 515
	Penaeus semisulcatus	Green tiger prawn	Kumaebi	11 133	2989
	Penaeus orientaris	Yellow sea prawn	Kouraiebi	3217	3204
	Metapenaeus monoceros	Speckled shrimp	Yoshiebi	43 908	24 383
	Pandalus kessleri	Northern prawn	Hokkaiebi	36	140
	Pandalus hypsinotus	Pink shrimp	Toyamaebi	542	301
Palinuridae	*Panulirus japonicus*	Japanese spiny lobster	Iseebi		10
Lithodidae	*Paralithodes brevipes*	Hanasaki king crab	Hanasakigani	127	20
Atelecyclidae	*Erimacrus isenbeckii*	Horsehair crab	Kegani	171	60

Appendix 3. Continued.

Family name	Scientific name	English name	Japanese name	Fingerling production ($\times 10^3$) ($\times 10^3$)	Fingerling released ($\times 10^3$)
Portunidae	*Scylla tranquebarica*	Mangrove crab	Nokogirigazami	1186	688
	Scylla oceanica	Green mangrove crab	Amimenokogirigazami	2	
	Portunus trituberculatus	Swimming crab	Gazami	59 178	30 985
	Portunus pelagicus	A kind of swimming crab	Taiwangazami	3356	1752
Grapsidae	*Eriocheir japonicus*	Hair crab	Mokuzugani	381	
Mollusca					
Haliotidae	*Sulculus diversicolo squatilus*	Ear shell	Tokobushi	412	447
	Sulculus diversicolor	A kind of ear shell	Fukutokobushi	2369	1966
	Nordotis discus	A kind of abalone	Kuroawabi	6235	3711
	Nordotis discus hannai	A kind of abalone	Ezoawabi	20 273	17 163
	Nordotis gigantea	A kind of abalone	Madakaawabi	15	35
	Nordotis gigantea sieboldii	A kind of abalone	Megaiawabi	3209	1755
Trochidae	*Tectus niloticus maximus*	Button shell	Sarasabatei	133	
Turubinidae	*Batillus cornutus*	Top shell	Sazae	2488	1971
Pteriidae	*Pinktada fukata*	Japanese pearl oyster	Akoyagai	139	10
Buccinidae	*Babylonia japonica*	Marine snail	Bai	133	177
Arcidae	*Scapharca broughtonii*	Ark shell	Akagai	4374	1871
Mytilidae	*Mytilus coruscus*	Native mussel	Igai	35	
Pectinidae	*Chlamys nobilis*	A kind of scallop	Hiougi	50	50
	Patinopecten yessoensis	Ezo giant scallop	Hotategai	2 896 561	2 960 325
Corbiculidae	*Corbicula japonica*	A kind of freshwater clam	Yamatoshijimi		21 309
Cardiidae	*Fulvia mutica*	Cockle	Torigai	1039	159
Tridacnidae	*Tridacna crocea*	Smooth giant clam	Himejyako		42
Veneridae	*Meretrix lusoria*	Common hard clam	Hamaguri	4200	8710
	Meretrix lamarckii	Shield clam	Chosenhamaguri	3475	3660
	Gomphina melanaegis	A kind of clam	Kotamagai		190
	Tapes philippinarum	Short-necked clam	Asari	42 569	10 930 117
Mactridae	*Mactra chinensis*	Hen clam	Bakagai	887	600
	Mactra carneopicta	A kind of hen clam	Ezobakagai	27	698
	Spisula sachalinensis	Northern clam	Ubagai	8957	9364
	Tresus keenae	Gaper	Mirukui	395	170
Tellinidae	*Peronidia venulosa*	A kind of sand clam	Saragai		192
Veneridae	*Solen strictur*	Razor shell	Mategai	40	40
Others					
Octopodidae	*Octopus vulgaris*	Octopus	Madako	25	41
Toxopneustidae	*Tripneustes gratilla*	Sea urchin	Silahige-uni	145	85
	Pseudocentrotus depressus	A kind of sea urchin	Aka-uni	2807	2948
Strongylocentrotidae	*Hemicentrotus pulcherrimus*	A kind of sea urchin	Bafun-uni	922	862
	Strongylocentrotus intermedius	A kind of sea urchin	Ezobafun-uni	56 244	54 636
	Strongylocentrotus nudus	A kind of sea urchin	Kitamurasaki-uni	11 553	15 120
Echinothuridae	*Anthocidaris crassispina*	A kind of sea urchin	Murasaki-uni		82
Stichopodidae	*Stichopus japonicus*	Sea cucumber	Manamako	2557	1692

Data from JASFA (1996)

Hydrobiologia **352**: 181–193, 1997.
Y.-S. Wong & N. F.-Y. Tam (eds), Asia–Pacific Conference on Science and Management of Coastal Environment.
©1997 *Kluwer Academic Publishers. Printed in Belgium.*

Human influence or natural perturbation in oceanic and coastal waters – can we distinguish between them?

Jarl-Ove Strömberg
Royal Swedish Academy of Sciences, Kristineberg Marine Research Station S-450 34 Fiskebäckskil, Sweden

Key words: marine biosphere, natural perturbation, human influence

Abstract

The review brings up several case studies on what is perceived as natural perturbations and/or human impacts on the marine biosphere. Case studies include polar seas, the green house effect, introduction of new species, pollution in Antarctic waters, pelagic fish stock fluctuations in oceanic waters, and eutrophication and pollution in temperate coastal systems. In coastal waters human impact is often obvious, but climatic fluctuations also influence the systems. In the open ocean the two factors are difficult to distinguish and some large scale fish stock fluctuations still need to be understood.

Introduction

Human influence on the marine environment occurs on many time and space scales and in many ways. Impact on a global scale may result from transport of pollutants via the atmosphere or via changes in the air chemistry both in the troposphere and the stratosphere. Example of the former is the spread of organochlorine contaminants into almost all marine ecosystems (Loganathan & Kannan, 1991) and of the latter the increase of greenhouse gases like carbon dioxide, methane, and chlorofluorocarbons (CFCs) leading to a possible climate change (Morel et al., 1990) and in the last case also a depletion in the stratospheric ozone layer (Morel, *loc. cit*, WMO, 1994). Such changes may have secondary effects on the marine physical, chemical and biological systems, and their effects occur over decadal or longer time-scales.

In coastal waters the impact of human activities may be much more immediate, particularly if bays or inlets with poor water circulation are considered. Here time scales can be from days to weeks and years.

An increasingly important factor to consider is the world population growth, which is mainly occurring in areas close to or on the coasts (e.g. Ohlin, 1992; Arrhenius, 1992). The effect of large population centers on coastal waters is thus of major concern. As a result

a wise use of coastal environments and their marine resources, i.e. coastal management, is needed.

Besides the human influence on the climate of the oceans we also have naturally occurring phenomena, most often triggered by atmospheric perturbations – whatever causes they may have. Examples of this are the so called ENSO phenomena (El Niño – Southern Oscillation) (e.g. Bjerknes, 1969; Brock, 1984; Quinn et al., 1987; Wuethrich, 1995), possible changes in the oceanic conveyor belt (Broecker, 1991; Broecker & Denton, 1990) and the North Atlantic Oscillation (Mann & Lazier, 1991). Such phenomena are normally cyclic, but the frequency in which they occur may vary or change in unilateral direction over centennial time scales.

The relative role of natural perturbations in the oceans vis-à-vis anthropogenically caused ones is most often difficult to distinguish and evaluate. A significant understanding of the dynamics of oceanic systems (whether physical, chemical or biological) and ocean – atmosphere interaction as well as long-term observations are most often required before the distinctions can be made.

The present paper gives some examples of both oceanic and coastal cases where natural and/or human influence may be the cause(s) of change and where often a clear conclusion on relative importance of such

182

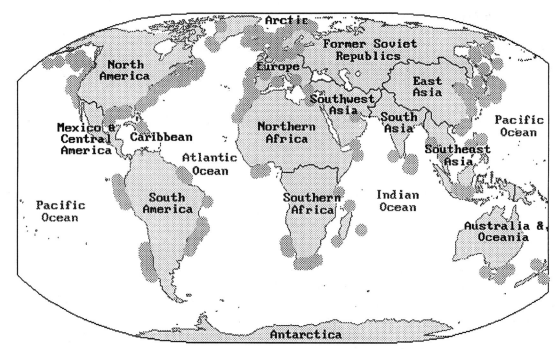

Areas of intensive fishing

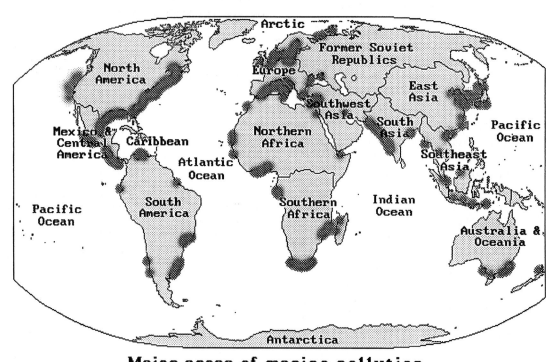

Major areas of marine pollution

Figure 1. A comparison of major fishing areas in the world (A) with areas of strong input of pollutants into the marine environment (B) shows a strong overlap over shelf seas, where most of the production in the ocean takes place.

causes is difficult to reach. Emphasis will be put on fluctuations affecting the marine biosphere with population dynamics of pelagic fishes as one oceanic example and of benthos in shallow waters as an example from coastal environments.

Case studies

Obviously all parts of the marine environment are exposed to natural fluctuations. These may be generated by long-term cyclic phenomena, like changes in the angle of the earth's axis (41 000 years) or the shape of the earth's path around the sun (100 000 years) (Broecker & Denton, 1990) or shorter ones like the 11-year solar activity cycle (Haigh, 1996; Robock, 1996). There are also shorter but irregular phenomena, like the already mentioned El Niño – Southern Oscillation events.

Human activities seem now to have influenced most parts of the marine environment. Most major fishing areas of the world, mainly in coastal and shelf waters, are close to major human population centers. Thus it is not only the fishing activity itself which influences the fish stocks, but it is also the impact of eutrophication and pollution from often close by landbased sources (Figure 1).

Polar Seas

Are there no parts of the ocean realm that has escaped human impact? Possibly this is the case in parts of the deep sea, where dumping has not yet taken place and where waste and pollutants has not yet had a chance to reach.

Polar seas are often regarded as being unaffected by human activities. Yet, in Antarctic circumpolar waters hunting of the big baleen whales from 1920's into the 1960's caused a dramatic reduction in the stocks of fin, blue, humpback and sei whales reducing their populations to a few percent of what they were at the turn of the century (Beddington & May, 1982) (Figure 2). There is little doubt that this had an effect on the major source of food, krill, of these whales, which in turn must have had an effect on other parts of the Antarctic ecosystem. Antarctic krill (*Euphausia superba*) has a central role in the marine ecosystem of the upper water column.

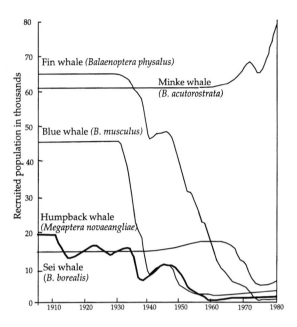

Figure 2. Populations of large balean whales declined dramatically in the 1930's to 1950's all over the world because of hunting and other anthropogenic disturbances. In the ocean area south of the equator, between 70° E and 130° E, the fin and blue whales decreased most, while the small minke whale increased in population size. Apparently hunting of sei whales became important only in the 1960's (from Beddington & May, 1982).

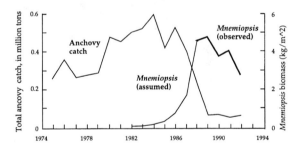

Figure 3. The comb yellyfish, *Mnemiopsis leidyi*, was probably introduced with bulk water to the Black Sea in 1982. Observations of biomass from 1982 in conjunction with records of anchovy catch indicates a strong decline in the Black Sea anchovy stock as the abundance of Mnemiopsis increased. Food competion has been implied to be the major forcing factor (from Zaika, 1994 and Anon, 1994).

The greenhouse effect

Depletion of the stratospheric ozone layer above the poles particularly during late winter–early spring, is another anthropogenic effect, which potentially may affect the upper ocean ecosystem of the Arctic and Antarctic, although this has not yet been clearly demonstrated in the field (Halpert et al., 1994; Voytek,

184

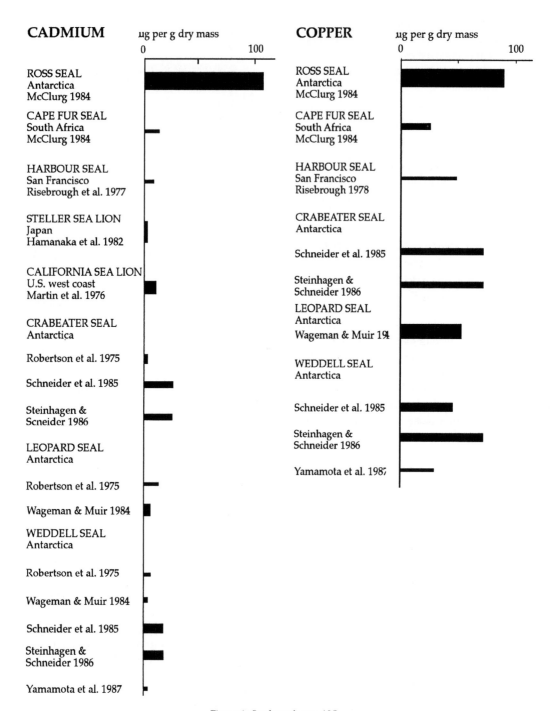

CADMIUM   µg per g dry mass   COPPER   µg per g dry mass

ROSS SEAL
Antarctica
McClurg 1984

CAPE FUR SEAL
South Africa
McClurg 1984

HARBOUR SEAL
San Francisco
Risebrough et al. 1977

STELLER SEA LION
Japan
Hamanaka et al. 1982

CALIFORNIA SEA LION
U.S. west coast
Martin et al. 1976

CRABEATER SEAL
Antarctica

Robertson et al. 1975

Schneider et al. 1985

Steinhagen &
Scneider 1986

LEOPARD SEAL
Antarctica

Robertson et al. 1975

Wageman & Muir 1984

WEDDELL SEAL
Antarctica

Robertson et al. 1975

Wageman & Muir 1984

Schneider et al. 1985

Steinhagen &
Schneider 1986

Yamamota et al. 1987

ROSS SEAL
Antarctica
McClurg 1984

CAPE FUR SEAL
South Africa
McClurg 1984

HARBOUR SEAL
San Francisco
Risebrough 1978

CRABEATER SEAL
Antarctica

Schneider et al. 1985

Steinhagen &
Schneider 1986

LEOPARD SEAL
Antarctica
Wageman & Muir 1984

WEDDELL SEAL
Antarctica

Schneider et al. 1985

Steinhagen &
Schneider 1986

Yamamota et al. 1987

Figure 4. See legend on p. 185.

1989). However, harmful effects of increased UV-B radiation on both phyto- and zooplankton have been well documented in the laboratory although some phytoplankton species seem to be able to protect themselves thanks to special pigments (Wängberg et al., 1996).

Chlorofluorocarbons (CFCs) which comprise one factor causing depletion of the stratospheric ozone

185

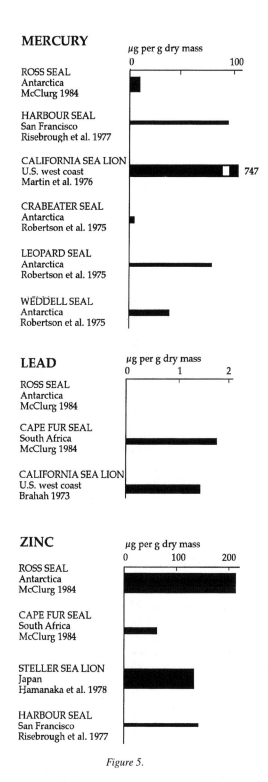

Figure 5.

Figures 4 and 5. Contents of cadmium, copper (4), mercury, lead and zinc (5) in seals from Antarctica (Ross, Weddell, crabeater and leopard seals) and from seas affected by human activities (Cape fur seal, Steller and California sea lions, harbour seals). For interpretation see the text. (From Strömberg et al., 1990; also references given in the Figure)

layer are also powerful greenhouse gases. Together with anthropogenically caused increases in atmospheric methane and above all carbon dioxide they have an impact on the global heat budget (Morel et al., 1990). In the last 39 years, or since measurements of atmospheric carbon dioxide started at the Maona Loa Climate Observatory in Hawaii in 1957, the concentration of this gas has increased from 315 ppm to about 355 ppm in 1995 (WMO, 1994), which is higher than at any time during the last 160 000 years as recorded from ice-cores from continental icecaps (Barnola et al., 1987). Natural events, like volcanic eruptions sending ash and other material into the upper troposphere and lower stratosphere, changes the solar transmission through the atmosphere. The eruption of Mt. Pinatubo in the Philippines may be the reason for a slightly diminished increase in the carbon dioxide level in the early 1990's (Kerr, 1994; WMO, 1994). The effects of the dramatic increase of greenhouse gases during the present century have so far only been demonstrated in a global mean temperature increase of about 0.5 °C, and surprisingly enough most of this change took place during the first half of the present century (e.g. Morel et al., 1990). No general change of temperature of the surface of the ocean since 1860 has been recorded (Morel et al., *loc.cit.*) , but there are indirect signs which could point in the direction of a temperature increase, e.g. the melting rate of shelf ice in the Antarctic (Overpeck, 1996), a shrinking sea ice cover in the Arctic (Johannessen et al., 1995), or a shift in the benthic fauna on the central Californian coast with some northern species disappearing and an influx of more southerly species (Barry et al., 1995). If this is a general trend, nobody can say at present, but it is worthwhile to be observant on such changes, which may be observed in both open ocean and coastal systems.

Introduction of new species

A different type of faunal and floral change may occur because of transport of bulk water by ships with the accidental introduction of new species into an area. A well-known example is the introduction of the comb jelly *Mnemiopsis leidyi* into the Black Sea and the dramatic effect this species is implied to have had on the recruitment of anchovy through food competition (Zaika 1994) (Figure 3). Another example is the accidental introduction of the brown alga *Sargassum muticum* in western Europe, probably via the dissemination of Japanese oyster sprat in the early to mid 1970's. This has become a major concern because of

its rapid spreading to other parts of Europe and the ecological consequences of its introduction (e.g. Critchley
et al. 1990; Jones & Farnham, 1973; Norton & Benson,
1983; Rueness 1989).

Pollution in Antarctic waters

Pollution is an increasing problem, not only in coastal
waters, but also in the open ocean. This is particularly noticeable in the Northern Hemisphere where
atmospheric transport of pollutants is a major pathway from land to open ocean waters. What is then the
situation in the most remote oceanic waters, the Southern Ocean? There is a poor coverage of the whole
area but the data available indicate that concentrations
of pesticides and other chlorinated hydrocarbons are
orders of magnitude lower than those in the Northern
Hemisphere. Thus, in this respect Antarctic waters are
the least contaminated in the world (Strömberg et al.,
1990).

However, studies of heavy metal concentrations in
Antarctic seals offer an interesting apparent contradic-
tion (Strömberg et al., *loc. cit.*). (Figures 4–5) The tissue level of copper is almost double as high in Antarctic
seals as in those from South Africa or the San Francisco Bay. The reason for this is obviously that the
major food source of the Antarctic seals is krill (e.g.
Laws, 1983), the blood of which contains copper as
the central atom in the hemocyanin molecule. Thus the
bioacculmulation of copper is natural. The food of the
other seals is more varied and consists mainly of fish,
in which the blood pigment is hemoglobin with iron as
the central atom.

Investigations of cadmium show very high concentrations in one Antarctic Seal, the Ross seal
(*Ommatophoca rossi*), while most others have low values. The most likely explanation for this is, that this
seal largely feeds on squid and and other deep living
animals, possibly also benthic ones which accumulate
cadmium from the sediment, which in its turn is rich
in this metal.

Other heavy metals, e.g. mercury and lead, show
the reverse picture, with much higher bioaccumulation
in species living in the vicinity of big human populations. Here a clear anthropogenic effect may be
inferred.

Finally, there is the case of zinc with relatively
high concentrations in both the Ross seal and some
more northern species. Interpretation is difficult, but
it may be assumed that both natural (Ross seal) and
anthropogenic causes (the other seals) are at work.

These examples show that it is imperative to know
about the biology and the habitats of the organisms
which are analyzed for various pollutants, before conclusions are made on the reasons for the pollution levels
encountered.

Stock fluctuations in the open ocean

The longest record of pelagic fish stock fluctuations
was presented by Baumgartner et al. (1992). The study
was based on counts of fish scales and otoliths from
the annually layered (varved) and unperturbed sediments in the deep, anoxic part of the Santa Barbara
Basin. It showed dramatic fluctuations in both Pacific
sardine and Northern anchovy from about year 300 to
present time. The biomass fluctuation of the Pacific sardine varied from almost nil up to about 16×10^6 metric
tons while the values of the anchovy were fluctuating
between 0.5 and up to 3.5×10^6 tons. The population
peaks of the two species did not coincide in most cases
(Figure 6). This is a remarkable demonstration of naturally occuring fluctuations, the reasons for which still
are under debate. However, contrary to initial belief, it
is obvious that fishing has had nothing to do with them.

In this context it is interesting to compare with
changes in catches during the present century of widely
separated stocks of one species of sardine, *Sardinops
sagax* (Kawasaki, 1991). (Figure 7). Three populations, the Far Eastern sardine in Japaneses waters, the
Californian sardine west of North America, and the
Chilean sardine west of South America, have long been
thought of as different species but are now regarded as
one. The co-variation in catches, reflecting abundance
and probably also production, is clear with an increase
in the 1920's leading to a peak in the 1930's and then
a drop in the 1940's. The catches were very small in
the 1950's and 1960's and then increased dramatically
in the 1970's to reach a peak in the early 1980's. The
catch went from almost none to nearly 6×10^6 tons
in less than ten years for both the Far Eastern and the
Chilean sardines.

The increase in the sardine populations must have
made a radical change in grazing rates in their respective areas. The build up of the populations most likely followed an increase in both phyto- and zooplankton, and the crashes of the sardine populations probably trailed similar events taking place in the plankton
communities. During the 1970's and 1980's there was
indeed an increase in the standing stock of phytoplankton (chlorophyll) in the Pacific Ocean north of Hawaii
(Venrick et al., 1987) and a doubling of the zooplank-

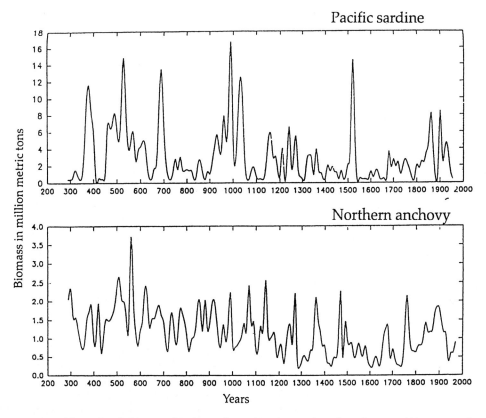

Figure 6. Reconstruction of fluctuations in biomass of Pacific sardine and Northern anchovy from about year 300 to present time. The record is based on analyses of fish scales and otholites from varved sediments in the Santa Barbara Basin (from Baumgartner et al., 1992).

ton in the northeast Pacific (Brodeur & Ware, 1992). In the southeast Pacific, off the Peruvian coast, the zooplankton population started to decline about 1972 and the anchovetta fishery crashed in 1974 (Carrasco & Lozano, 1989.) It is surprising that the Chilean sardine did not suffer in the same way as the Peruvian anchovetta. However, it has been demonstrated that population peaks in sardine and anchovy or anchovetta normally do not co-occur. This has been shown in Japanese waters (Kondo, 1991) (Figure 8) and the Southeast Pacific (Avaria, 1985) (Figure 9).

How can the covariation (teleconnections) in the dynamics of the widely separated sardine populations be explained? Since most of the Pacific Ocean is affected in some way, the reason must be large-scaled, possibly a climatic shift causing changes in wind stress and paths of ocean currents. Kasai et al. (1992) suggested changes in wind field and Kobayashi & Kuroda (1991) looked at availability of spawning grounds and shifts in the path of the Kurishio current southeast of Japan. Could ENSO-events explain such changes? There is as yet no clear picture of this, and the fluctuations of

the Chilean sardine and Peruvian anchovetta do not indicate a connection (Sharp & McLain, 1993). The anchovetta stock was high and the sardine stock low during the late 1950's, 1960's and early 1970's in spite of several ENSO events. However, after the 1971–72 El Niño the former crashed and the latter increased, and during the 1982–83 El Niño this trend was strengthened.

Did fishing cause the changes? Sharp & McLain (*loc. cit.*) concluded that fishing could contribute to the dramatic stock changes, but that the natural causes were the overriding ones. Looking at the very long-term records of Baumgartner et al. (1992), this conclusion seems logical, and Rothschild (1991, 1993) reached the same conclusion. There is still the possibility that ENSO events influence the climate of the North Pacific and the air pressure oscillations in that part of the ocean. However, interrelation is still unclear.

In the Atlantic Ocean there are also fluctuations in clupeid stocks. The catches of herring on the Swedish west coast are well documented (Höglund, 1972) and show a periodicity with peaks at about 100 years' inter-

188

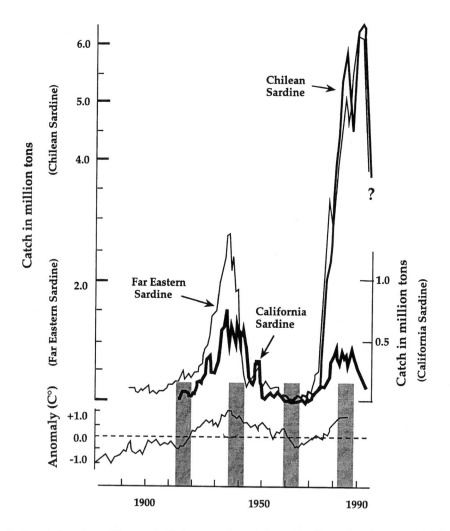

Figure 7. Records of catch from three different and widely separated populations of Pacific sardine show a remarkable covariation. The fluctuations may be compared with the temperature anomaly from the North Pacific which shows a similar change (bottom graph) and with ENSO events (stippled bars) where there seems to be no clear relationship with fish catch fluctuations. (From Kawasaki, 1991 and Sharp & McLain, 1993).

val. The periods lasted for more than 50 years each time and had a marked socio-economic impact on the coastal society. The coupling of a population decline of the North Sea herring during the 1960's and 1970's to a similar trend in the lower trophic levels was proposed by Aebischer et al. (1990).

Long-term studies of plankton in the northeast Atlantic made within the Continuous Plankton Recorder program indeed supports the role of lower trophic levels. Dickson et al. (1988) found that an increase in the wind field and a shift to a northerly direction in the 1970's coincided with a marked decrease in phytoplankton and zooplankton standing stocks. The

indices declined by a factor of about one standard deviation during the 1970's and 1980's, and if the 1950's and 1960's are included, the decline was almost two standard deviations (Figure 10).

The north Atlantic oscillation with a low pressure south of Iceland forces the westerly winds in a northerly direction in the northeast Atlantic, while during the other extreme a high pressure north of the Azores causes the westerlies to the north, with a main direction from the northwest in the northeast part of the ocean (Mann & Lazier, 1991).

The causes of the events related to clupeid and plankton standing stocks (and possibly production) are

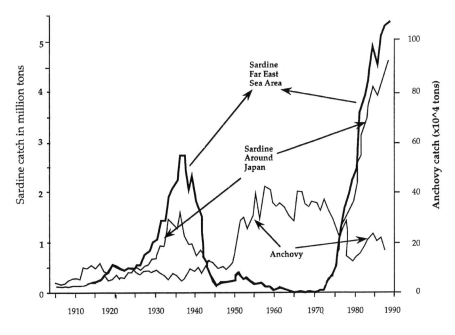

Figure 8. Graphs of annual catch of Japanese sardine and anchovy from 1905–1988 clearly show that populations of the two species peak during different periods (from Kondo, 1991).

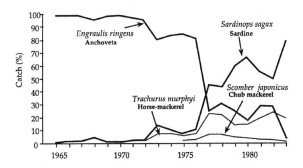

Figure 9. Annual catches of pelagic fish in percentage contribution in the south-east Pacific 1965–1983. The drop in anchoveta catch began in 1972 but the major decrease occurred in 1976, when the catch of the Chilean sardine started to increase (from Avaria, 1985).

indeed complex. If wind field is implicated, its manifestation in effects on the stocks is not understood. It could be wind-driven advection or a change in the relative advantage of predator and prey owing to small-scale turbulent flow (Rothschild, 1991). It could also be a direct link between stronger winds, a deeper mixed layer and poorer conditions for phytoplankton production. This in turn would affect secondary and tertiary production.

It is argued (Anon., 1994) that the observed dynamics of small pelagic fish stocks is due to high- or low-frequency climatic variability. High frequency causes interannual variation in recruitment. Low frequency causes changes in recruitment, growth, abundance, and distribution on a decadal (or multidecadal) scale. On top of this, comes man's fishing activities affecting abundance.

Eutrophication and pollution in a coastal system

The west coast of Sweden borders towards the Skagerrak, which is in open communication with the North Sea and the northeast Atlantic. The coastal water is strongly stratified with a surface layer coming from the Baltic Sea, an intermediate layer from the North Sea and in the deep Atlantic water from north of the British Isles (e.g. Svansson, 1975; Rodhe, 1996). Pollution may enter into the Baltic water from sources in the Baltic or as the water passes northward along the west coast of Sweden and the east coast of Denmark. It may also reach the Swedish coast via the Jutland current from the North Sea. Transfer of pollutants via atmospheric transport from surrounding countries in western and central Europe into the Kattegat-Skagerrak surface water is also of major importance (e.g. Preston, 1992; North Sea Task Force, 1993 a, b).

The coastal ecosystem is stressed because of eutrophication (Rosenberg et al., 1996) which has

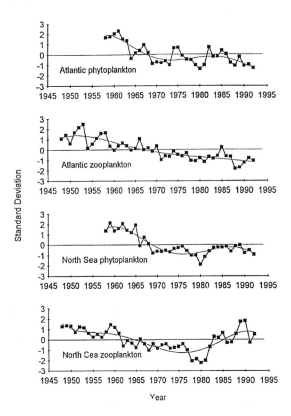

Figure 10. Between 1950 and 1990 the biomass of both phyto- and zooplankton decreased in the north-east Atlantic with about two standard deviations. The trend was the same in the North Sea for phytoplankton while the zooplankton recovered in the 1980's. (From Dickson *et al.*, 1988, and by courtesy of the late John Gamble.)

been documented for more than 20 years. This causes increase of benthic biomass in well ventilated deeper areas (e.g. Josefson, 1990; Rosenberg et al., 1987) and decrease in number of species and sometimes lack of oxygen in less well ventilated deep areas, like in fjords (e.g. Josefson & Rosenberg, 1988; Andersson, 1996). In shallow bays the effect is seen in excessive growth of filamentous algae (Pihl et al., 1995; Pihl et al., 1996).

Pollutants like PCBs (polychlorinated biphenyls) and DDT are present in the ecosystem, although at relatively low levels (Ahnoff & Josefsson, 1975; Blomkvist et al., 1992) as are heavy metals (e.g. Cato, 1983; Dave & Nilsson, 1994).

Effects of eutrophication on the planktonic system seem to vary from year to year, partly depending on variation in precipitation and run off from land each year (e.g. Lindahl et al., 1993) but the system is also heavily dependent on water exchange with the outer Skagerrak (Lindahl & Hernroth, 1983, 1988).

The changes taking place in the ecosystems on the Swedish west coast are thus easy to blame on pollution and eutrophication, i.e. man's impact. Is this then the whole truth? Recent analyses of the last 10 years of data on changes in benthic communities along the coast at depths shallower than 50 m, at about 50 m and at 100 m have revealed a striking co-variation of total softbottom macrofaunal abundance (Tunberg & Nelson, *pers. comm.*) and the North Atlantic Oscillation Index (Mann & Lazier, 1991) (Figure 11). This clearly indicates the close connection between climate variability over an ocean basin and coastal faunal response whether this is in shallow or deep waters. On top of such natural variability there is the effect of human activities, but one has to be careful to estimate the relative role of the two factors. It also shows that long term series of data are needed to have a chance to bring in the natural variations. Preferable monitoring activities should not be shorter than 20 years.

Conclusion

The complex interaction between natural variability and effects of human activities calls for careful and long-term investigations. In many cases impact of man's activity is obvious, e.g. direct pollution in coastal waters, spreading of radioactive isotops, DDT, PCBs, and CFCs which have no natural sources, overfishing on stocks with limited abundance and distribution, but most often – and certainly when dealing with the open ocean and long-term changes – the picture becomes complex. Above all natural changes on decadal and multidecadal time-scales and their causes are difficult to discern. To understand the complex picture, data from large-scale, international investigations are needed. Some programs are already in operation through the World Climate Research Program (WCRP), such as the World Ocean Circulation Experiment (WOCE) and the Tropical Ocean and the Global Atmosphere (TOGA), or through projects of the International Geosphere-Biosphere Programme (IGBP), the Scientific Committee on Oceanic Research (SCOR), and the Intergovernmental Oceanographic Commission (IOC), e.g. the Joint Global Ocean Flux Study (JGOFS) and the Global Ocean Ecosystem Dynamics project (GLOBEC). Besides data collection the efforts are directed towards modelling with the final aim of being able to predict future developments under different scenarios (Anon., 1990; Anon., 1997).

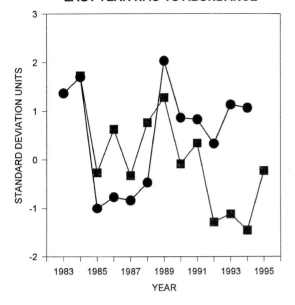

A

LAG1 YEAR NAO VS ABUNDANCE

Such research programs will feed their results into a Global Ocean Observing System (GOOS), which parallels the Global Climate Observing System (GCOS), and which will be responsible for the long-term monitoring activities.

National efforts will play an absolutely crucial role in these activities, and in the end the final results will depend on the quality assurance of the data, i.e. how well the data collection is made and controlled, and how the data are analyzed (Spellerberg, 1991).

Acknowledgments

I am greatly indepted to many collegues, particularly within the GLOBEC scientific community, who have helped in supplying information about pertinent literature and observations, e.g. Jürgen Alheit, Lars Hernroth, Mats Olsson, Brian Rothschild, and Jan Thulin. Special mention should be given to Björn Tunberg and Walter G. Nelson who allowed me to use unpublished material.

References

Aebischer, N. J., J. C. Coulson & J. M. Colebrook, 1990. Parallel long-term trends across four marine trophic levels and weather. Nature 347: 753–755.

Ahnoff, M. & B. Josefsson, 1975. Polychlorinated biphenyls (PCB) in Göta River water. Ambio 4: 172–174.

Andersson, L., 1996. Trends in nutrient and oxygen concentrations in the Skagerrak-Kattegat. J. Sea Res. 35: 63–71.

Anon, 1990. The International Geosphere-Biosphere Programme: a study of global change. The initial core projects. IGBP Rep. 12: 242 pp.

Anon, 1994. International GLOBEC small pelagic fishes and climate change program,. Report of the first planning meeting. GLOBEC Rep. 8: 1–72.

Anon, 1997. GLOBEC Science Plan. GLOBEC Report 9, IGBP Report 40, IGBP and SCOR: 82 pp.

Arrhenius, E., 1992. Population, development, environmental disruption – an issue on efficient natural-resource management. Ambio 21: 9–13.

Avaria, S., 1985. Efectos de El Niño en las pesquerías del Pacífico Sureste. Invest. Pesq. Chile 32: 101–116.

Barnola, J. M., D. Raynaud, Y. S. Korotkevich & C. Lorius, 1987. Vostok ice core provides a 160 000 year record of atmospheric CO_2. Nature 329: 408–414.

Barry, J. P., C. H. Baxter, R. D. Sagarin & S. E. Gilman, 1995. Climate-related. Long-term faunal changes in a California rocky intertidal community. Science 267: 672–675.

Baumgartner, T. R., A. Soutar & V. Ferreira-Bartrina, 1992. Reconstruction of the history of pacific sardine and northern anchovy popuations over the past two millennia from sediments of the Santa Barbara Basin, California. Calif. Coop. Oceanic Fish. Invest. Rep. 33: 24–40.

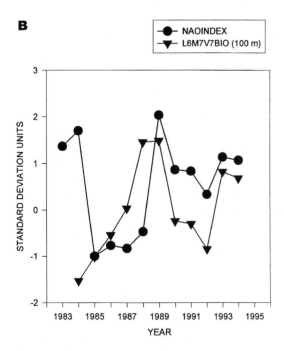

Figure 11. Covariation between the North Atlantic Oscillation Index (NAOIndex) and the total abundance of the benthic soft-bottom macrofauna off the Swedish west coast in waters shallower than 50 m (A) and at 100 m (B). (By courtesy of B. G. Tunberg and W. G. Nelson.)

Beddington, J. R. & R. M. May, 1982. The harvesting of interacting species in a natural ecosystem. Sci. Am., Nov. 1982: 42–49.

Bjerknes, J., 1969. Atmospheric teleconnections from the equatorial Pacific. Monthly Weather Rev. 97 (Jan.): 163–172.

Blomkvist, G., A. Roos, S. Jensen, A. Bignert & M. Olsson, 1992. Concentrations of sDDT and BCB in seals from Swedish and Scottish waters. Ambio 21: 539–545.

Bond, G. C., 1995. Climate and the conveyor. Nature 377: 383–384.

Brock, R. C., 1984. El Niño and world climate: piecing together the puzzle. Environment 26 (April): 14–20, 37–39.

Brodeur, R. D. & D. M. Ware, 1992. Long-term variability in zooplankton biomass in the subarctic Pacific Ocean. Fish. Oceanogr. 1: 32–38.

Broecker, W. S., 1991. The great ocean conveyor. Oceanogr. 4: 79–89.

Broecker, W. S. & G. H. Denton, 1990. What drives glacial cycles? Sci. Am. 262: 43–50.

Carrasco, S. & O. Lozano, 1989. Seasonal and long-term variations of zooplankton volumes in the Peruvian Sea 1964–1987: 82–85. In Pauly, D., P. Muck, J. Mendo & I. Tsukayama (eds), The Peruvian upwelling ecosystem: Dynamics and Interactions, Callao, Peru, 1987 ICLARM Conf. Proc. 18: 483 pp.

Cato, I., 1983. Tungmetallbelastningen i västerhavets sediment. Medd. Havsfiskelaboratoriet Lysekil 292: 12–34 (English summary).

Critchley, A. T., W. F. Farnham, T. Yoshida & T. A. Norton, 1990. A bibliography of the invasive alga Sargassum muticum (Yendo) Fensholt (Fucales; Sargassaceae). Bot. Mar. 33: 551–562.

Dave, G. & E. Nilsson, 1994. Sediment toxicity in the Kattegat and Skagerrak. J. aquat. Ecosyst. Health 3: 193–206.

Dickson, R. R., P. M. Kelly, J. M. Colebrook, W. S. Wooster & D. H. Cushing, 1988. Winds and production in the eastern Atlantic. J. Plankton Res. 10: 151–169.

Haigh, J. D., 1996. The impact of solar variability on climate. Science 272: 981–984.

Halpert, M. S., G. D. Bell, V. E. Kousky & C. F. Ropelewski (eds), 1994. Fifth annual climate assessment 1993. Climate Analysis Center, Camp Springs, Md, 111 pp.

Höglund, H., 1972. On the Bohuslän herring during the great herring fishery period in the eighteenth century. Inst. Mar. Res., Lysekil (Sweden), Biol. Rep. 20: 1–86.

Johannessen, O. M., M. Miles & E. Bjørgo, 1995. The Arctic's shrinking sea ice. Nature 376: 126–127.

Jones, E. B. G. & W. F. Farnham, 1973. Japweed: new threat to British coast. New Scientist 60: 394–395.

Josefson, A. B., 1990. Increase of benthic biomass in the Skagerrak-Kattegat during the 1970s and 1980s. Mar. Ecol. Prog. Ser. 66: 117–130.

Josefson, A. B. & R. Rosenberg, 1988. Long-term soft-bottom faunal changes in three shallow fjords, west Sweden. Neth. J. Sea Res. 22: 149–159.

Kasai, A., M. J. Kishi & T. Sugimoto, 1992. Modeling the transport and survival of Japanese sardine larvae in and around the Kuroshio current. Fish. Ocean. 1: 1–10.

Kawasaki, T., 1991. Long-term variability in the pelagic fish populations. In Kawasaki, T., S. Tanaka, Y. Toba & A. Taniguchi (eds). Long-term variability of pelagic fish populations and their environment. Pergamon Press: 47–60.

Kerr, R. A., 1994. Did Pinatubo send climate-warming gases into a dither? Science 263: 1562.

Kobayashi, M. & K. Kuroda, 1991. Estimation of main spawning grounds of the Japanese sardine from a viewpoint of transport condition of its eggs and larvae. In Kawasaki, T., S. Tanaka, Y. Toba & A. Taniguchi (eds). Long-term variability of pelagic fish populations and their environment. Pergamon Press: 109–116.

Kondo, K., 1991. Interspecifik relation between Japanese sardine and anchovy populations that reflects the essential mutual relation between fluctuation mechanisms of the two species based on 'organism-environment' coupling. In Kawasaki, T., S. Tanaka, Y. Toba & A. Taniguchi (eds). Long-term variability of pelagic fish populations and their environment. Pergamon Press: 129–134.

Laws, R., 1983. Antarctica: a convergence of life. New Scientist, Sept. 1983: 608–616.

Lindahl, O. & L. Hernroth, 1983. Phyto-zooplankton community in coastal waters of western Sweden – An ecosystem off balance? Mar. Ecol. Prog. Ser. 10: 119–126.

Lindahl, O. & L. Hernroth, 1988. Large-scale and long-term variation in the zooplankton community of the Gullmar fjord, Sweden, in relation to advective processes. Mar. Ecol. Prog. Ser. 43: 161–171.

Lindahl, O., G. Persson & H. Olsson, 1993. Eutrofiering av svenska kustområden samt omgivande hav: tillstånd, utveckling, orsak och verkan. Swedish Environmental Protection Agency, Solna. Report 4151: 1–85.

Loganathan, B. G. & K. Kannan, 1991. Time perspectives of organochlorine contamination in the global environment. Ambio 22: 582–584.

Mann, K. H. & J. R. N. Lazier, 1991. Dynamics of marine ecosystems. Biological-physical interactions in the oceans. Black. Sci. Publ., Boston: 466 pp.

Morel, P., P. Hulm & N. Meith, 1990. Global climate change, a scientific review presented by the World Climate Research Programme (WCRP). WMO Secretariat, Geneva, 35 pp.

North Sea Task Force, 1993a. North Sea quality status report, 1993. Oslo and Paris Commissions, London: 1–132.

North Sea Task Force, 1993b. North Sea subregion 8. Assessment Report 1993 State. State Pollution Control Authority, Oslo: 1–79.

Norton, T. A. & M. R. Benson, 1983. Ecological interaction between the brown seaweed Sargassum muticum and its associated fauna. Mar. Biol. 75: 169–177.

Ohlin, G., 1992. The population concern. Ambio 21: 6–9.

Overpeck, J., 1996. Warm climate surprises. Science 271: 1820–1821.

Pihl, L., I. Isaksson, H. Wennhage & P.-O. Moksnes 1995. Recent increase of filamentous algae in shallow Swedish bays: effects on the community structure of epibenthic fauna and fish. Neth. J. aquat. Ecol. 29: 349–358.

Pihl, L., G. Magnusson, I. Isaksson & I. Wallentinus, 1996. Distribution and growth dynamics of ephemeral macroalgae in shallow bays on the Swedish west coast. J. Sea Res. 35: 169–180.

Preston, M. R., 1992. The interchange of pollutants between the atmosphere and oceans. Mar. Pollut. Bull. 24: 477–483.

Quinn, W. H., V. T. Neal & S. E. Antunez de Mayolo, 1987. El Niño occurrences over the past four and-a-half centuries. J. geophys. Res. 92: 449–461.

Robock, A., 1996. Stratospheric control of climate. Science 272: 972–973.

Rodhe, J., 1996. On the dynamics of the large-scale circulation of the Skagerrak. J. Sea Res. 35: 9–21.

Rosenberg, R., J. S. Gray, A. B. Josefson & T. H. Pearson, 1987. Petersen's benthic stations revisited. II. Is the Oslofjord and eastern Skagerrak enriched? J. exp. mar. Biol. Ecol. 105: 219–251.

Rosenberg, R., I. Cato, L. Förlin, K. Grip & J. Rodhe, 1996. Marine environment quality assessment of the Skagerrak – Kattegat. J. Sea Res. 35: 1–8.

Rothschild, B. J., 1991. On the causes for variability of fish populations – the linkage between large and small scales. In Kawasaki, T., S. Tanaka, Y. Toba & A. Taniguchi (eds). Long-term variability of pelagic fish populations and their environment. Pergamon Press: 367–376.

Rothschild, B. J., 1993. Fishstock fluctuations as indicators of multidecadal fluctuations in the biological productivity of the ocean. In Beamish, R. J. (ed.), Climate change and northern fish populations, Can. Spec. Publ. Fish. aquat. Sci. 121: 203–211.

Rueness, J., 1989. *Sargassum muticum* and other Japanese introduced macroalgae: biological pollution of European coasts. Mar. Pollut. Bull. 20: 173–175.

Sharp, G. D. & D. R. MaLain, 1993. Fisheries, El-Niño-Southern Oscillation and upper-ocean temperature records: an eastern Pacific example. Oceanography 6: 13–22.

Spellerberg, I. F., 1991. Monitoring ecological change. Cambridge University Press, 334 pp.

Strömberg, J. O., L. G. Anderson, G. Björk, W. N. Bonner, A. C. Clark, A. L. Dick, W. Ernst, D. W. S. Limbert, D. A. Peel, J. Priddle, R. I. L. Smith & D. W. H. Walton, 1990. State of the marine environment in Antarctica. UNEP Regional Seas Reports and Studies No. 129, 34 pp.

Svansson, A., 1975. Physical and chemical oceanography of the Skagerrak and the Kattegat. I. Open sea conditions. Fishery Board of Sweden, Institute of Marine Research, Rep. 1: 1–88.

Venrick, E. L., J. A. McGowan, D. R. Cayan & T. L. Hayward, 1987. Climate and chlorophyll *a*: Long-term trends in the central north Pacific Ocean. Science 238: 70–72.

Voytek, M. A., 1989. Ominous future. Under the ozone hole: assessing biological impacts in Antarctica. Environmental Defense Fund, Inc., Washington, D.C., 69 pp.

Wängberg, S.-Å., J.-S. Selmer, N. G. A. Eklund & K. Gustavson, 1996. UV-B effects on nordic marine ecosystem – a literature review. TemaNord 1996: 515, Nordic Council of Ministers, Copenhagen 1996, 45 pp.

World Meteorological Organization, 1994. WMO Statement on the status of the global climate in 1993. WMO-No. 809, 20 pp.

Wuethrich, B., 1995. El Niño goes critical. New Scientist (Febr.): 32–35.

Zaika, V. E., 1994. The drop of anchovy stock in the Black Sea: result of biological pollution? FAO Fish. Rep. 495: 78–83.

Hydrobiologia **352**: 195–200, 1997.
Y.-S. Wong & N. F.-Y. Tam (eds), Asia–Pacific Conference on Science and Management of Coastal Environment.
©1997 *Kluwer Academic Publishers. Printed in Belgium.*

Toxic events in the northwest Pacific coastline of Mexico during 1992–1995: origin and impact

José Luís Ochoa, Arturo Sánchez-Paz, Ariel Cruz-Villacorta, Erick Nuñez-Vázquez &
Arturo Sierra-Beltrán
The Center for Biological research, CIBNOR, Box 128, La Paz, baja California Sur, 23000, México

Key words: Marine biotoxins, red times, harmful algal blooms, domoic acid, okadaic acid, TTX

Abstract

Previously considered as toxin-free, the Baja California Peninsula has witnessed several toxic algal blooms during the past three years. Apparently these 'red-tide' phenomena's outbreaks are not linked to any human related activity. This may just reflect better detection and training. Such events may be periodical and natural rather than induced. The most common types of marine toxins have been detected along the coast of the Peninsula and neighboring waters by mouse bioassay and chromatographic techniques. These are: Tetrodotoxin (TTX), Amnesic Shellfish Poison (ASP), Paralytic Shellfish Poisons (PSP), Diarrhetic Shellfish Poisons (DSP) and even Ciguatera (CFP), which are related to the presence of organisms of *Prorocentrum* sp. and *Alexandrium* sp. groups, and the diatom *Pseudonitzschia* sp. among others. There are also some indications about different kinds of TTX in the puffer fish of the region, and reasons to believe that we are facing a quite different pattern in toxic components, since PSP toxic potency (defined as the number of mouse units per gram (MU/g) of shellfish meat) is very high in spite of low dinoflagellates cell density registered. The ecological and social impact of the above has been considerable, with mass deaths of shellfish, seagulls, dolphins and turtles, and even some human casualties. The locally registered toxicity records: PSP found in one single fanshell reaches to 23 000 MU/100 g of tissue as determined by the mouse bioassay and, on a different event, two persons killed after ingesting puffer fish fillet. The largest reservoir of commercial marine organisms in Mexico is precisely the Northwest coast of the country and important plans for building large harbors and develop aquaculture areas are in progress. Therefore, a monitoring program is essential for an adequate management of such resources. Considering the large extension of the Peninsula (about 1600 km) and, at this time, the lack of efficient communication means and scarce population, the implementation of such monitoring programs presents a big challenge.

Introduction

Outbreaks of mass shellfish deaths on commercial fishing grounds during 1991 in Bahía Concepción (BACO) and 1992 in Bahía Magdalena (BAMA), in the state of Baja California Sur, México, as well as a mass mortality of fish, birds and sea mammals along the coast of *B. Magdalena* in 1992, highlighted the need for the implementation of a program for monitoring 'red tide' occurrences and their impact. To obtain a picture of the prevalence and distribution of marine biotoxins, and to delimit the areas at risk, two sampling programs were carried out at Bahía Concepción,

and Bahía Magdalena, where massive mortalities of Catarina scallop, *Argopecten ventricosum* (Sowerby, 1842) had occurred. The nature and amount of toxins in marine bivalves, and the identification and quantification of phytoplankton species present in the area, were determined. In addition, in this report we describe other algal blooms, of different nature and magnitude, that occurred in different areas along the coast of the Peninsula during the past three years. All this demonstrates the need for a widespread monitoring program along the Peninsula.

Description of sites studied

The Northwest Pacific coast of Mexico extends from Latitude 23° to 33° N, and from Longitude 106° to 117° W, and includes the Peninsula of California, the second largest in the world (about 1600 km length), numerous islands and the Eastern coast of the Gulf of California, making a total coast line well above 8000 km.

The sampling program so far implemented was concentrated into two areas along the Peninsula: On the Pacific, Bahía Magdalena (BAMA), from Lat. 24° 15' to 25° 15' N and from Long. 111° 30' to 112° 15' W, which is an area of about 1390 km² constituted by a series of coastal channels of low depth. And, on the Gulf of California, Bahía Concepción (BACO), from Lat. 26° 30' to 26° 55' N and from Long. 111° 40' to 112° 00' W, with an area of about 400 km2, is a shallow basin with the maximum depth at 30 m in the widest portion. Other sites where important 'red tides' phenomena have been recently observed include Ojo de Liebre at Lat. 27° 45' N and Long. 114° 14' W, Vizcaino at Lat. 27° 45' N and Long. 113° 41' W, San Hipolito at Lat. 27° 00' N, and Long. 114° 00' W, Alijos Rocks at Lat. 24° 57' N, and Long. 115° 45' W, Cabo San Lucas at Lat. 22° 30' N, and Long. 110° 30' W, La Paz Bay, located at Lat. 24° 15' N and Long. 110° 15' W, Loreto at Lat. 25° 45' N, and Long. 111° 25' W, and San Felipe at Lat. 31° 15' N, and Long. 114° 45' W (Figure 1).

Material and methods

Frequency of sampling

The frequency of sampling at BACO and BAMA was variable due to the high cost of sampling activities. During the first part of 1992, samples were collected each two weeks, then monthly for the rest of the year. During 1993, samples were taken each three months, during the peak of each season (winter, spring, summer and autumn). In 1994 samples were collected monthly for the first 6 months, and then every 4 months in 1995. The rest of the samples were obtained when a blooming event or its sequel was present.

Biological data

Phytoplankton species and density were estimated using Uthermöl's method (1958).

Toxicological data

The toxins were deterrnined by mouse bioassay according to their nature employing several different procedures: PSP and TTX by the AOAC method (1984); DSP by the Yasumoto method (1978); ASP by using a modification of the AOAC method proposed by Iverson (1989); Ciguatoxin (CTX) by an adaptation of the method of Lewis (1992); and cyanobacterial toxins (Hepatotoxins) by the method described by Carmichael (1986). Animals were handled according to UFAW guidelines (1976). Toxin analysis was carried out by means of High Pressure Liquid Chromatography (HPLC) techniques described for each group of toxins: PSP by Thielerth (1991) and Lawrence (1991) methods, TTX by Yasumoto (1985), DSP by Shen (1991), ASP by Quilliam (1995), and CTX by Legrand (1992).

Meteorological data

Meteorological and oceanographic data were obtained from satellite images provided by the Climate Diagnostics Bulletin/NOM, National Weather Service Forecast Office. USA.

Chemicals

Domoic acid (DA) and okadaic acid (OA) were from Sigma Chemicals Co., St. Louis, Mo, USA. Dinophysistoxin-1 (DTX-1), saxitoxin (SAX), dc-saxitoxin (dc-SAX) and neo-saxitoxin (neo-SAX) were donated by the Measurements and Testing Programme, Commission of the European Communities. As qualitative standards, cultures of the cyanobacteria *Aphanizomenon flos-aquae* (Hearey & Jaworsky, 1977) for SAX and neo-saxitoxin and, the dinoflagellate *Alexandrium tamarense* (Balech, 1985) for gonyautoxin-2 (GTX-2) and gonyautoxin-3 (GTX-3) were used.

Results

Sampling at Bahía Concepción (BACO) and Bahía Magdalena (BAMA)

The sampling program for marine biotoxins on the Northwest region of Mexico, confirmed that BACO presents annual cycles of toxicity of the PSP type, mainly during spring but some times a second increase in toxicity is evident during autumn. Previous data

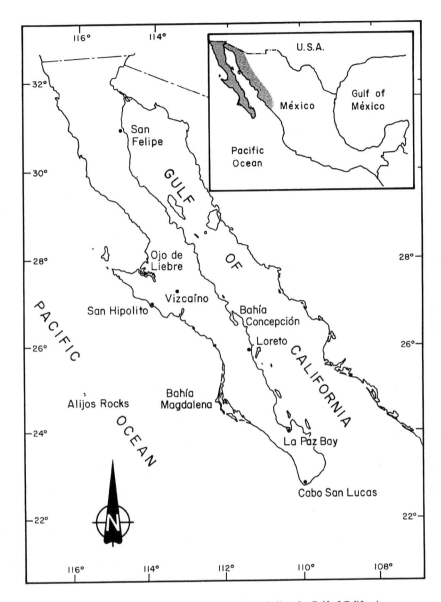

Figure 1. Northwest Pacific coast of Mexico, including the Gulf of California.

obtained in 1992 drew our attention to the fact that high toxicity it is often evident at BACO (caused mainly by PSP carbamates rather than by PSP type B and C), while there is not an evident increase of the phytoplanktonic communities (algal blooms or 'red-tide') (Sierra-Beltrán et al., 1996). During May 1992, an extremely high PSP toxicity value was recorded for a single fanshell, *Pinna rugosa* (Sowerby, 1835) which showed as much as 23 000 MU/100 g. Also, although not quantified, the presence of DSP toxins, okadaic acid (OA) and dynophysistoxin-1 (DTX-1), has been detected in 1992, 1994 and 1995 in shellfish at BACO, with apparently higher values, according to the mouse bioassay, occurring together with the spring peak of PSP toxicity (Figure 2).

At BAMA, on the other hand, only low PSP toxicity was observed in shellfish during spring, barely reaching the detection level of the mouse bioassay, while no DSP has been detected. The ASP-mouse bioassay however, suggested the presence of domoic acid (DA) in shellfish of BAMA during winter in 1994 and 1995.

198

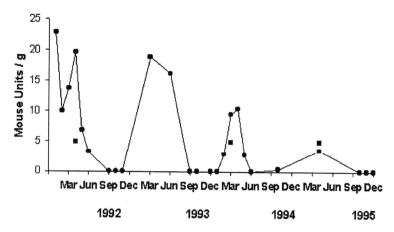

Figure 2. Cycles of shellfish toxicity at BACO as detected by mouse bioassay. Circles: quantified levels of PSP. Squares: time where DSP activity was detected.

Other toxic events

The presence of cyanobacterial toxins in this region has been confirmed with an outbreak bloom of *Oscillatoria erythraea* (= *Trichodesmium erythraeum*) (Ehrenberg, 1830) in La Paz Bay on April 1993. In this case, the mouse bioassay allowed us to observe the hepatotoxicity of water samples containing the cyanobacteria. Interestingly, one month later, a similar but non-toxic event occurred at BACO and later, in 1995, we were again able to isolate toxic strains of such cyanobacteria, without blooming, from the water body of BACO. This finding suggests that toxicity is not an intrinsic property of the microorganism and its conversion from a non-toxic strain into a toxic one remains a mystery, this could be an issue of life cycle as well as strain.

In a different episode, on May 1993, the crew of a fishing boat became ill after eating grouper fish (*Serranidae* sp. and *Labridae* sp.) caught close to Alijos Rocks, 480 km offshore from Bahía Magdalena in Baja California Sur. The symptoms indicated ciguatera poisoning. Extraction from the fish flesh remains with organic solvents yielded a toxic extract which induced liver enlargement without bleeding and intestine swelling due to liquid accumulation, as well as shrinking of the spleen, in inoculated animals. All the mice suffered severe diarrhea leading to considerable weight loss (Lechuga-Devéze & Sierra-Beltrán, 1995). HPLC analysis of hexane washed extracts showed a peak with a mobility resembling that of ciguatoxin (Legrand, *loc. cit*), notwithstanding that no internal standard toxin was available and therefore the identification remains uncertain. Additional fish poisoning

events resembling ciguatera during the last three years at different locations in the Peninsula let us to propose the presence of this toxin in the region. In all cases, poisoning has been associated with consumption of livers from big groupers and snappers (*Serranidae* and *Lutjanidae*), while the livers of young or small ones appear innocuous. Extracts obtained from such large specimens were highly toxic to mice, producing diarrhea, dizziness, breathing difficulties and sometimes death by respiratory arrest.

Globally a more common fish poisoning is known to occur after consumption of the puffer fish. Four of the five species found in the Baja California Peninsula (Bullseye puffer *Sphaeroides annulatus* (Jenyns, 1843); lobeskin puffer *Sphaeroides lobatus* (Steindachner, 1870); Guineafowl puffer *Arothron meleagris* (Bloch & Schneider, 1801) black and golden phases; sponed sharpnose puffer *Canthigaster punctatissima* (Günther, 1870) and *Sphaeroides nsp.*) show high levels of toxicity at least in one organ (liver, skin, flesh, mucus, gonads or guts). In *A. meleagris* (golden phase), for example, a decreasing gradient of toxicity was observed from mucus>flesh>guts>gonads>liver, while in *Sphaeroides* nsp. (a new and unclassified organism in the region) the pattern was liver>gonads/gut>mucus>flesh. In general, the toxins present are TTX, 4-epiTTX and 4,9-anhydroTTX, TTX being the dominant toxin, followed by 4,9-anhydroTTX and 4-epiTTX scantily represented. The exception is the mucus of *A. meleagris*, in which 4,9-anhydroTTX is the dominant toxin, followed by TTX and 4-epiTTX.

Table 1. Events associated with marine toxins in the Northwest of Mexico during 1991–1996.

Time	Place	Organism	Species affected	Toxin determined
1991/Nov	Bahia Concepcion	?	Several tons of shellfish loss	Not done
1992/Jan & Apr (2 times)	Ojo de Liebre	?	Several tons of shellfish loss	Not done
1992/Oct	Bahia Magdalena	Dinoflagellate	Dolphins, sea lions, sea birds, fish & turtles loss	Not done
1992/Apr	La Paz	*Oscillatoria erythraea*	None	Hepatotoxins
1993/May	Bahia Concepcion	*Oscillatoria erythraea*	None	No toxic
1993/May	Alijos Rocks	Serranidae & Labridae fish	7 Humans ill, no casualties	Ciguatoxins *
1994/Apr & Jun (2 times)	La Paz (Lohmann, 1908)	*Mesodinum rubrum*	None	No toxic
1994/Jun	San Hipolito	*Gymnodinium sanguineum* (Hirasaka, 1922)	Fish & sea birds loss	No PSP
1995/Jan	San Felipe	?	Dolphins, whales, sea birds loss	Not done
1995/Jun	Vizcaino	*Sphaeroides* sp.	2 Humans dead	Not done
1995/Oct	La Paz	?	Fish loss (Tetrodontiforme)	Highly potent liposoluble toxin in sampled shellfish
1996/Jan	Atil, Sonora	*Microcystis*/LPPB	Fish loss (*Oreochromis* sp.)	Mucus & scum
1996/Jan	Cabo San Lucas	*Pseudonitzchia* sp.	>150 Brown pelicans loss	Domoic acid **
1996/Feb	Loreto	*Noctiluca scintillans* (Ehrenberg, 1834) & *Pseudonitzchia* sp.	None	No toxic/Ammonia production
1996/Feb	Sta. Ma. del Oro, Nayarit	Cyanobacteria LPPB	Fish loss (*Oreochromis* sp.)	Oxygen depletion
1996/Mar	Cabo San Lucas	Cyanobacteria LPPB & *Chatonella* sp.?	Bentonic fish & fan corals loss	? Probably a secondary infection

* Lechuga-Devéze & Sierra-Beltran, (1995).
** Sierra-Beltran et al. (1997).

A summary of 'red-tide' events, and/or mass mortalities of fish, sea-birds, and other marine animals is shown in Table 1. It may be concluded that this phenomenon is a common and natural one in the area and, therefore, fishing practices and aquaculture activities should take into consideration the risk involved in consuming contaminated organisms: a permanent sampling program needs to be established.

Discussion

We have been able to demonstrate that at Bahía Concepción (BACO), in the Peninsula of Baja California (one of the selected places for the development of aquacultural sites for shellfish), the periodic appearance of PSP toxicity imposes a risk factor that calls for a permanent monitoring program in this area. It is noteworthy that in spite of the high level of toxicity in some shellfish samples, no cases of human poisoning have been reported. This could be explained by the local custom of eating only the callus organ (abductor muscle) and not the viscera of the organisms, where most of the toxin usually concentrates. On the other hand, it is difficult to assess how many of the 300 000 diarrhea cases registered annually in the Southern State of Baja California were due to DSP, since we lack the resources for a correct diagnosis. The knowledge of the exposure rate to these toxins should be highlighted, since the number of cancer patients in the region is well above the national average index and is well known that DSP toxins possess tumor inducing properties (Fujiki et al., 1990).

The results obtained from the sampling at Bahía Magdalena (BAMA) are quite different and of less concern than those of Bahía Concepción (BACO). Bahía Magdalena has been already considered an ideal site for aquaculture exploitation and huge projects are underway. Since PSP seems only a sporadic event, and no DSP toxicity has been detected so far, the site appears suitable for aquaculture practice. Yet, it should be borne in mind that the presence of domoic acid has been detected recently during winter months, and although toxicity values are well below the risk limit; a continuous monitoring of such phenomena is recommended (Sierra-Beltrán et al., 1997).

We have also concluded that 'red tides' are not an isolated phenomenon in the Baja California Peninsula. The events appear to present some periodicity and are specially frequent in cold months, in which temperature, marine currents, and nutrients favor the blooms. There is no evidence that such blooms are due to agriculture or man-related activities: they are most likely of natural occurrence and should be considered difficult to prevent or avoid.

Acknowledgments

TTX HPLC chromatography was developed in the laboratory of Prof. Yasumoto under the supervision of Dr M. Yotsu-Yamashita.

References

AOAC, 1984. Official methods of analysis. 14th edn. AOAC, Arlington (VA), Sects. 10.086–18.092.

Carmichael, W. W., 1986. Isolation, culturing and toxicity testing of toxic freshwater cyanobacteria (blue-green algae). In Silow, Y. G. (ed.), Fundamental Research in Homogeneous Cathalysis. Gordon & Breach Publ., New York 3: 1249–1262.

Fujiki, I., M. Suganuma, H. Suguri, S. Yoshizawa, K. Takagi, M. Nakayasu, M. Ojika, K. Yamada, T. Yasumoto, R. Moore & T. Sugimura, 1990. New tumor promoters from marine natural products. In Hall, S. & G. Strichartz (eds), Marine Toxins: Origin, Structure, and Molecular Phammacology. Am. Chem. Soc. Press. Washington, D.C. 418: 232–240.

Iverson, F., J. Truelove, E. Nera, L. Tryphonas, J. S. Campbell & E. Look, 1989. Domoic acid poisoning and mussel associated intoxication: preliminary investigation into the response of mice and rats to toxic mussel extracts. Food. Chem. Toxicol. 27: 377–384.

Lawrence, J. F. & C. Menard, 1991. Determination of paralytic shellfish poisons by pre-chromatographic oxidation and HPLC. In Fremy, J. M. (ed.), Proceedings of Symposium on Marine Biotoxins, CNEVA, Paris: 127–129.

Legrand, A. M., M. Fukui, P. Cruchet, Y. Ishibashi & T. Yasumoto, 1992. Characterization of ciguatoxins from different fish species and wild Gambierdiscus toxicus. In Tosteson, T. R. (ed.), Proceedings of the Third International Conference on Ciguatera Fish Poisoning. Polysciences, Quebec: 25–32.

Lewis, R. J., M. Sellin, R. Street, M. J. Holmes & N. C. Gillespie, 1992. Excretion of ciguatoxin from moray eels (murenidae) of the central Pacific. In Tosteson, T. R. (ed.), Proceedings of the Third International Conference on Ciguatera Fish Poisoning. Polysciences, Quebec: 131–143.

Lechuga-Devéze, C.& A. P. Sierra-Beltrán, 1995. Documented case of ciguatera on the mexican Pacific coast. Nat. Toxins 3: 415–418.

Quilliam, M. A., M. Xie & W. R. HArdstaff, 1995. Rapid Extraction and cleanup for liquid chromatographic determination of domoic acid in unsalted seafood. J. AOAC 78: 543–554.

Shen, J. L., G. Ganzlin & B. Luckas, 1991. HPLC determination of DSP toxins. In Fremy, J. M. (ed.), Proceedings of Symposium on Marine Biotoxins, CNEVA, Paris: 101–106.

Sierra-Beltrán, A. P., M. L. Morquecho-Escamilla, C. Lechuga-Devéze & J. L. Ochoa, 1996. PSP monitoring program at Baja California Sur, Mexico. In Yasumoto, T. (eds), Harmful and Toxic Algal Blooms. IOC/UNESCO, Paris: 105–108.

Sierra-Beltrán, A. P., M. Palafox-Uribe, J. Grajales-Montiel, A. Cruz-Villacorta & J. L. Ochoa, 1997. Sea bird mortality at Cabo San Lucas, Mexico: evidence that toxic Pseudonitzchia sp. is spreading. Toxicon. 35.

Thielert, G., K. Peters, I. Kaiser & B. Luckas, 1991. HPLC determination of PSP toxins. In Strategies for Food Quality Control and Analytical Methods in Europe. Proceedings of EURO FOOD CHEM 6. Hamburg: 816–820.

Yasumoto, T. & T. Michishita, 1985. Fluorometric determination of tetrodotoxin by high performance liquid chromatography. Agricult. Biol. Chem. 49: 3077–3080.

Yasumoto, T., Y. Oshima & M. Yamaguchi, 1978. Occurrence of a new type of shellfish poisoning in the Tohoku district. Bull. Jap. Soc. Sci. Fish. 44: 1249–1255.

UFAW, 1976. Handbook for the Care and Management of Laboratory Animals. Union of Federations of Animal Welfare, Washington D.C. 575 pp.

Utermöl, H., 1958. Mitt. int. Ver. Theor. Angew. Limnol. 2: 1–38.

Hydrobiologia **352**: 201–206, 1997.
Y.-S. Wong & N. F.-Y. Tam (eds), Asia–Pacific Conference on Science and Management of Coastal Environment.

A biological survey of ballast water in container ships entering Hong Kong

K. H. Chu[1], P. F. Tam[1], C. H. Fung[1] & Q. C. Chen[2]
[1]*Department of Biology, The Chinese University of Hong Kong, Hong Kong*
[2]*South China Sea Institute of Oceanology, The Chinese Academy of Sciences, Guangzhou, China*

Key words: ballast water, container ship, introduced species, exotic species, Hong Kong

Abstract

The role of ballast water in the introduction of exotic species has recently received extensive attention. The aim of this study is to assess the importance of ballast water discharge as a vector for the introduction of exotic species into Hong Kong waters. Twelve ballast water samples were collected from 5 container ships entering Hong Kong between June 1994 and October 1995. The ballast water originated from ports on both sides of the Pacific Ocean. At least 81 species from 8 animal phyla and 5 protist phyla were found. Most of the major marine taxonomic groups were represented and many planktonic larval stages were included. Species richness in the ballast tanks decreased with the age of ballast water. Copepoda was the most diverse and abundant taxonomic group. The density of calanoid and cyclopoid copepods decreased with the age of ballast water, but that of harpacticoid copepods did not change significantly with time. Bivalve, crustacean, polychaete and ascidian larvae from ballast water samples were observed to settle in laboratory culture tanks. The mussel *Mytilopsis sallei* which was introduced to Hong Kong in 1980, was one of the bivalves that settled readily. Results of this study indicate that ballast water can be a major source for the introduction of exotic species to Hong Kong waters. Regulatory guidelines on the discharge of ballast water should be established.

Introduction

The introduction of exotic species has long been recognized to be a serious environmental problem (Elton, 1958). The introduction of rabbit to Australia and of Dutch elm disease to North America are two noticeable examples. While the invasion of exotic terrestrial organisms have been widely studied, extensive attention has not been paid to the invasion of aquatic organisms until the introduction of zebra mussels (*Dreissena* sp.) to the Great Lakes (Hebert et al., 1991; Griffiths et al., 1991; Nalepa & Schloesser, 1993) and of toxic dinoflagellates to Australia (Hallegraeff & Bolch, 1991, 1992), both of which have resulted in great economic loss. The release of ballast water from ships is known to be a major vector for the introduction of exotic aquatic species (Carlton, 1985 review). In addition to the examples cited earlier, ballast water discharge is believed to be responsible for the recent introduction of many exotic aquatic organisms, including plankton

(Cordell et al., 1992), benthic invertebrates (Buttermore et al., 1994; Gosliner, 1995), fish (Baltz, 1991; Pratt et al., 1992), fish parasites (Cone et al., 1994), and bacteria such as epidemic cholera (McCarthy & Khambaty, 1994). Risk assessment of ballast water discharge has been undertaken and regulatory guidelines have been considered and implemented in many countries (Hutchings, 1992; Locke et al., 1993; MacLeod, 1995).

Hong Kong is one of the busiest ports in the world and its coastal waters are highly susceptible to the invasion of exotic species. Morton (1987) reviewed the introductions of marine organisms to Hong Kong. No biological studies have been conducted on the ballast water in ships entering Hong Kong, in spite of the fact that similar studies have been conducted elsewhere (Williams et al., 1988; Hallegraeff & Bolch, 1992; Carlton & Geller, 1993; Kelly, 1993). The objective of this study is to investigate the diversity and abundance of organisms in the ballast water in container

202

ships entering Hong Kong and to assess the possibility of exotic species being successfully introduced into local waters. Container ships were chosen for this study because their large volume of ballast water would likely carry a fauna of higher diversity.

Materials and methods

A total of 12 ballast water samples were obtained between June 1994 and October 1995 from 5 container vessels under the NYK Line (Hong Kong) Ltd. All the vessels were of the same type with a tonnage of 43 kilotons and were travelling between the coasts of North America and Asia. The ballast water samples originated from the coastal waters of California, Washington, British Columbia, Japan and Taiwan, with some being mixtures from more than one of the above sites. The total capacity of ballast water of the vessels varied from 8800 to to 13 300 tonnes. The ballast tanks sampled were totally or partially full, containing 180–1000 m^3 of water to a depth of 1.5 to 12 m.

Samplings were made a few hours after the container ships arrived Hong Kong and docked in the Kwai Chung Container Port. In each sampling trip samples from one or two ballast tanks were taken from the ship. No sediment samples were collected because of the difficulties involved. In some cases, the two tanks contained water from the same origin ballasted on the same date and they were treated as a single sample. Ballast water samples were collected using a 80 μm plankton net. From each ballast tank, a sample for qualitative analysis was collected by repetitive tows in the tank to a total depth of about 20 m. Duplicate samples for quantitative analysis were collected by taking vertical tows from the bottom of tank at a speed of about 0.5 m s^{-1}. The quantitative samples were preserved in 90% ethanol. Temperature, salinity and dissolved oxygen of the water in the tank were measured. These parameters were also measured in water at the shipside in order to compare with those in the tanks.

In the laboratory, samples for qualitative analysis were transferred to rectangular culture tanks containing about 7 litres of 20 μm filtered seawater. In some cases, a mud bed was provided for the settlement of planktonic larvae. The mud was collected locally and frozen at −20 °C for 1 month before use. The cultures were usually maintained for several weeks. The diatom *Chaetoceros gracilis* at about 10^5 cells ml^{-1} and a formulated diet (Artificial Plankton-Rotifer, Ocean Star International Inc., Snowville, Utah, USA) at about 0.1 g l^{-1}

was added regularly to feed the organisms. Organisms in the tanks were sampled with a 75 μm sieve for observation and identification under an inverted microscope. Some organisms including those settled in the cultured tanks were preserved in 10% formalin or 90% ethanol for further taxonomic identification. Densities of copepods in quantitative samples were counted under the microscope.

Results and discussion

Age of the ballast water ranged from 1 day to 1 year. Temperature and salinity of water in the ballast tanks ranged from 24 to 31 °C and 29 to 39‰, respectively. Temperature of ballast water was always within 3 °C of the water at the side of the ship (23–29 °C). Salinity of water at the shipside ranged from 18 to 35‰. Lower salinity was caused by heavy rain as many samples were collected in the rainy season. Dissolved oxygen level of the ballast water and the water at the shipside ranged from 4.8 to 10.7 mg l^{-1} and 3.0 to 7.0 mg l^{-1}, respectively. Dissolved oxygen level from the shipside was often lower, but always within 3.5 mg l^{-1} of the corresponding ballast water samples.

Ballast water samples contained a wide variety of organisms including at least 81 species from 8 animal phyla and 5 protist phyla (Table 1). This was a conservative estimate as many planktonic stages of invertebrates could not be identified to the species level. Most of the major marine taxonomic groups were represented. Species richness of the ballast water sample was found to decrease with the age of the ballast water (Figure 1). As expected, the species diversity in a ballast tank depends very much on the location and seasonality of the water ballasted. Given the inherent differences in diversity when the water was ballasted, the present study shows a general decrease in species richness as the ballast water aged. Although similar trends have been documented by other investigators (Williams et al., 1988; Carlton, pers. comm.), our samples encompassed a much greater range of age of ballast water. The decrease in species richness most likely resulted from the adverse conditions in the ballast tanks for planktonic organisms, such as the absence of light, reduced food availability and the fluctuating physical and chemical variables, noticeably water temperature (Carlton, 1985). Nevertheless, a few species noticeably diatoms, protozoans and copepods, were able to survive under these conditions for up to a year in our most aged sample. This phenomenon may be related to the

Hydrobiologia **352**: 201–206, 1997.
Y.-S. Wong & N. F.-Y. Tam (eds), Asia–Pacific Conference on Science and Management of Coastal Environment.
©1997 *Kluwer Academic Publishers. Printed in Belgium.*

A biological survey of ballast water in container ships entering Hong Kong

K. H. Chu[1], P. F. Tam[1], C. H. Fung[1] & Q. C. Chen[2]
[1]*Department of Biology, The Chinese University of Hong Kong, Hong Kong*
[2]*South China Sea Institute of Oceanology, The Chinese Academy of Sciences, Guangzhou, China*

Key words: ballast water, container ship, introduced species, exotic species, Hong Kong

Abstract

The role of ballast water in the introduction of exotic species has recently received extensive attention. The aim of this study is to assess the importance of ballast water discharge as a vector for the introduction of exotic species into Hong Kong waters. Twelve ballast water samples were collected from 5 container ships entering Hong Kong between June 1994 and October 1995. The ballast water originated from ports on both sides of the Pacific Ocean. At least 81 species from 8 animal phyla and 5 protist phyla were found. Most of the major marine taxonomic groups were represented and many planktonic larval stages were included. Species richness in the ballast tanks decreased with the age of ballast water. Copepoda was the most diverse and abundant taxonomic group. The density of calanoid and cyclopoid copepods decreased with the age of ballast water, but that of harpacticoid copepods did not change significantly with time. Bivalve, crustacean, polychaete and ascidian larvae from ballast water samples were observed to settle in laboratory culture tanks. The mussel *Mytilopsis sallei* which was introduced to Hong Kong in 1980, was one of the bivalves that settled readily. Results of this study indicate that ballast water can be a major source for the introduction of exotic species to Hong Kong waters. Regulatory guidelines on the discharge of ballast water should be established.

Introduction

The introduction of exotic species has long been recognized to be a serious environmental problem (Elton, 1958). The introduction of rabbit to Australia and of Dutch elm disease to North America are two noticeable examples. While the invasion of exotic terrestrial organisms have been widely studied, extensive attention has not been paid to the invasion of aquatic organisms until the introduction of zebra mussels (*Dreissena* sp.) to the Great Lakes (Hebert et al., 1991; Griffiths et al., 1991; Nalepa & Schloesser, 1993) and of toxic dinoflagellates to Australia (Hallegraeff & Bolch, 1991, 1992), both of which have resulted in great economic loss. The release of ballast water from ships is known to be a major vector for the introduction of exotic aquatic species (Carlton, 1985 review). In addition to the examples cited earlier, ballast water discharge is believed to be responsible for the recent introduction of many exotic aquatic organisms, including plankton

(Cordell et al., 1992), benthic invertebrates (Buttermore et al., 1994; Gosliner, 1995), fish (Baltz, 1991; Pratt et al., 1992), fish parasites (Cone et al., 1994), and bacteria such as epidemic cholera (McCarthy & Khambaty, 1994). Risk assessment of ballast water discharge has been undertaken and regulatory guidelines have been considered and implemented in many countries (Hutchings, 1992; Locke et al., 1993; MacLeod, 1995).

Hong Kong is one of the busiest ports in the world and its coastal waters are highly susceptible to the invasion of exotic species. Morton (1987) reviewed the introductions of marine organisms to Hong Kong. No biological studies have been conducted on the ballast water in ships entering Hong Kong, in spite of the fact that similar studies have been conducted elsewhere (Williams et al., 1988; Hallegraeff & Bolch, 1992; Carlton & Geller, 1993; Kelly, 1993). The objective of this study is to investigate the diversity and abundance of organisms in the ballast water in container

Table 1. Number of species and frequency of occurrence of different taxa found in 12 ballast water samples from container ships entering Hong Kong in 1994-95. +++, ++ and + refer to the occurrence of a taxon in >70, 30-70 and <30% of the samples, respectively.

Taxon	No. of species found	Frequency of occurrence
Phylum Arthropoda		
Subphylum Crustacea		
Class Ostracoda	1	+
Class Copepoda		
Order Harpacticoida	7	+++
Order Calanoida	>16	++
Order Cyclopoida	9	++
Class Cirripedia	>2	++
Class Malacostraca	>3	++
Subphylum Chelicerata	1	+
Phylum Mollusca		
Class Gastropoda	>1	++
Class Bivalvia	>2	++
Phylum Annelida		
Class Polychaeta	>3	++
Class Oligochaeta	1	+
Phylum Nematoda	>2	++
Phylum Platyhelminthes	2	++
Phylum Cnidaria		
Class Hydrozoa	2	+
Class Scyphozoa	2	+
Phylum Chaetognatha	1	+
Phylum Chordata		
Class Ascidiacea	1	+
Phylum Chrysophyta	>13	+++
Phylum Pyrrophyta	5	++
Phylum Rhizopoda		
Foraminifera	3	++
Phylum Actinopoda		
Radiolaria	>2	+
Phylum Ciliophora	3	++
Total	>81	

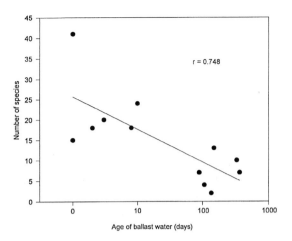

Figure 1. Number of species against the age of ballast water for 12 plankton samples. The line represents linear regression line fitted by the least square method. The correlation coefficient, *r*, is significant ($P=0.005$).

coida and the age of ballast water was not significant ($P>0.05$). Because of the absence of light and algal growth, the decrease in calanoid and cyclopoid density with increasing age of ballast water is likely to be the result of food limitation. On the other hand, harpacticoid copepods that graze on bacteria (Hardy, 1978) could be supported by continual bacterial production for several months in the ballast tanks (Carlton, 1985).

Of the copepod species identified in the samples, only one calanoid species, *Pseudodiaptomus marinus* Sato, has not been reported in Hong Kong (Chen, 1982). Yet as the species is found in the coasts of Japan and China, it is possible that it is also present in waters around Hong Kong. The individuals, found in two independent samples in the present study, originated from Taiwan where the species was found (C. T. Shih, pers. comm.). This species has been recognized to be introduced to other areas in the Pacific Ocean as well as the Indian Ocean, and the most plausible route appeared to be ballast water transport (Jones, 1966; Grindley & Grice, 1969; Orsi & Walter, 1991). Many species of *Pseudodiaptomus* are neritic but euryhaline and hardy, and thus are believed to survive in long journeys in ballast tanks (Grindley & Grice, 1969). The recent introduction of *P. inopinus* (Burckhardt) to North America was also believed to be via ballast water (Cordell et al., 1992).

Larvae of a number of invertebrates were observed to settle in laboratory culture tanks. These include the rock oyster *Saccostrea cucullata* (Born), the mussel *Mytilopsis sallei* (Recluz), the barnacle *Bal-*

large volume of water in the ballast tanks of container ships, which is likely to be less subject to fluctuation of temperature and other abiotic factors.

Copepoda represented by at least 32 species is the most abundant taxonomic group. The relationship between the density of the three groups of copepods, Calanoida, Cyclopoida and Harpacticoida, and the age of ballast water is shown in Figure 2. While the density of Calanoida and Cyclopoida was found to be negatively correlated with the age of ballast water, the correlation coefficient between the density of Harpacti-

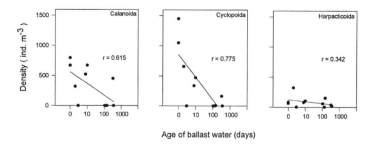

Figure 2. Density of calanoid, cyclopoid and harpacticoid copepods against the age of ballast water for 11 plankton samples. One of the 12 samples was not analysed quantitatively as the depth of water in the tank (1.5 m) was too shallow to allow accurate analysis. The lines represent linear regression lines fitted by the least square method. The correlation coefficient, r, is significant for Calanoida ($P = 0.044$) and Cyclopoida ($P = 0.005$), and not significant for Harpacticoida ($P = 0.304$).

anus amphitrite amphitrite Darwin, the shrimp *Acetes chinensis* Hansen, unidentified species of the crab *Nanosesarma* and the polychaete *Boccardia*, and an unidentified ascidian. *Acetes chinensis* is believed to be found in Hong Kong waters. The grapsid crab *Nanosesarma minutum* (de Man) has been reported in the seashore of Hong Kong (Morton & Morton, 1983). The individual found in the present study originated from Taiwan and might represent one of the six species reported there (Fukui et al., 1989). *Balanus amphitrite amphitrite* and *Saccostrea cucullata* are also found in the seashore of Hong Kong (Morton & Morton, 1983). *Balanus amphitrite amphitrite* is a cosmopolitan species (Henry & McLaughlin, 1975; Newman & Ross, 1976). It has been reported as an introduced species in San Francisco Bay, California (Carlton, 1979), and Korea (Kim, 1992).

The mussel *Mytilopsis sallei* is an introduced species in Hong Kong waters (Morton, 1980, 1989; Huang & Morton, 1983). The first record of the Central American mussel *Mytilopsis sallei* in Asia was from Visakhapatnam harbour of India (Ganapati et al., 1971). It was first found in Hong Kong in 1980 (Morton, 1980) and was believed to be introduced by a Vietnamese refugee boat from ports close to Hong Kong (Huang & Morton, 1983). The present study suggests that ballast water discharge is an alternative route of introduction. The ballast water in which *Mytilopsis sallei* was found was a mixture of water from Seattle, Washington and Keelung, Taiwan. We believe that the mussel originated from Keelung as it was introduced to Taiwan in the seventies (Chang, 1985). This species has wide environmental tolerance and its rapid growth, fast maturation and high fecundity allow it to be transported easily (Morton, 1981, 1989).

Species of the polychaete *Boccardia* are widely distributed in many parts of the world. Larvae of Spionidae were frequently found to survive in ballast waters (Howarth, 1981, cited in Carlton, 1985 review). The unidentified species we found came from either Japan or the west coast of North America where *Boccardia* species were found from Washington to southern California and beyond (Ricketts & Calvin, 1968). *Boccardia* is not included in species lists given by Shin (1977, 1982) and Mak (1982), but since the polychaetes in Hong Kong are poorly known, we cannot be certain that the *Boccardia* sp. we found is an exotic species.

Conclusion

Results from this survey demonstrate rich diversity and high abundance of biological organisms in ballast water of container vessels entering Hong Kong. Since many of these organisms could be cultured in local water and some larvae have successfully settled and developed, we believe that ballast water is a major source of invasion of exotic species into Hong Kong waters. Regulatory guidelines on the discharge of ballast water, such as mid-ocean exchange and treatment of ballast water before discharge (MacLeod, 1995) should be implemented to minimize the risk of exotic species introduction.

Acknowledgments

We are grateful to the NYK Line (Hong Kong) Ltd.'s administration for access to the ballast tanks of the container ships. We would like to thank A. L. C. Chan,

K. C. Cheung, M. K. Cheung, R. W. T. Foo, F. L. S. Leong and Y. C. Tam for assistance in the collection of samples, and B. R. Morton (University of Hong Kong), F. X. Li (Xiamen University) and P. Y. Qian (Hong Kong University of Science and Technology) for identifying some of the organisms. We also thank J. C. Carlton (Williams College), T. C. Lan (Taiwan), C. T. Shih (National Taiwan University) and C. K. Wong (Chinese University of Hong Kong) for their valuable discussion of the work. C. K. Wong also gave constructive comments on the manuscript. The research was supported by a grant from the Centre for Environmental Studies, The Chinese University of Hong Kong.

References

Baltz, D. M., 1991. Introduced fishes in marine systems and inland seas. Biol. Conserv. 56: 151–177.

Buttermore, R. E., E. Turner & M. G. Morrice, 1994. The introduced northern Pacific seastar *Asterias amurensis* in Tasmania. Mem. Queensland Mus. 36: 21–25.

Carlton, J. T., 1979. Introduced invertebrates of San Francisco Bay. In Canomos, T. J. (ed.), San Francisco Bay: The Urbanized Estuary. Pacific Division, American Association for the Advancement of Science, San Francisco (CA): 427–444.

Carlton, J. T., 1985. Transoceanic and interoceanic dispersal of coastal marine organisms: the biology of ballast water. Oceanogr. mar. Biol. ann. Rev. 23: 313–371.

Carlton, J. T. & J. B. Geller, 1993. Ecological roulette: The global transport of nonindigenous marine organisms. Science 261: 78–82.

Chang, K. M., 1985. Newly introduced false mussel to Taiwan (Bivalvia: Dreissenidae). Bull. Malacol. R.O.C. 11: 61–67.

Chen, Q. C., 1982. The marine zooplankton of Hong Kong. In Morton, B. S. & C. K. Tseng (eds), The Marine Flora and Fauna of Hong Kong and Southern China. Vol. 1. Hong Kong University Press, Hong Kong: 789–799.

Cone, D., T. Eurell & V. Beasley, 1994. A report of *Dactylogyrus amphibothrium* (Monogenera) on the gills of European ruffe in western Lake Superior. J. Parasitol. 80: 476–478.

Cordell, J. R., C. A. Morgan & C. A. Simenstad, 1992. Occurrence of the Asian calanoid copepod *Pseudodiaptomus inopinus* in the zooplankton of the Columbia river valley. J. Crust. Biol. 12: 260–269.

Elton, C. S., 1958. The Ecology of Invasions by Plants and Animals. Chapman & Hall, London: 181 pp.

Fukui, Y., K. Wada & C. H. Wang, 1989. Ocypodidae, Mictyridae and Grapsidae (Crustacea: Brachyura). J. Taiwan Mus. 42: 225–238.

Gosliner, T. M., 1995. Introduction and spread of *Philine auriformis* (Gastropoda: Opisthobranchia) from New Zealand to San Francisco Bay and Bodega Harbor. Mar. Biol. 122: 249–255.

Griffiths, R. W., D. W. Schloesser, J. H. Leach & W. P. Kovalak, 1991. Distribution and dispersal of the zebra mussel (*Dreissena polymorpha*) in the Great Lakes region. Can. J. Fish. aquat. Sci. 48: 1381–1388.

Grindley, J. R. & G. D. Grice, 1969. A redescription of *Pseudodiaptomus marinus* Sato (Copepoda, Calanoida) and its occurrence at the Island of Mauritius. Crustaceana 16: 125–134.

Ganapati, P. N., M. V. Lakshmana Rao & R. Nagabhushanam, 1971. On *Congeria sallei* Recluz, a fouling bivalve mollusc in the Visakhapatnam harbour. Curr. Sci. 40: 409.

Hallegraeff, G. M. & C. J. Bolch, 1991. Transport of toxic dinoflagellate cysts via ships' ballast water. Mar. Pollut. Bull. 22: 27–30.

Hallegraeff, G. M. & C. J. Bolch, 1992. Transport of diatom and dinoflagellate resting spores in ships' ballast water: implications for plankton biogeography and aquaculture. J. Plankton Res. 14: 1067–1084.

Hardy, B. L. S., 1978. A method for rearing sand-swelling harpacticoid copepods in experimental conditions. J. exp. mar. Biol. Ecol. 34: 143–149.

Hebert, P. D. N., C. C. Wilson, M. H. Murdoch & R. Lazar, 1991. Demography and ecological impacts of the invading mollusc *Dreissena polymorpha*. Can. J. Zool. 69: 405–409.

Henry, D. P. & P. A. McLaughlin, 1975. The barnacles of the *Balanus amphitrite* complex (Cirripedia, Thoracica). Zool. Verh., Leiden 141: 3–254.

Huang, Z. G. & B. Morton, 1983. *Mytilopsis sallei* (Bivalvia: Dreissenoidea) established in Victoria Harbour, Hong Kong. Malacol. Rev. 16: 99–100.

Hutchings, P., 1992. Ballast water introductions of exotic marine organisms into Australia – current status and management options. Mar. Pollut. Bull. 25: 5–8.

Jones, E. C., 1966. A new record of *Pseudodiaptomus marinus* Sato (Copepoda, Calanoida) from brackish waters of Hawaii. Crustaceana 10: 316–317.

Kelly, J. M., 1993. Ballast water and sediments as mechanisms for unwanted species introductions into Washington State. J. Shellfish Res. 12: 405–410.

Kim, I. H., 1992. Invasion of foreign barnacles into Korean waters. Korean J. syst. Zool. 8: 163–175. (in Korean with English abstract).

Locke, A., D. M. Reid, H. C. van Leeuween, W. G. Sprules & J. T. Carlton, 1993. Ballast water exchange as a means of controlling dispersal of freshwater organisms by ships. Can. J. Fish. aquat. Sci. 50: 2086–2093.

MacLeod, L., 1995. Risks associated with the uncontrolled discharge of ships' ballast and the legislative controls available in Scotland. J. CIWEN 9: 173–178.

McCarthy, S. A. & F. M. Khambaty, 1994. International dissemination of epidemic *Vibrio cholerae* by cargo ship ballast and other nonpotable waters. Appl. envir. Microbiol. 60: 2597–2601.

Mak, P. M. S., 1982. The coral associated polychaetes of Hong Kong, with special reference to the serpulids. In Morton, B. S. & C. K. Tseng (eds), The Marine Flora and Fauna of Hong Kong and Southern China. Vol. 2. Hong Kong University Press, Hong Kong: 595–617.

Morton, B., 1980. *Mytilopsis sallei* (Recluz) (Bivalvia: Dreissenacea) recorded from Hong Kong: An introduction by Vietnamese refugees? Malacol. Rev. 13: 90–92.

Morton, B., 1981. The biology and functional morphology of *Mytilopsis sallei* (Recluz) (Bivalvia: Dreissenacea) fouling Visakhapatnam Habour, Andhra Pradesh, India. J. moll. Stud. 47: 25–42.

Morton, B., 1987. Recent marine introductions into Hong Kong. Bull. mar. Sci. 41: 503–513.

Morton, B., 1989. Life-history characteristics and sexual strategy of *Mytilopsis sallei* (Bivalvia: Dreissenacea), introduced into Hong Kong. J. Zool., Lond. 219: 469–485.

Morton, B. & J. Morton, 1983. The Sea Shore Ecology of Hong Kong. Hong Kong University Press, Hong Kong: 350 pp.

Nalepa, T. F. & D. W. Schloesser (eds), 1993. Zebra Mussels: Biology, Impacts, and Control. Lewis Publishers, CRC Press, Boca Raton (FL), 810 pp.

Newman, W. A. & A. Ross, 1976. Revision of the babanomorph barnacles including a catalogue of the species. San Diego Soc. nat. Hist. Mem. 9: 1–108.

Orsi, J. J. & T. C. Walter, 1991. *Pseudodiaptomus forbesi* and *P. marinus* (Copepoda: Calanoida), the latest copepod immigrants to California's Sacramento-San-Joaquin estuary. Bull. Plankton Soc. Jap., Special vol.: 553–562.

Pratt, D. M., W. H. Blust & J. H. Selgeby, 1992. Ruffe, *Gymnocephalus cernuus*: newly introduced in North America. Can. J. Fish. aquat. Sci. 49: 1616–1618.

Ricketts, E. F. & J. Calvin, 1968. Between Pacific Tides. 4th ed. Stanford University Press, Stanford, California: 614 pp.

Shin, P. K. S., 1977. A quantitative and qualitative survey of the benthic fauna of the territorial waters of Hong Kong. M. Phil. Thesis, University of Hong Kong: 195 pp.

Shin, P. K. S., 1982. Some polychaetous annelids from Hong Kong waters. In Morton, B. S. & C. K. Tseng (eds), The Marine Flora and Fauna of Hong Kong and Southern China. Vol. 1. Hong Kong University Press, Hong Kong: 161–172.

Williams, R. J., F. B. Griffiths, E. J. Van der Wal & J. Kelly, 1988. Cargo ship ballast water as a vector for the transport of non-indigenous marine species. Estuar. coast. Shelf Sci. 26: 409–420.

Hydrobiologia **352**: 207–218, 1997.
Y.-S. Wong & N. F.-Y. Tam (eds), Asia–Pacific Conference on Science and Management of Coastal Environment.
©1997 *Kluwer Academic Publishers. Printed in Belgium.*

Decision-making processes in ecological risk assessment using copper pollution of Macquarie Harbour from Mt. Lyell, Tasmania, as a case study

J. R. Twining & R. F. Cameron
Australian Nuclear Science and Technology Organisation, PMB 1, Menai, 2234

Key words: Ecological risk assessment, copper, marine, estuarine

Abstract

Ecological risk assessment is increasingly being used to make decisions on the acceptability of industrial processes, as well as on the appropriate approach to take with remediation of contaminated sites. In this approach, the risks and costs must first be determined before decisions can be made.

In principle, the procedure for undertaking an ecological risk assessment for a site with existing contamination is fairly straight forward. Probability distributions are obtained for the concentration of the contaminant of concern and for the biological and/or structural impacts likely to occur in the affected habitat. The degree of overlap between these distributions determines the risk from the contaminant to the habitat. With water-borne contamination, the level of assessment can vary from a simple comparison with water quality criteria, through site specific literature surveys, to laboratory and field studies depending on the importance of the environment, the concentration and perceived nature of the contaminant, the resources available, and the likely benefit from the process to be developed.

However, a number of uncertainties make this process more difficult. These include the lack of a standard methodology, availability of appropriate data and agreed definitions of acceptable risk. Thus several arbitrary or considered decisions need to be made before and during such an assessment.

This paper discusses the application of an ecological risk assessment of copper pollution in Macquarie Harbour, Tasmania, using data from long-term monitoring of waters and literature searches on lethal and sub-lethal effects of copper in marine and estuarine environments. This study is part of a much larger program established to determine best methods for the remediation of the Mt. Lyell copper mine site as well as the freshwater and marine habitats downstream. The results of the assessment indicated that there was at present a probability greater than 0.9 of the occurrence of anodic stripping voltametry-labile copper water concentrations harmful to 5% of all species. For total dissolved copper the probability was higher than 0.98. The upper value of total dissolved copper in Macquarie Harbour that encompassed 90% of the probable concentrations would need to be reduced by a factor of approximately 30, without the inclusion of any additional application factors, to achieve (sub-lethal) protection for 95% of species.

Introduction

Ecological risk assessment is increasingly being used to make decisions on the acceptability of industrial activities, as well as on the appropriate approach to take with remediation of contaminated sites. It provides a framework for comparing the ecological effects with the acceptability criteria while including the uncertainty in determination of the risk parameters. Remediation strategies can then be assessed in terms of their poten-tial for reducing risks as a function of cost. The use of ecological risk in this context is still developing, but the increasing use of risk based decision-making in other areas would imply that it will become increasingly important.

In principle, the assessment process is straightforward. A probability distribution for the concentration of the contaminant of concern is first derived by on-site monitoring or by some assessment of the source terms and dispersion. A distribution is also determined for the biological and/or structural impacts likely to occur in the affected habitat. The degree of overlap between these distributions determines the risk from the contaminant to the habitat. With water-borne contamination, the level of assessment will increase as we move from simple comparisons with water quality criteria, through site specific literature surveys, to laboratory and field studies. The choice of which comparison to use depends on the importance of the environment, the concentration and perceived nature of the contaminant, the resources available, and the likely benefit from the process to be developed.

However, the process is made more difficult by the lack of key data and the need to incorporate the uncertainties which exist in data, in modelling and in representation of the actual environmental conditions. Some appreciation of the underpinning ecological structure and function of the environment in question must also be available in order to set minimal requirements for remediation. The degree of recovery to and beyond that level will depend on: the extent and severity of pollution; available resources in relation to treatments required and, from that, the *a priori* decisions as to what is adequate or acceptable; the natural resilience of the various systems; and time.

The case study

Due to the mining of copper for a period in excess of 100 years, heavy pollution has occurred around the mine sites at Queenstown, Tasmania, as well as within the affected habitats downstream in the Queen and King Rivers and Macquarie Harbour (Figure 1). In addition to the problems that have arisen from the extraction and smelting processes of the past, acid rock drainage and consequent leaching of residual copper and other metals from the large waste rock heaps at the site continue to contribute to the presently poor condition of the affected area.

The area has a naturally high rainfall (2500 mm a^{-1}). Dissolved organics, particularly humic substances, persist in the fresh water of the system and give rise to values of 2–7 mg DOC l^{-1} in mid-salinity Harbour waters.

Currently, the Mt Lyell Remediation Research and Demonstration Program is determining the extent of the pollution off site and its severity in terms of ecological impact. As part of the larger program, the aim of this risk assessment is to evaluate the probability of harm to aquatic populations in the marine environment downstream. This will be done by determining the degree of overlap between the distribution of measured concentrations of copper in water samples from Macquarie Harbour and the distribution of concentrations of copper reported in the literature to have significant effects on biota in similar environments. By comparing these distributions, the probabilities of exceeding critical values of copper in the environment relevant to selected end-points, such as proportional lethality to a prescribed range of species across trophic levels, can be determined within set confidence limits. We can then assess the generic risk that copper, in waters of specific Harbour habitats, presents to biota likely to inhabit those regions.

The water quality distributions can also be used to compare with site specific ecotoxicological data yet to be determined for algae, crustaceans and fish. As these values will be based on actual Macquarie Harbour waters, they will give a better indication of any synergistic or antagonistic influences on copper toxicity when compared with the predicted effects from the literature.

Data and assessment methods

Macquarie Harbour water monitoring data

Monitoring data for various stations within Macquarie Harbour were provided by Dr Lois Koehnken of Tasmanian Department of Environment and Land Management in Hobart. The data comprised a comprehensive but incomplete (for a variety of reasons) set over the period from May 1993–August 1995 at approximately 3 month intervals. The incompleteness was mainly due to poor weather or low water levels at the time of sampling. Quality assurance checks on the electronic transfer of the information indicated that the data arrived safely. Missing data were ignored. Stations sampled on only one or a few occasions were excluded for general consistency between dates.

The data were arranged by analysis type, i.e. anodic stripping voltametry-labile copper (ASV), total dissolved copper (hereafter referred to as dissolved) (μg l^{-1} or ppb) and particulate copper (mg l^{-1} or ppm). The ASV and dissolved copper values were determined after filtration (0.45 μm), whilst the particulate copper was determined from that retained on the filter. For each station, samples were taken at the surface, at mid-water and immediately above the sediment. The mid-water samples were taken at the point at which 20 ppt salinity was measured in the profile. This repre-

209

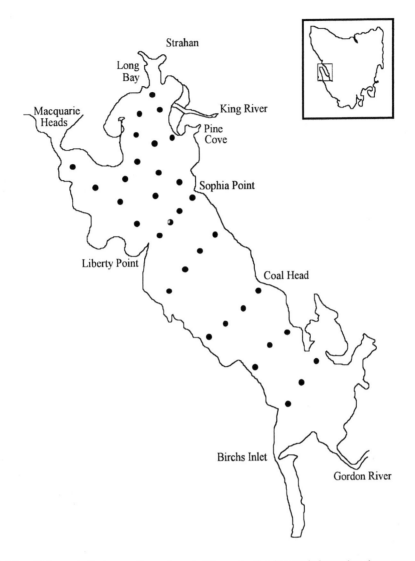

Figure 1. Map of Macquarie Harbour on the west coast of Tasmania, showing regularly monitored water sampling sites.

sented the middle of the salt wedge boundary between the deeper, more dense, sea water and the shallower, less dense, river water.

Mid-water data were selected for modelling. This selection coincided with the choice of this salinity for the ecotoxicological studies carried out within the project. The assumption was that the marine species to be tested could tolerate these salinity conditions and also that copper input from the fresh waters would be both more concentrated and more soluble, and hence bioavailable, under these conditions than in the deeper, saltier waters. It was thus inferred that these conditions would give the most conservative assessment of copper toxicity in the Harbour. Within this category, ASV and dissolved copper were selected for distribution analyses. Dissolved copper represents the upper extreme of measurable copper likely to be toxic. On the other hand, ASV would more closely represent the bioavailable fraction, but by its nature this measure will still tend to overestimate toxicity (see following discussion) and as such is still an ecologically conservative estimate. Total copper, the sum of dissolved and particulate copper, was also derived to compare with the ANZECC (1992) guideline values.

210

Biotic data

Literature values were taken from the recent marine-specific review by Ahsanullah et al. (1995) and the more comprehensive pre-1980 review by Hodson et al. (1979). Only criteria specific data were selected, that is, marine or estuarine, LC or LD values for lethal end-points and Lowest or No Observable Effect Concentrations or EC values for sub-lethal effects. Algal toxicity data from experimental studies carried out in full nutrient media, containing compounds that absorb or complex copper, reduce copper toxicity and thus underestimate it's effect (Stauber, pers comm). These data were therefore excluded. Given the resource constraints on the study, no other quality criteria, such as listed in Emans et al. (1993), were applied to reject literature data. Identical data from both review sources were included once only.

The data were arranged by broad taxonomic group, ie. Algae, Crustaceans, Fish, and combinations of all Marine Invertebrates and All Marine taxa. Significant variability was encountered within each criterion. For example: exposure periods for experimental tests ranged between 2–720 hours; salinity varied between 5–36 ppt; temperature ranged from 5–35 °C; and the end-points ranged across all life-stages encompassing metabolic activity, physiological and behavioural responses, growth rates and mortality. The data within each category were separated into lethal (LD_{50} and LC_{50}) and sub-lethal (EC_{50}; LOEC and NOEC) criteria. In some categories not every criterion was available.

Despite the fact that some species were more heavily represented than others and that the endpoints within the different effect categories varied considerably, all individual results were included separately to ensure that the total variability inherent in the data was included. This was to allow for a more accurate uncertainty estimate when calculating the critical values.

Statistical analysis

When relatively large sets of data are available (usually 8 or more sets), it is possible to derive a species sensitivity model from fitting a distribution function to the frequency distributions of the test data and deriving a criterion using a prescribed percentile of that distribution (OECD, 1992). In using this approach, it is assumed that the species have some distribution of sensitivity with a few very sensitive species and many moderately sensitive species. This approach is

not, however, able to account for interactions between species and is based on an assumption that differences in toxicity only arise because of differences in sensitivities between species. In general, the models have been developed to arrive at a concentration which protects 95% of the aquatic genera.

In deriving these models, the major issues are the choice of the distribution, the choice of the protection level and the statistical evaluation of the uncertainties due to the limited number of toxicity data. The former issue relates to the error in fitting the actual data by distributions and the latter to using samples to estimate population parameters.

In deriving values for protection of the ecosystem, there are various approaches in current use. The method of Stephan et al., (1985) estimates a Final Chronic Value (FCV) to protect 95% of the species, based on the geometric means of the species chronic values. A triangular distribution is used to estimate the FCV from the 4 lowest genus means. In the method of Aldenberg and Slob (Aldenberg & Slob, 1993), an extrapolation factor, T, is estimated and used to derive a concentration expected to be harmful to no more than 5% of the community. This method takes into account the uncertainty caused by using estimated parameters to represent a population and is based on the use of a log-logistic distribution. The method of Wagner and Lokke (Wagner & Lokke, 1991) is similar to Aldenberg and Slob in estimating a parameter to protect p% of the community, but these authors base their results on a lognormal distribution.

In comparisons made with the various methods (OECD, 1992), it was noted that:
- the choice of distribution did not have a marked effect on the protection level calculated.
- the level of confidence of the estimate (from 50% to 95%) altered the protection level by around a factor of six. (for relatively small numbers of test results).
- larger differences arose when the number of data were small.
- the use of acute values divided by 100 gave similar results to using the NOEC values divided by a factor of 10.

Thus the choice of a distribution should be determined by the data themselves, especially where larger data sets are available. The distributions differ in how extreme the tailing is at the lower and upper ends. Clearly, a lowest level of zero is a requirement.

In general, the use of chronic values is more appropriate as a protection level, for species exposed to pollutants over a long period. LC$_{50}$ values are, however, appropriate to use for uncommon exposures of short duration, where some level of mortality would be acceptable. Since large kills may not be acceptable, it may be desirable to use a lower percentage than 50 in the LC criterion. This may be a better approach than application of an arbitrary factor to the LC$_{50}$ value. The difficulty with use of NOEC or LOEC values is that measurement of the effect is more difficult because of the variability in magnitude of the type of effect chosen and the greater variability among individual members of the species.

Most approaches use the 95% level as the appropriate protection level for most species. It is, however, justifiable to use a lower level for distributions of algae and bacteria in consideration of the higher diversity, functional redundancy, highly variable structure and low public concern for these communities. Generally, there is no consensus on what level of confidence should be applied to the protection level with some groups favouring the 95% level and others the 50% value. Those groups favouring the 95% level, calculate the bounds using methods based on random sampling. These can add uncertainty since the data set variability is often systematic rather than random. The US EPA does not have a provision for specifying this level of confidence and hence its methods should be compared with the 50% levels used by others. Because of the emphasis on determining a parameter at the lower end of the data (the 5% level), some censoring of the very high values is also considered appropriate. This will prevent distortion of the 5% level, from either being too high or too low.

In this study, preliminary data analysis showed that the water concentrations and subsets of the biological effect data were biased towards higher copper concentrations. In some cases this was extreme. Because of this, the data were assumed to be log-normally distributed and geometric means and standard deviations were derived. This form of distribution is typical for data of this nature (e.g. Kooijman, 1987). Probability distribution functions were generated using these statistics within the STATISTICA software package (Statsoft Inc., 1994). The goodness of fit of each derived log-normal distribution was determined using the Kolmogorov–Smirnov one sample test or the Chi-Squared test (Steel & Torrie, 1981) at a significance level of 5%. The extreme high values mentioned earlier did not allow an adequate fit to the log-normal model. Thus, these values, which can be considered as outliers, were excluded in order to achieve at statistically significant

goodness of fit for the biota distributions. This action will make the assessed risk more conservative as the species most tolerant of copper pollution have been excluded in favour of more sensitive taxa.

Assuming that 5% of the representative population could be affected (i.e. a protection level of 95%), the critical hazardous copper concentrations (HC$_{5\%}$) for each of the subsets of biota distributions were derived (Wagner & Lokke, 1991). The 95% and 50% confidence intervals around these estimates were also determined as per Aldenberg & Slob (1994). These values were then imposed on the distributions generated for the water sample copper concentrations (ASV and dissolved) to determine the prevailing probability of exceeding the critical water concentrations and also the degree by which water concentrations would need to be reduced in order to achieve the nominated degree of protection.

Results and discussion

Copper water concentrations

The selected water concentration monitoring data were observed for any seasonal and other temporal trends in their maximum, minimum and average values at stations for each sampling period (Figures 2a, b and c). Despite the occurrence of occasional high values, that may reflect sediment disturbance or increased pollutant inflow from the King River due to storm activity, there were no persistent patterns over the period of monitoring. These observations imply that copper concentrations, in this specific compartment of the areas affected by pollution from Mt Lyell, are currently relatively constant. On this basis, all further comparisons in this report used data combined from all sampling times.

Comparison with water quality guidelines

Figure 3 shows the degree of overlap between measured Macquarie Harbour copper concentrations and the ANZECC (1992) guideline values of total copper (dissolved plus particulate) for the protection of marine ecosystem health. None of the measured total copper concentrations were less than the guideline value of 5 ppb (0.7 on the log$_{10}$ scale) which is commonly taken as the default regulatory limit. Even dissolved copper (the typical measure of environmental copper concentrations in water) and ASV copper (a value more close-

212

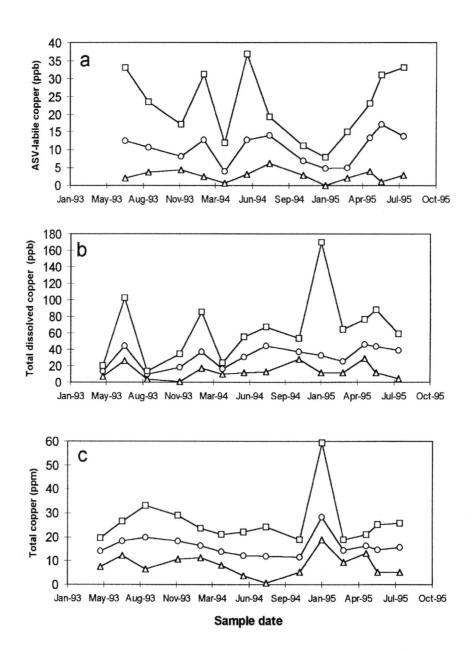

Figure 2. Maximum (□), average (○) and minimum (△) copper concentrations in mid-salinity Macquarie Harbour waters at each sampling period as measured by: (a) anodic stripping voltametry-labile copper; (b) total dissolved copper (0.45 μm) (both in μg l^{-1}); and (c) total copper (dissolved plus filterable) (mg l^{-1}).

ly approximating bioavailable concentrations) were in excess of the guideline value most of the time. More than 95% of the measured dissolved copper values and 75% of the ASV values were greater than the guideline at all times.

A reduction in total copper levels within the mid-salinity waters of Macquarie Harbour by 4 orders of magnitude is required to achieve levels that are beneath the guideline value at least 90% of the time.

However, total Cu is not a good measure of environmental hazard as most of the measured metal is not biologically available and as such will not directly contribute to toxic effect. Indirect contribution is possible depending upon the degree to which the particle

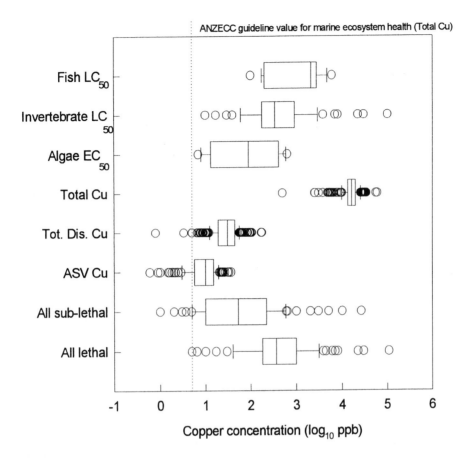

Figure 3. Box plots of measured copper concentrations at mid-salinity depths of selected Macquarie Harbour sampling stations (see text) and sub-sets of raw ecotoxicity data (lethal and sub-lethal) from the literature in relation to the ANZECC guideline copper concentration for marine ecosystem health (dotted line). The boxes extend from the 25th–75th percentiles with the median as a mid-line. The capped bars indicate the 10th and 90th percentiles and symbols indicate data outside these values.

associated copper can be mobilised into a bioavailable form. Exceptions to this generalisation include toxicity from particulate copper to members of the ecological community that are filter feeders, detritivors that ingest particles containing copper, and plants that use the copper bearing particulates as a nutrient substrate. These exposure pathways can be especially significant if the affected taxa include keystone species.

Biological data

It is generally assumed that the concentrations for each of the biological end-points used in the data sorting will decrease in the order $LC_{50} > LD_{50} > EC_{50} \geq LOEC > NOEC$. Chronic NOECs were found to be 10–30 times lower than acute median lethal values on average by Hendricks (1995) when studying organic toxicants. This general pattern could

be observed in the raw data of our current study, particularly where results for a single species or within closely related taxa were examined. However, this was not always found to be the case as some of the observed sub-lethal criteria were less sensitive than others and there were wide ranging degrees of tolerance between species. That is, some very tolerant organisms showed no or low observed response to very high concentrations of copper (high NOEC/LOEC) whilst some extremely sensitive organisms died at low concentrations or exposures (low LC_{50}, LD_{50}).

In Figure 3 the measured copper concentrations in Macquarie Harbour waters are compared with subsets from the literature data (Ahsanullah et al., 1995; Hodson et al., 1979) indicating the biological effect of copper. Both acute and chronic, lethal and sub-lethal parameters are represented. The plots of All lethal and All sub-lethal data include information in addition to that given for the sub-sets at the top of the page. Given the high copper loads in the Harbour at present it is unlikely that any sensitive taxa persist in that environment unless local populations of relatively fecund species have undergone gradual selective adaptation to the copper levels to which they have been exposed since mining began.

The available literature data cover several trophic levels. The information density varies between these levels but the discrepancies are minor. Also, the biological effects within any category occur over orders of magnitude differences in copper concentrations. From these observations it is apparent that the data have provided a representative spread of effect levels for both sensitive and insensitive species across most trophic levels and, as such, they are providing a reasonable basis for ecosystem scale assessment.

The boxplots representing all lethal and sub-lethal data show a substantial overlap (Figure 3). However, sub-lethal effects may be seen to generically occur at copper concentrations an order of magnitude lower than those observed for lethal effects.

The probability distribution functions of combined taxa lethality data (LC_{50} and LD_{50} values) and sub-lethality data (EC_{50}, LOEC and NOEC values) are shown in Figures 4a and b respectively. The parameters of these distributions were chosen to give the best 'goodness-of-fit' to the raw data. The most sensitive comprehensive subset, algal EC_{50} data, could not be adequately fitted by a log-normal model.

From the raw data, the copper concentrations likely to be hazardous to 5 and 10% of the biota at the given end-point are given in Table 1. Also given are the parameters derived from the fitted distributions. These are the $HC_{5\%}$ and the lower limits of the 50% and 95% uncertainty, or confidence, ranges about this critical value.

Comparison of the water and biota distributions

Predominantly, the literature data refer to soluble or dissolved copper concentrations, particularly when dealing with determination of lethal end-points. Hence total copper concentrations (dissolved + particulate) provide a poor basis for comparison. Field data in particular refer mainly to dissolved copper concentrations. Experimental data are generally concerned with ionic copper species and therefore more closely correspond to the ASV values.

Thus, for comparison of likely toxic effect, dissolved copper will give the upper limit to possibly toxic

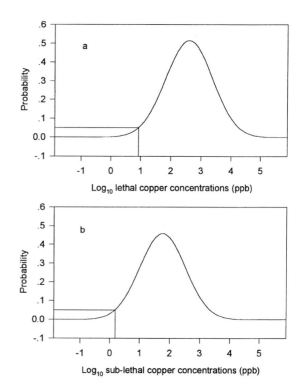

Figure 4. Probability distribution functions of copper toxicity data for (a) lethal and (b) sub-lethal end-points taken from the literature. Extreme (high) values were excluded to allow statistically adequate fit of the distributions. The intercepts indicate the copper concentrations below which only 5% of species are predicted to show a response.

copper concentrations in Macquarie Harbour. However, Macquarie Harbour waters are known to have a very high complexation capacity, predominantly from the levels of organics input from the surrounding fresh water catchments. From this, it is reasonable to refer to the ASV copper distribution for a more realistic appraisal of ecological risk. This assessment will still be conservative as the copper measured by ASV will include species such as carbonates that are non-toxic (Hunt, 1987) and copper that is moderately bound to some organic ligands within the water column (Batley, pers comm). These components of the measured copper are not considered to be biologically available. The cumulative probability distributions of ASV and dissolved copper in Macquarie Harbour mid-salinity water samples are shown in Figure 5.

In Table 2 are shown the probabilities that the measured concentrations of copper in Macquarie Harbour waters currently exceed the values that would protect 95% of species from lethal or sub-lethal effects. The

Table 1. Critical values of copper concentrations (ppb) that have lethal and sub-lethal affects on biota. Outliers were high values that were removed from the data sets to allow for statistically adequate model fitting.

Taxonomic group	Criteria	Raw data		Fitted distribution		
		5%	10%	HC$_{5\%}$ value	50% conf. value	95% conf.
All	lethal	30	60			
All-outliers	lethal	17	40	21.1	20.8	9.4
All	sub-lethal	4.9	5			
All-outliers	sub-lethal	3.7	5	2.1	2.1	0.9
algae	sub-lethal	15	15			
invertebrates	lethal	17	40	23	21	5.1
fish	lethal	100	200	93.8	84.9	18.1

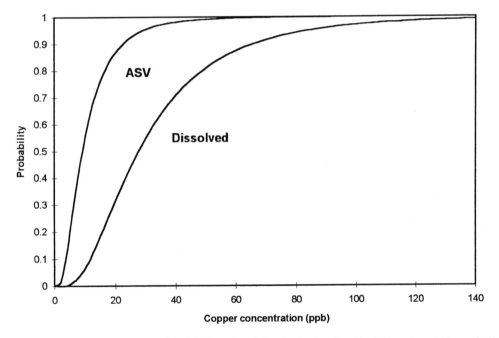

Figure 5. Cumulative probability distributions of ASV-labile and total dissolved copper in mid-salinity waters of Macquarie Harbour. The curves indicate the probability of measuring a concentration less than any specified value, based on monitoring data.

high outliers were excluded from the raw data across all species.

From the sub-lethal values, the 0.90 probability concentration of total dissolved copper in Macquarie Harbour would need to be reduced by a factor of approximately 30 to achieve adequate protection based solely on this criteria without application factors. To protect 95% of the algal species 90% of the time a reduction in ASV copper concentrations by a factor of approximately 2 is required.

When making assessments of the degree of reduction in copper required to achieve this type of end-point, the degree of dissolution from sediment must also be taken into account in addition to reductions in input via the King River.

The lethal parameters for invertebrates and fish are included in Table 1. The raw data are also shown in Figure 3. It can be seen that invertebrates are relatively comparable to the generic lethal data and, as such, sub-sets of crustacean data may be a reasonable surrogate for more extensive biological data in other comparisons. The fish data are relatively insensitive when compared with the end-points derived for combined taxa lethal data or any sub-lethal parameters.

216

Table 2. Current probabilities of exceeding the critical values to protect 95% of species.

Water quality	Lethal Effects		Sub-lethal effects	
	ASV	Diss. Cu	ASV	Diss. Cu
5% value from raw data for all species	0.19	0.76	0.90	~1.0
5% value (at 50% confidence level) from fitted distribution for all species	0.12	0.66	0.98	~1.0
5% value for Algae from raw data			0.24	0.82

Factors that affect the relative degree of safety implicit in the risk analysis

The following decisions and other factors provide inherent conservatism (safety) to the risk assessment.
- Use of measured values that overestimate the biologically available copper water concentrations.

The labile copper measured by ASV as well as the dissolved concentrations include some chemical species that are less, or not, bioavailable. These chemical species are less toxic than the free ionic form of copper.
- Use of mid-salinity water quality data maximised the measured copper concentration in the marine waters of Macquarie Harbour and, hence, the perceived risk to biota.
- Use of all water measurements rather than averages for any period.

There is little likelihood of all sites being simultaneously contaminated to high levels. The average values at any particular sampling period (Figure 2) indicate a distribution about a factor of two lower than the maximum values measured at the same time. This provides an additional safety factor in the assessment given that mobile species will be at an advantage in that they may avoid or move out of highly contaminated zones to other depths or locations, and widely distributed species will be able to recolonise affected areas.
- The data from the literature studies represent effects to proportions of individuals within populations rather than to populations as a whole. Higher concentrations would be required to affect all individuals within the tested populations.
- Probable bias in the literature data towards sensitive species.

Research workers will tend to select species that are most likely to show a significant response to any test. Sensitive species are also likely to have been chosen for testing or monitoring on the basis of their relative response in field surveys. In addition, the exclusion of extremely tolerant species from the biota data to achieve normal distributions has biased the data towards more sensitive taxa.
- The use of laboratory studies to estimate environmental risk.

Most controlled laboratory studies constrain the experimental parameters to minimise variability. Many natural water quality parameters that reduce toxicity (e.g. complexation capacity) are thereby excluded from these studies. Hence, this may lead to an overestimate of toxic effect when the results of laboratory studies are applied to natural systems.
- The use of the lower limit of the uncertainty estimate of the critical hazardous concentration is inherently conservative. Emans et al. (1993) also found that the Aldenberg & Slob (1993) model based on single-species data, tended to overestimate toxic effect when compared with relevant field multi-species studies.
- The likelihood of the occurrence of tolerant populations of species within the Harbour brought about by over a century of natural selection pressure.

The following factors are of unknown significance or could contribute to an estimate of greater risk from copper to biota in Macquarie Harbour:
- The magnitude of significant water quality parameters.

This study has looked solely at copper concentrations in the mid-salinity habitat of Macquarie Harbour. There has been no specific attempt in this study to address the other habitat parameters that can influence copper toxicity. These include possible additive or synergistic affects from other toxic materials (eg zinc) and antagonistic effects such as complexation by organic ligands or the formation of non-toxic metal species.
- the keystone species for ecological sustainability have not been identified.

At present, too little is known of the local biological communities, either within Macquarie Harbour or in similar habitats unaffected by the pollution from Mt Lyell. As such the keystone or indicator species have yet to be adequately identified for the overall study. The successful identification and re-occurrence of these species within Macquarie Harbour is certain to be one of the criteria for success of the overall remediation process.

- The impact of copper concentrations in the upper water layer habitat of Macquarie Harbour has not been addressed.

Very high levels of copper are present in the less dense river water suspended above the saline wedge within the Harbour. The copper levels are well in excess of the ANZECC guidelines for fresh waters. The impact on euryhaline, migratory or fresh water species could be significant at the measured concentrations. Any assessment of this habitat should include toxicity to water fowl including bioaccumulation pathways.

- Bioaccumulation pathways and their associated risk, to other biota or humans, have not been addressed.
- Quality criteria have not been applied to the selection of literature data used for the models. Exclusion of data will lead to changes in the probability distributions but the significance of these changes cannot be assessed at present.
- Sediment effects.

Estimations on the degree of copper concentration reductions required will need to include an assessment of the likely remobilisation of sediment bound copper as well as reductions in riverine input. It must also be recognised that some keystone organisms, environmentally critical to the remediation, may occupy a benthic habitat. As such, these species will be at risk from current and future sedimentary copper.

Conclusions

When compared with the ANZECC (1992) water quality guidelines, copper loads in Macquarie Harbour are too high by at least four orders of magnitude. This preliminary comparison justifies the need for a more comprehensive risk assessment.

When alternate, less restrictive, criteria are used to compare concentrations of copper in Macquarie Harbour water with literature data on the biological effects of copper in marine systems, the monitored water concentrations still exceed the critical hazard levels using both lethal and sub-lethal end-points. Based on these more realistic evaluations the prevailing total dissolved copper water concentrations have a probability of close to 100% of exceeding the sub-lethal critical limit and of 66% of exceeding the critical limit for lethality.

To achieve water concentrations that have an adequately low probability (i.e. 10%) of exceeding the critical sub-lethal limit across all taxa, a reduction in total dissolved copper water concentrations by a factor of at least 30 is required. For algae, important as the autochthonous primary producers of the ecosystem and the most sensitive taxonomic group, the reduction of the ASV-labile copper concentration in water that is required to protect 95% of species is an approximate factor of only 2 based on the available literature.

The risk assessment is reasonably conservative for a variety of reasons. Predominant amongst these are that a relatively low risk of hazard (5%) was chosen as the critical assessment level; that sub-lethal end-points were considered; and that bioavailable copper was over estimated.

Factors that may contribute to risk but which have not been addressed in this study include the possible presence of metals such as zinc that could influence the overall toxicity of harbour waters. It is also imperative that a biological survey be undertaken to identify, if possible, potential keystone species in equivalent environments with particular note of any filter feeders or benthic species that may be affected by the high concentration of copper in the Harbour sediment.

Acknowledgments

The authors would like to express their gratitude to Drs Ahsanullah, Stauber and Koehnken for providing access to their data. Mr C. Rehberg helped in data input and sorting. Both he and Mr J. Perera provided some of the figures used in this report.

References

Ahsanullah, M., T. M. Florence & J. L. Stauber, 1995. Ecotoxicology of copper to marine and brackish water organisms. Report to the Mt Lyell Remediation Research and Development Program (Project 9). CSIRO Inv. Rep. CET/IR402.

Aldenberg, T. & W. Slob, 1993. Confidence limits for hazardous concentrations based on logistically distributed NOEC toxicity data. Ecotoxicol. Envir. Safety 25, 48–63.

ANZECC, 1992. 'Australian Water Quality Guidelines for Fresh and Marine Waters'. Australian and New Zealand Environment and Conservation Council secretariat. Government Printing Office, Canberra.

Emans, H. J. B., E. J. van de Plassche, J. H. Canton, P. C. Okkerman & P. M. Sparenburg, 1993. Validation of some extrapolation methods used for effect assessment. Envir. Toxicol. Chem. 12: 2139–2154.

Hendricks, A. J., 1995. Modeling response of species to microcontaminants: Comparative ecotoxicology by (sub)lethal body burdens as a function of species size and partition ratio of chemicals. Ecotoxicol. Envir. Safety 32: 103–130.

218

Hodson, P. V., U. Borgmann & H. Shear, 1979. Toxicity of copper to aquatic biota. Ch 11. In Nriagu, J. O. (ed.), 'Copper in the Environment' Pt 2. Health Effects. J. Wiley & Sons, New York.

Hunt, D. T. E., 1987. Trace metal speciation and toxicity to aquatic organisms: A review. TR 247, Water Research Centre: Environment, Marlow, U.K., 51 pp.

Kooijman, S. A. L. M., 1987. A safety factor for LC_{50} values allowing for differences in sensitivity between species. Wat. Res. 21: 269–276.

OECD, 1992. Report of the workshop on the extrapolation of laboratory aquatic toxicity data to the real environment. OECD Environmental Monographs No. 59. Organisation for Economic Co-operation and Development. Paris. OCDE/GD (92)169.

Steel, R. G. D. & J. H. Torrie, 1981. 'Principles and procedures of statistics. A biometrical approach' (2nd ed.). Magraw-Hill International.

Wagner, C. & H. Lokke, 1991. Estimation of ecotoxicological protection levels from NOEC toxicity data. Wat. Res. 25: 1237–1242.

Statsoft Inc., 1994. STATISTICA for Windows. Release 5. Tulsa, Oklahoma, USA.

Stephan, C. E., D. I. Mount, D. J. Hanson, J. H. Gentile, G. A. Chapman & W. A. Brungs, 1985. Guidelines for deriving numeric National Water Quality Criteria for the protection of aquatic organisms and their uses, PB85-227049. US EPA, Duluth, Minnesota, USA.

Hydrobiologia **352**: 219–230, 1997.
Y.-S. Wong & N. F.-Y. Tam (eds), Asia–Pacific Conference on Science and Management of Coastal Environment.
©1997 *Kluwer Academic Publishers. Printed in Belgium.*

Spatial variations of size-fractionated Chlorophyll, Cyanobacteria and Heterotrophic bacteria in the Central and Western Pacific

Nianzhi Jiao[1] & I-Hsun Ni[2]
[1] *Institute of Oceanology, Chinese Academy of Sciences, Qingdao, China*
[2] *Department of Biology, Hong Kong University of Science & Technology, Hong Kong*

Key words: Size-fraction, Chlorophyll, Cyanobacteria, Heterotrophic bacteria, Pacific

Abstract

Geographic and vertical variations of size-fractionated (0.2–1 μm, 1–10 μm, and >10 μm) Chlorophyll *a* (Chl.a) concentration, cyanobacteria abundance and heterotrophic bacteria abundance were investigated at 13 stations from 4°S, 160°W to 30°N, 140°E in November 1993. The results indicated a geographic distribution pattern of these parameters with instances of high values occurring in the equatorial region and offshore areas, and with instance of low values occurring in the oligotrophic regions where nutrients were almost undetectable. Cyanobacteria showed the highest geographic variation (ranging from 27×10^3 to $16,582 \times 10^3$ cell l^{-1}), followed by Chl.a (ranging from 0.048 to 0.178 μg l^{-1}), and heterotrophic bacteria (ranging from 2.84×10^3 to 6.50×10^5 cell l^{-1}). Positive correlations were observed between nutrients and Chl.a abundance. Correspondences of cyanobacteria and heterotrophic bacteria abundances to nutrients were less significant than that of Chl.a. The total Chl.a was accounted for 1.0–30.9%, 35.9–53.7%, and 28.1–57.3% by the >10 μm, 1–10 μm and 0.2–1 μm fractions respectively. Correlation between size-fractionated Chl.a and nutrients suggest that the larger the cell size, the more nutrient-dependent growth and production of the organism. The ratio of pheophytin to chlorophyll implys that more than half of the >10 μm and about one third of the 1–10 μm pigment-containing particles in the oligotrophic region were non-living fragments, while most of the 1–10 μm fraction was living cells. In the depth profiles, cyanobacteria were distributed mainly in the surface layer, whereas heterotrophic bacteria were abundant from surface to below the euphotic zone. Chl.a peaked at the surface layer (0–20 m) in the equatorial area and at the nitracline (75–100 m) in the oligotrophic regions. Cyanobacteria were not the principle component of the picoplankton. The carbon biomass ratio of heterotroph to phytoplankton was greater than 1 in the eutrophic area and lower than 1 in oligotrophic waters.

Introduction

Phytoplankton, cyanobacteria and heterotrophic bacteria are the principal components of the basic communities in oceanic environments: their dynamics are of crucial importance to the understanding of the structures and functions of marine ecosystems, especially with respect to material cycling and climate changes (Cho & Azam, 1990; Krupatkina, 1990; SCOR, 1990). The Pacific ocean is the largest water body in the world; it can be divided into two divisions according to the nutrient availability: the eutrophic region and the oligotrophic region. The central equatorial area is a typical

upwelling-driven eutrophic area, and coastal areas constitute another kind of eutrophic area due to the input of nutrients from the land. However, the majority of the open ocean is made up of oligotrophic waters. It is, therefore, of interest to investigate the spatial dynamics of the aforementioned microorganisms from the central Pacific to coastal areas and to determine the controlling factors for their standing biomass.

There has been a large number of investigations on the geographic distribution and temporary variation of phytoplankton around the world's oceans, but most of them only dealt with whole phytoplankton assemblages. Size fractionated biomass measurements are still rare especially in the Pacific Ocean (Chavez, 1989; Bouteiller et al., 1992). Furthermore, glass fiber filters (Whatman GF/F), rather than Nuclepore filters, were often used, allowing some picoplankton species to escape collection and thus leading to an underestimation of Chlorophyll *a* (hereafter, Chl.a) concentration in subtropical and transition waters by a factor of two to four (Dickson & Wheeler, 1993). Also, autotrophic and heterotrophic organisms are often studied separately. In addition, most of the previous studies were conducted over relatively small areas. Basinwide investigations however are limited because either they involve time differences which limit data comparability or because they require a heavy outlay. The present study is therefore aimed at (1) displaying the geographic variation based on the synchronous measurement of size-fractionated chlorophyll *a* concentration, cyanobacteria and heterotrophic bacteria abundance on a transregional scale from the Central Pacific to the Northwest Pacific; and (2) outlining the vertical profiles of these three indices for microorganisms in typical eutrophic and oligotrophic areas, so as to understand their general distribution pattern and the controlling mechanism for that distribution pattern.

Description of study sites and sampling techniques

This study was conducted during a cruise of the Hakuho Maru of Tokyo University in November 1993. There were 13 sampling sites (Figure 1, with Coordinates in Table 1) from transect lines distributed from the equatorial central Pacific (4° S, 160° W) to the Northwest Pacific (30° N, 42° E). Samples for our spatial dynamic study were collected approximately 10 am daily from the surface water (about 5 m) of Stations 1–12 with an on-board sampling bump. Samples for vertical distribution study were collected from the central equatorial upwelling area (Station 2) at 0, 20, 50, 75, 100 and 150 m, as well as the oligotrophic area (Station 0) at 0, 20, 25, 50, 75, 100, 150, 200 and 300 m with 10-liter Niskin bottles attached to a CTD rosette system (General Oceanics). Sampling depth were determined according to CTD fluorescence profiles monitored by SeaTech Fluorometer.

Materials and methods

Size fractionation of microorganisms

Natural phytoplankton assemblages were divided into 3 size categories: picoplankton (hereafter, Pico; 0.2–1 μm), nanoplankton (Nano; 1–10 μm) and netplankton (Net; >10 μm) respectively. In order to reduce the systematic error from repeated filtering with Nuclepore filters of different pore sizes, size fractionation was performed by filtering respective 1000 ml, 500 ml, and 100 or 200 ml water sample directly onto corresponding 10 μm Nuclepore filter at gravity, and 1 μm and 0.2 μm Nuclepore filters in a vacuum of less than 0.03 MPa respectively. The Nano fraction would, therefore, be obtained by subtracting the >10 μm fraction (Net) from the >1 μm fraction (Net + Nano) and the Pico fraction by subtracting the >1 μm fraction from the >0.2 μm fraction (Net + Nano + Pico). Duplicate measurements were done for each sample. Usually the coefficients of variation are less than 10%. Size fractionation of microorganisms was performed at St. 2 through St. 12.

Determination of pigments

Chl.a and pheophytin (Phe.) in the filter samples were extracted by N, N-Dimenthylformamide which has been shown to be more efficient than acetone (Suzuki & Ishimaru, 1990), and then measured on board by fluorometric techniques using a Turner Design model-10AU fluorometer which was calibrated with commercial Chl.a (Sigma).

Cell counting

Aliquots of 30–50 ml water were subsampled into polycarbonate tubes, which were then fixed with formalin at a final concentration of 2% and stored at 4 °C in a refrigerator, for subsequent numerical counting of cyanobacteria and heterotrophic bacteria. Microscopic examination of these samples was performed within 4 weeks of the fixing at which time the stored samples were filtered onto 0.2 μm black Nuclepore at a vacuum of less than 0.03 MPa. Cyanobacteria were counted by epifluoreoscence method (Wood et al., 1985) with a Zeiss epifluoreoscence microscope set equipped with BP450-490, FT510, and LP515 light filters. Heterotrophic bacteria were stained with 4,6-diamidinno-2-phenylindole (DAPI) (Porter & Feig, 1980) and counted by epifluorescence microscope with

Table 1. Location of 13 stations with contributions of size-fractionated Chlorophyll a to total Chlorophyll a.

Station No.	Longitude	Latitude	Total Chl.a (μg l^{-1})	Net Chl.a (%)	Nano Chl.a (%)	Pico Chl.a (%)
0	160°00′W	5°00′N	0.098	–	–	–
1	160°00′W	4°00′S	0.154	–	–	–
2	159°00′W	0°	0.178	30.89	41.01	28.10
3	165°00′W	3°00′N	0.103	6.80	35.92	57.28
4	171°00′W	6°00′N	0.074	10.81	39.19	50.00
5	176°30′W	8°58′N	0.048	14.58	45.83	39.58
6	177°14′E	12°08′N	0.069	8.69	53.62	37.68
7	171°12′E	15°08′N	0.079	6.33	48.10	45.57
8	165°00′E	18°16′N	0.077	7.79	46.75	45.45
9	159°00′E	21°07′N	0.079	8.86	48.10	43.04
10	153°43′E	24°23′N	0.067	8.96	44.78	46.29
11	147°06′E	27°00′N	0.105	0.90	50.94	48.11
12	141°36′E	29°18′N	0.127	11.03	45.67	43.30
mean			0.102	10.51	45.45	44.04

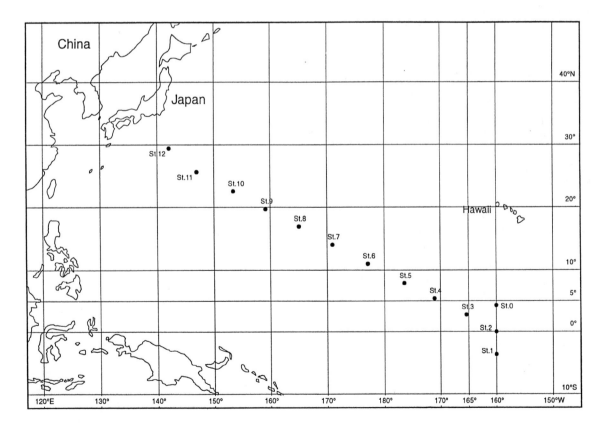

Figure 1. Locations of sampling sites (see Table 1 for coordinates). Stations 0 and 2 were designed especially for vertical variation study.

222

a FT390, and LP395 filter set. Heterotrophic bacteria were divided into 'long-shaped' and 'round shaped' groups according to whether the ratio of long axis of the cell to short axis of the cell was obviously greater than 2.

Determination of nutrients

Nitrate, nitrite, ammonia and phosphate concentrations were measured with a Technicon II autoanalyzer according to the method described by Parsons et al. (1984). Nutrient detection limits were 0.08, 0.01, 0.01 and 0.02 μmol l^{-1} for nitrate, nitrite, ammonia and phosphate respectively.

Results

1. Geographic variations of nutrients

The geographic distribution of temperature and salinity are displayed in Figure 2. The sea surface temperature was relatively low and the salinity relatively high in the equatorial area which indicated the existence of upwelling. The geographic variations of such nutrients as nitrate, nitrite, ammonia, and phosphate are shown in Figure 3. The univariate statistics (mean, standard deviation and range) for each variable among 12 stations are summarized in Table 2. Along the cruise from St. 1 to St. 12. nitrate varied from undetectable to 2.30 μmol l^{-1} with two peaks occurring at St. 2 and around St. 12. St. 2 is within the equatorial upwelling area where new nitrogen, especially nitrate, (Dugdale & Goering, 1967) is upwelled to the surface layer from below the thermocline (about 100 m, refer to Figure 7a). Here nitrate had the highest concentration among all the investigation sites. The secondary peak at St. 12. was apparently due to the land nutrient input since there was an inshore-direction increasing trend from St. 10 to St. 12. The concentrations of nitrite and phosphate followed the same pattern as for nitrate, ranging from undetectable to 0.59 μmol l^{-1} and from undetectable to 0.47 μmol l^{-1} respectively. Ammonia, a typical regenerated nitrogen (Dugdale and Goering, 1967), only ranged from undetectable to 0.06 μmol l^{-1}; its geographic coefficients of variation were less significant than that of the other three nutrients investigated.

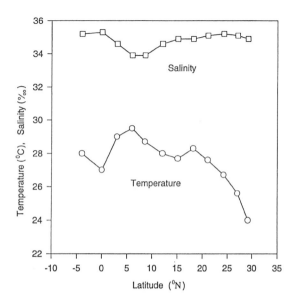

Figure 2. Geographic variations of surface temperature and salinity. Refer to Figure 1 and Table 1 for the corresponding latitude for each station.

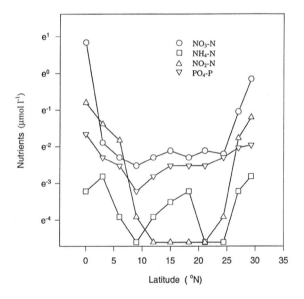

Figure 3. Geographic variations of surface nitrate, nitrite, ammonia and phosphate.

2. Geographic variations of pigments

The geographic variations of total and size-fractionated Chl.a are shown in Figure 4a. For total Chl.a, the peak values occurred at St. 1, 2 and 12, the former two stations are in the equatorial area, and the latter is in an offshore area. The lowest value occurred at St. 5 where the levels of all the nutrients were lowest. Mea-

Table 2. Univariate statistics of temperature, salinity, nitrate, ammonia, phosphate, total and fractionated (Net, Nano, Pico) Chlorophyll a, cyanobacteria, total bacteria and long-shaped heterotrophic bacteria for stations 1 to 12.

	Temp °C	Salinity ‰	NO$_3$ μ mol l^{-1}	NH$_4$ μ mol l^{-1}	PO$_4$ μ mol l^{-1}	Total Chl.a μg l^{-1}	Net Chl.a μg l^{-1}	Nano Chl.a μg l^{-1}	Pico Chl.a μg l^{-1}	Cya Bact ×10^7 cell l^{-1}	Total Bact ×10^8 cell l^{-1}	Long Bact ×10^8 cell l^{-1}
Mean	27.51	34.80	0.38	0.03	0.14	0.10	0.01	0.04	0.04	0.30	4.01	0.37
SD	1.53	0.47	0.64	0.02	0.16	0.04	0.01	0.01	0.01	0.47	1.18	0.12
Min	24.00	33.90	UD#	UD#	UD#	0.05	0.00	0.02	0.02	0.00	2.84	0.11
Max	29.50	35.30	2.30	0.06	0.47	0.18	0.06	0.07	0.07	1.66	6.50	0.52
CV*	0.06	0.01	1.68	0.67	1.14	0.40	1.34	0.36	0.35	1.61	0.29	0.33

∗ CV: Coefficient of variation
UD: nutrients undetectable

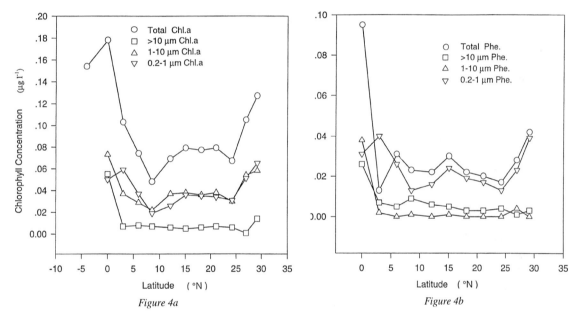

Figure 4a

Figure 4b

Figure 4. Geographic variations of surface chlorophyll *a* (4a) and pheophytin (4b).

surements of the three Chl.a size-fractions followed the same geographic distribution trend as the total, but with different variabilities. The coefficients of variation for the Net, Nano and Pico fractions were 1.34, 0.36 and 0.35 respectively (Table 2). Regarding contributions of different fractions to the total Chl.a, the Net fraction was on average responsible for only 10.53% of the total, the remaining contribution (approximately 90%) was shared by Nano and Pico fractions (Table 1). The Net fraction also had the most geographic variability in terms of its relative contribution (%), which peaked where the nutrients were relatively abundant. Phe. (Figure 4b) had a geographic variation pattern similar to that of Chl.a, but the variations in different

size fractions were quite different. The ratio of Phe. to Chl.a (Table 2) indicates that almost half of the Net fraction and about 1/3 of the Pico fraction were non-living organic particles, whereas most of the particulate matter in the Nano fraction was living organisms.

Significant correlations between Chl.a and nutrients were observed (Figures 5a, 5b and 5c). Nitrate was significantly correlated to the larger fractions (Net and Nano), but not to the smaller fraction (Pico) (Figure 5a). On the other hand, ammonia correlated better with the smaller fractions (Pico and Nano) than with the larger fraction (Net) (Figure 5b). Among all the nutrients, phosphate correlated significantly with total Chl.a as well as each of the three fractions (Figure 5c).

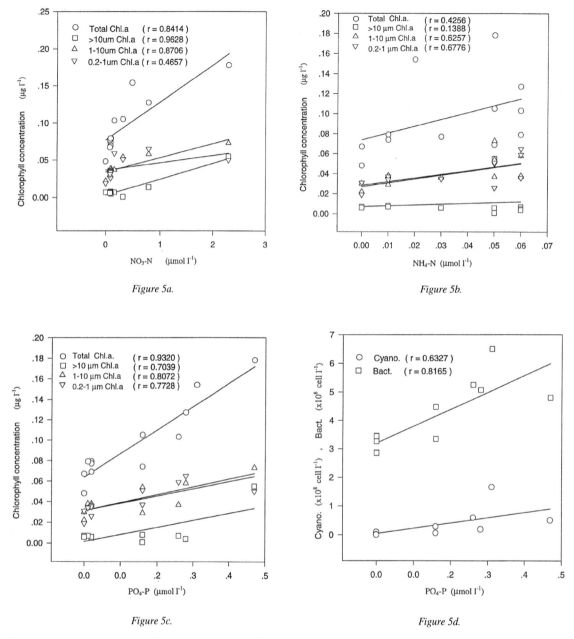

Figure 5a.

Figure 5b.

Figure 5c.

Figure 5d.

Figure 5. Correlations between chlorophyll *a* and nitrate (5a), chlorophyll *a* and ammonia (5b), chlorophyll *a* and phosphate (5c), and Correlation between cyanobacteria, heterotrophic bacteria and phosphate (5d).

3. *Geographic distribution of cyanobacteria and heterotrophic bacteria*

The abundance of cyanobacteria varied from the order of magnitude of 10^5 cell l^{-1} in the vast oligotrophic region (St. 5 through St. 11) to the order of magnitude of 10^7 cell l^{-1} in the equatorial area (Figure 6a).

The coefficient of variation for geographic distribution was as high as 1.61 (Table 2) which was the highest among all the biological parameters investigated. The abundance of heterotrophic bacteria, however, was less geographically variable ranging from 2.85×10^8 to 6.50×10^8 cell l^{-1} with a mean of 4.0×10^8 cell l^{-1} and a coefficient of variation of 0.29 (Figure 6b).

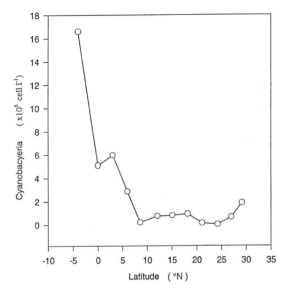

Figure 6a.

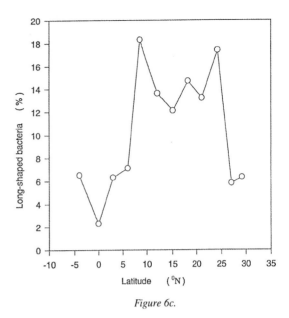

Figure 6c.

Figure 6. Geographic variations of cyanobacteria (6a), heterotrophic bacteria (6b) and the relative abundance of the long-shaped bacteria (6c).

4. Vertical distribution of nutrients

St. 0 is a typical oligotrophic site and St. 2 is a typical equatorial upwelling eutrophic site. The sea surface temperature at St. 2 was slightly lower than at St. 0, but the thermoclines at both stations were almost identical (Figure 7a). The vertical distribution of nitrate, and to a lesser extent, phosphate were characterized by a sharp variation at the thermocline (Figure 7b). In the oligotrophic region, nutrients commonly found in the euphotic zone were very limited. Phosphate was about 0.16 μmol l^{-1}, and nitrate and ammonia were essentially undetectable. Below the thermocline, nitrate increased rapidly and reached a level of 30 to 35 μmol l^{-1}, at a depth of 200 to 300 m. Phosphate also increased slightly with increasing depth, but ammonia was uniform throughout the water column. The vertical distribution of nutrients in the equatorial area was quite different from that in the oligotrophic region. First, all the nutrients were relatively abundant in the euphotic zone, the average concentrations of nitrate, ammonia and phosphate being 4.2, 0.04 and 0.52 μmol l^{-1} respectively. Second, the depth profiles for nitrate and phosphate were less drastic since the upwelling brought 'new nutrients' to the upper layer which reduced the differences between the layers above and below the thermocline. On the other hand,

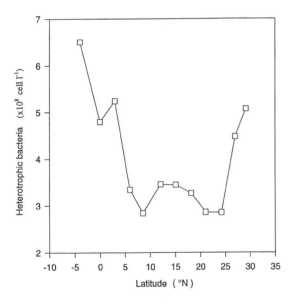

Figure 6b.

Among the heterotrophic bacteria, most of them were round-shaped while about 10% of them were long-shaped (possibly bacillus and vibrio). The percentages of the long-shaped bacteria were higher in the oligotrophic area than in the equatorial region (Figure 6c). Correlation analyses between cyanobacteria, heterotrophic bacteria and nutrients indicated that only phosphate correlated significantly with these microorganisms (Figure 5d).

ammonia was undetectable below the euphotic zone, this demonstrated the typical depth profile for regenerated nitrogen.

5. Vertical distribution of chlorophyll a

Figure 7c shows the vertical distribution of Chl.a in both the oligotrophic region and the equatorial region. With regard to the former case, the concentration of Chl.a in the upper layer (<50 m) was very low, and it decreased slowly with increasing depth from 0.091 μg l^{-1} at the surface to 0.074 μg l^{-1} at a depth of 50 m, then increased abruptly to 0.32 μg l^{-1} over the thermocline, after which it again decreased gradually to undetectable at 200 m. In the equatorial region, the vertical distribution pattern of Chl.a was much more simple. Chl.a peaked at the surface layer (<50 m) with a high value of 0.043 μg l^{-1}, and then gradually decreased to less than 0.02 μg l^{-1} at a depth of 150 m.

6. Vertical distribution of cyanobacteria and heterotrophic bacteria

The depth profile of cyanobacteria was similar to that of Chl.a, however, cyanobacteria distributed shallower than did Chl.a at both oligotrophic sites and equatorial sites (Figure 7d). The maximum abundance of cyanobacteria was about 3×10^6 cell l^{-1} in both the oligotrophic region and the equatorial area, but at different depths. The vertical distribution pattern of heterotrophic bacteria was similar to those of Chl.a in both the equatorial region and the oligotrophic area, but heterotrophic bacteria distributed much deeper than that of Chl.a. Below 200 m, Chl.a was undectable, and heterotrophic bacteria was still abundant. The maximum abundance of heterotrophic bacteria was about 20×10^6 cell l^{-1} both in the oligotrophic region and in the equatorial area.

Discussion

Factors controlling phytoplankton abundance

For the geographic distribution, correlation analysis (Figures 5a, 5b and 5c) indicated that phytoplankton abundance was basically controlled by nutrient availability. For the vertical distribution, in the oligotrophic area, the low phytoplankton biomass (as shown by Chl.a concentration) in the upper layer was apparently due to a lack of nutrients. Given the same nutri-

Table 3. The ratio of Pheophytin to Chlorophyll a.

Station No.	Total Chl.a (%)	Net Chl.a (%)	Nano Chl.a (%)	Pico Chl.a (%)
2	53.4	47.3	52.1	62.0
3	12.6	100.0	5.4	67.8
4	41.9	62.5	0.0	70.3
5	47.9	128.6	4.5	68.4
6	31.9	100.0	0.0	61.5
7	38.0	100.0	2.6	66.7
8	28.6	50.0	UD#	54.3
9	25.3	42.9	UD#	50.0
10	25.4	66.7	UD#	41.9
11	26.7	100.0	7.4	45.1
12	33.1	75.0	0.0	60.0
Mean	33.16	79.36	9.00	58.91

UD — Pheophytin was undetectable.

ent conditions, the decrease of Chl.a in this layer was believed to be due to a decrease in the availability of light. The sudden increase of Chl.a at the thermocline layer, though partially due to higher Chl.a content per cell (light- adaption), was no doubt resulted principally from the abundant supply of nutrients. The subsequent decrease was again caused by the limitation of the availability of light. On the other hand, in the equatorial upwelling area, with sufficient nutrient supply, the phytoplankton depth profile was basically controlled by light availability. In comparison with other normal eutrophic waters, Chl.a in equatorial Pacific waters was still low due to lack of soluble iron as suggested by Martin et al. (1994).

The size structure of phytoplankton

Since the size structure of microorganisms as well as species composition is of crucial importance to the energy flow pathway, it has become one of the focuses of recent studies in marine productivity. Previous research has revealed that picoplankton (usually 0.2–2 μm) accounts for about 40–80% of the total phytoplankton in terms of both biomass and productivity in oligotrophic regions (Grandinger et al., 1993; Gomes et al., 1993; Odate & Fukuchi, 1994) and about 30–50% in coastal waters (Jiao & Wang, 1994; Tata et al., 1994). In the present study, the Pico fraction (<1 μm) accounted for 45% of the total Chl.a, and the Net fraction accounted for only 10% of that total. This supports the results of a comparative study which found that surface Chl.a in the North Pacific measured with 0.2 μm

227

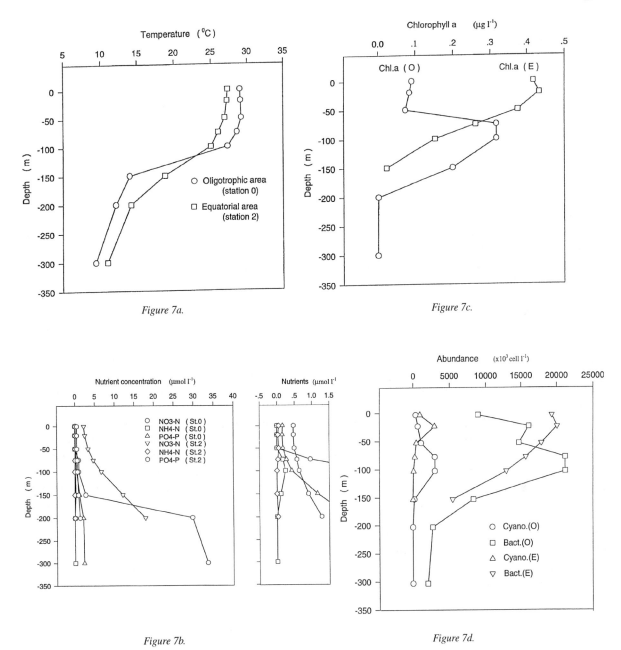

Figure 7a.

Figure 7c.

Figure 7b.

Figure 7d.

Figure 7. Vertical distribution of temperature (7a), nutrients (7b), Chlorophyll *a* (7c), cyanobacteria and heterotrophic bacteria (7d) in equatorial (St. 2) and oligotrophic (St. 0) areas.

Nuclepore was up to four times higher than that measured with Whatman GF/F filters which are extensively used for measurement of Chl.a (Dickson & Wheeler, 1993). With respect to the common concept for size fractionation, picoplankton (0.2–2 μm) would be even more dominant and netplankton (20–200 μm) might be less significant in the pelagic ocean since our size frac-

tionation was based upon 1 μm and 10 μm Nuclepore filters. In fact, an inverse relationship between percentage of picoplankton and total Chl.a was observed, and netplankton showed a decreasing trend with increasing oligotrophic conditions. This confirmed the idea that larger phytoplankton species are associated with eutrophic waters, whereas smaller microorganisms are dominant in oligotrophic waters (Chisholm, 1992). Moreover, the correspondence of fractionated Chl.a to nitrate and to ammonia were in inverse order: the significance of the correlation coefficient for nitrate formed a Net>Nano>Pico sequence, but for ammonia it formed a Pico>Nano>Net sequence. This suggests that larger phytoplankton cells prefer new nitrogen to regenerated nitrogen whereas smaller cells prefer regenerated nitrogen to new nitrogen. Moreover, nitrate is relatively abundant in the equatorial region and ammonia is relatively dominant in the oligotrophic region. Therefore, it can be concluded that new production would be based mainly on the larger fractions and regenerated production mainly on the smaller fractions, and that the f-ratio (new production/primary production) would be much higher in eutrophic areas than in oligotrophic areas.

The significance of cyanobacteria

During the past decade, cyanobacteria have been the subject of many studies, and they were regarded as forming the dominant component of picoplankton (Krupatkina, 1990). In some instances, cyanobacterial abundance was actually quite high (e.g., Northern Indian Ocean, Veldhuis & Kraay, 1993). But in the Northern Pacific Ocean, the present results for surface distribution and vertical profile do not support the point that cyanobacteria is the dominant component of the picoplankton. The abundance of cyanobacteria was 1–3 orders of magnitude lower than that of heterotrophic bacteria, and probably the 1 or 2 order of magnitude lower than that of *prochlorococcus*, a newly discovered unicellular procaryote with cell size of about 0.6 μm (Chisholm et al. 1992) which we were not able to examine its abundance but included its biomass in terms of Chl.a into pico fraction. Also, biomass of cyanobacteria would account for very small part of autotrophs in terms of carbon which can be calculated by using the conversion factor of 250 fgC per cell for cyanobacteria (Kana & Glibert, 1987), and 71 for phyto-C/Chl.a. (Campbell & Nolla, 1994) (Table 4). This result coincides with the conclusion of a flowcytometric investigation at the ALOHA station (22°45'N, 158°W) (Campbell & Nolla, 1994). Furthermore, the variability of cyanobacteria abundance is the largest among all the three parameters through Station 1–12 (Table 2). From the geographic distribution and the vertical profile, it can be speculated that the abundance of cyanobacteria was limited by either the availability of light or the availability of nutrients.

The role of heterotrophs and autotrophs

Whether bacteria, phytoplankton, or detritus dominate the food for grazers in the euphotic zone is likely to influence the food web structure, nutrient cycling and sinking flux (Cho & Azam, 1990). In the pelagic ocean, detritus is relatively insignificant. Bird & Kalff (1984) reported a positive correlation between bacterial abundance and Chl.a and demonstrated that from eutrophic to mesotrophic waters, the level of bacteria decreased less slowly than that of Chl.a. In this study, we also found a significant positive correlation between bacteria and Chl.a both for geographic distribution and vertical distribution within the euphotic zone. This suggests a dependence of heterotrophic bacteria on phytoplankton for dissolved organic carbon (DOC) supply and a dependence of phytoplankton on heterotrophic bacteria for nutrient supply. In view of their variability, heterotrophic bacteria are more independent of environmental factors than autotrophic organisms. The biomass ratio of heterotrophic bacteria to phytoplankton in terms of carbon (Table 4) is less than 1 in the equatorial eutrophic area and greater than 1 in the oligotrophic areas. This supports the idea that food webs in eutrophic waters are dominated by the primary producer biomass, while, food webs in oligotrophic waters are dominated by the decomposers (Dortch & Packard, 1989). Cho & Azam (1990) reported that the bacteria biomass in oligotrophic waters is commonly 2–3 times greater than the phytoplankton biomass, and it is a key link in food web structure and nutrient cycling pathways. In comparison, the present data for surface waters show less predominance of heterotrophic bacteria over phytoplankton (Table 4).

Campbell & Nolla (1994) pointed out that it is impossible to discriminate *Prochlorococcus* from heterotrophic bacteria when samples are labeled with DAPI, because the chlorophyll fluorescence of the former is too dim. Epifluorometric bacteria counts, therefore, provide an estimate of total bacteria that overstates (by about 31%) the importance of heterotrophic bacteria. On the other hand, bacteria stained with DAPI were reported to be underestimated (by about 26%) compared to those stained with acridine orange (AO) (Suzuki et al., 1993). Considering these two points, the positive and negative bias in the present study seem to be balanced. However, since microscopy is the routine method for the assessment of microorganisms, and since DAPI is preferred to AO because DAPI stained cells are more clearly visible and longer lasting, then it is necessary to directly compare different methods

Table 4. Carbon biomass ($< mu$g C 1^{-1}) for total phytoplankton assemblage with its size-fractions, cyanobacteria, and heterotrophic bacteria.

Station No.	Total	Net	Nano	Pico	Cyano bacteria	Hetero bacteria	Heterotropher/ Total phytoplankton
1	10.78	–	–	–	4.15	13.00	1.20
2	12.46	3.85	5.11	3.50	1.27	9.58	0.76
3	7.21	0.49	2.59	4.13	1.49	10.48	1.45
4	5.18	0.56	2.03	2.59	0.72	6.68	1.29
5	3.36	0.49	1.54	1.33	0.05	5.68	1.69
6	4.83	0.42	2.59	1.82	0.19	6.90	1.42
7	5.53	0.35	2.66	2.52	0.20	6.88	1.24
8	5.39	0.42	2.52	2.45	0.08	6.52	1.21
9	5.53	0.49	2.66	2.38	0.03	5.72	1.03
10	4.69	0.42	2.10	2.17	0.00	5.70	1.21
11	7.35	0.07	3.78	3.57	0.17	8.94	1.21
12	8.89	0.98	4.06	4.55	0.47	10.12	1.13
Mean	6.76	0.77	2.87	2.81	0.73	8.02	1.24
SD	2.70	1.04	1.03	0.99	1.18	2.36	0.22

involved in fluorescence microscopy techniques. For a better understanding of the structures and functions of microplanktonic communities in oceanic waters, scientists need to conduct further studies on the smallest autotrophic prokaryotes (*Prochlorococcus*, 0.6 μm) and eukaryotes (*Ostreococcus tauri*, 0.70–0.97 μm; Countiers et al., 1994) by the application of flowcytometry.

Conclusions

From this study, the following conclusions can be drawn:

1. The standing biomasses of phytoplankton, cyanobacteria and heterotrophic bacteria were relatively high in the equatorial area and the offshore area, but quite low in the oligotrophic area. Vertical profiles of Chl.a, cyanobacteria and heterotrophic bacteria peaked in the upper layer of the euphotic zone (<50 m) in the equatorial area, but at the bottom layer of the euphotic zone (around 100 m) in oligotrophic waters.
2. The total amount of chlorophyll a was mainly contributed by nanoplankton and picoplankton; netplankton was responsible for a very small part (only about 10%). More than half of the >10 μm and about one third of the 0.2–1 μm pigment-containing particulate organic matter were likely to be non-living particles whereas most of the 1–10 μm pigment-containing particles were phytoplankton.

3. Except for the equatorial area, the primary production in the open ocean seemed to be limited mainly by nutrient availability. Correspondences of microorganisms to NO_3-N were more significant in larger fractions, but the opposite was true in correspondences of smaller fractions to NH_4-N.

4. Cyanobacteria were the most variable organisms and seemed not to be the principle component of the picoplankton as previously expected. Heterotrophic bacteria were the most stable organisms in oceanic waters, but they were not always the dominant microorganisms. The carbon biomass ratio of heterotrophs to phytoplankton was less than 1 in the equatorial eutrophic area and greater than 1 in other oligotrophic waters.

Acknowledgments

We are greatly indebted to Professor I. Koike for his valuable suggestions and offering the use of his facilities both in the lab and in the field. We thank the KH-93-4-cruise scientists especially Dr K. Furuya and Dr T. Saino for their kind assistance in the field. We are grateful to Mr T. Usui for his help in analyzing the nutrients. We also thank the captain and crew of Hokaho Maru for their assistance in the field.

230

References

Bird, D. F. & F. Kalff, 1984. Empirical relationship between bacterial abundance and chlorophyll concentration in fresh and marine waters. Can. J. Fish. aquat. Sci. 41: 1015–1023.

Bouteiller, L. A., J. Blanchott & M. Rodier, 1992. Size distribution pattern of phytoplankton in the West Pacific: Toward a generalization for the tropical open ocean. Deep Sea Res. 39: 805–823.

Campbell, L. & H. A. Nolla, 1994. The importance of *Prochlorococcus* to community structure in the central North Pacific Oceano. Limnol. Oceanogr. 39: 954–961.

Chisholm, S. W., 1992. Phytoplankton size. In Falkowski, G. & A. D. Woodhead (ed.), Primary Productivity and Biogeochemical Cycles in the Sea. Plenum Press, New York: 213–237.

Chisholm, S. W., R. J. Olson, E. R. Zettler, R. Goericke, J. Waterbury & N. Welshmeyer, 1988. A novel free-living prochlorophyte abundant in the oceanic euphotic zone. Nature 334: 340–343.

Chavez, F. P., 1989. Size distribution of phytoplankton in the central and eastern tropical Pacific. Global Biochem. Cycles, 3: 27–35.

Cho, B. C. & F. Azam, 1990. Biogeochemical significance of bacterial biomass in the ocean's euphotic zone. Mar. Ecol. Prog. Ser. 63: 253–259.

Countiers, C., A. Vaquer, M. Troussellier & J. Lautier, 1994. The smallest eukaryotic organism. Nature 370: 255.

Dickson, M. L. & P. A. Wheeler, 1993. Chlorophyll concentration in the North Pacific: Does a latitudinal gradient exist? Limnol. Oceanogr. 38: 1813–1818.

Dortch, Q. & T. T. Packard, 1989. Differences in biomass structures between oligotrophic and eutrophic marine ecosystems. Deep-Sea Res. 36: 223–240.

Dugdale, R. C. & J. J. Goering, 1967. Uptake of new and regenerated forms of nitrogen in primary productivity. Limnol. Oceanogr. 12: 196–206.

Gomes, H. D. R., J. I. Goes & A. H. Parulekar, 1993. Size-fractionated biomass, photosynthesis and dark carbon dioxide fixation in a trophic oceanic environment. J. Plankton Res. 14: 1307–1329.

Grandinger, R., T. Weisse & T. Pillen, 1993. Significance of pico-cyanobacteria in the Red Sea and the Gulf of Aden. Bot. mar. 35: 245–250.

Jiao, N.-Z. & R. Wang, 1994. Size-fractionated biomass and production of microplankton in the Jiaozhou Bay. J. Plankton Res. 16: 1609–1625.

Kana, T. M. & P. M. Glibert, 1987. Effect of irradianaces up to 2000uEm-2S on marine *Synechococcus* WH7803. 1. Growth, pigmentation and cell composition. Deep-Sea Res. 34: 479–495.

Krupatkina, D. K., 1990. Estimates of primary production on oligotrophic waters and metabolism of picoplankton: A review. Mar. Microb. Food Web. 4: 87–101.

Martin, J. H., K. H. Coale, K. S. Johnson, S. E. Fitzwater, R. M. Gordon, S. J. Tanner, C. N. Hunter, V. A. Elrod, J. L. Nowicki, T. L. Coley, R. T. Barber, S. Lindley, A. J. Watson, K. Van Scoy, C. S. Law, M. I. Liddicoat, R. Ling, T. Stanton, J. Stockel, C. Collins, A. Anderson, R. Bidigare, M. Ondrusek, M. Latasa, F. J. Millero, K. Lee, W. Yao, J. Z. Zhang, G. Fredrich, C. Sakamoto, F. Chavez, K. Bick, Z. Kobler, R. Green, P. G. Falkowski, S. W. Chisholm, F. Hoge, R. Swift, J. Yungle, S. Turner, Pl. Nightinggale, A. Hatten, P. Liss & N. W. Tindale, 1994. Testing the iron hypothesis in ecosystems of the equatorial Pacific, Nature 371: 123–129.

Odate, T. & M. Fukuchi, 1994. Surface distribution of picoplankton along the first leg of the JARE-33 cruise, from Tokyo to Freemantle, Australia. Bull. Plankton Soc. Japan 41: 93–104.

Parsons, T. R., Y. Maita & C. M. Lalli, 1984. A Manual of Chemical and Biological Method for Seawater Analysis. Pergamon Press: 3–122.

Porter, K. G. & Y. S. Feig, 1980. The use of DAPI for identifying and counting aquatic microflora. Limnol. Oceanogr. 25: 943–948.

SCOR, 1990. Joint Global Ocean Flux Study (JGOFS) science plan. JGOFS report. 5: 1–51.

Suzuki, R. & T. Ishimaru, 1990. An improved method for the determination of phytoplankton chlorophyll using N, N-Dimethylformamide. J. Oceanogr. Soc. Japan 46: 190–194.

Suzuki, M. T., E. B. Sherr & B. F. Sherr, 1993. DAPI direct counting underestimates bacterial abundance and average cell size compared to AO direct counting. Limnol. Oceanogr. 38: 1566–1570.

Tata, K., K. Matsumoto, M. Tata & T. Ochi, 1994. Size distribution of phytoplankton community in Hiroshima Bay. Kagawa Daigaku Nogakubu Gakujutsu Hokoku 46: 27–35.

Veldhuis, M. J. W. & G. W. Kraay, 1993. Cell abundance and fluorescence of picoplankton in relation to growth irradiance and availability in the Red Sea. Neth. J. Sea Res. 31: 135–145.

Wood, A. M., P. K. Horan, K. Muirhead, D. A. Phinney, C. M. Yentsch & J. B. Waterbury, 1985. Discrimination between types of pigments in marine *Synechococcus spp.* by scanning spectroscopy, epifluoroscence microscopy and flow cytometry. Limnol. Oceanogr. 30: 1303–1315.

Hydrobiologia **352**: 231–240, 1997.
Y.-S. Wong & N. F.-Y. Tam (eds), Asia–Pacific Conference on Science and Management of Coastal Environment.
©1997 *Kluwer Academic Publishers. Printed in Belgium.*

Reproductive function state of the scallop *Mizuhopecten yessoensis* Jay from polluted areas of Peter the Great Bay, Sea of Japan

Marina A. Vaschenko[1], Iraida G. Syasina[1], Peter M. Zhadan[2] & Lyubov A. Medvedeva[3]
[1]*Institute of Marine Biology, Far East Branch of Russian Academy of Sciences, Vladivostok 690041, Russia*
[2]*Pacific Oceanological Institute, Far East Branch of Russian Academy of Sciences, Vladivostok 690041, Russia*
[3]*Moscow State University, Moscow 119899, Russia*

Key words: bivalve mollusc, scallop, pecten, reproduction, gametogenesis, embryogenesis, pollution

Abstract

The morphology of gonads and development of offspring of the scallop *Mizuhopecten yessoensis* sampled from six stations in Peter the Great Bay (Sea of Japan) were studied. The retardation of gametogenesis, oocytes resorption, autolysis of spermatozoa and their phagocytosis were observed in the gonads of the scallops from polluted sites. The number of hermaphrodites was about 6% against 0.3–0.4% in the scallop populations from clean areas. In the offspring development, a decrease in fertilisation success, diminution in percent of normal trochophores, D-veligers, veligers, and retardation of larval growth were recorded. The scallop populations inhabiting polluted areas of Peter the Great Bay seem to be incapable of normal reproduction. Development of offspring was a more sensitive index of disturbance of the reproductive function than morphology of scallop gonads. Analysis of the offspring development of common species of marine invertebrates is suggested to be used as a sensitive indicator of adverse environmental conditions.

Introduction

Marine animals inhabiting polluted coastal waters near urbanised areas may be affected by a complex of toxic substances of domestic and industrial origin. Pollution of the marine environment represents a stress whose effects on organisms can be measured at different levels of organisation, i.e., biochemical, cytological, physiological (Bayne et al., 1985). Physiological responses have an important advantage in assessing of organism's health because they represent integrated biochemical and cellular responses to the changes in the environment (Widdows, 1985). From an ecological perspective, one of the most important physiological responses of the organisms to pollution is the impairment of their reproduction (Bayne et al., 1981; Malakhov & Medvedeva, 1991; Vaschenko & Zhadan, 1993; Krause, 1994). Disturbance of the reproductive function of marine species which are sensitive to pollution may be one of the mechanisms that causes negative changes in community structure, abundance, diversi-

ty and composition. To date, nevertheless, only a few studies have explored the *in situ* effects of polluted coastal waters on the gonad state, fecundity and development of offspring in marine invertebrates (Koster & Van den Biggelaar, 1980; Dixon & Pollard, 1985; Lee, 1986; Sunila, 1986; Pekkarinen, 1991; Vaschenko et al., 1992, 1993, 1994).

The scallop *Mizuhopecten yessoensis* (Jay) is widely distributed in coastal waters of Peter the Great Bay (Sea of Japan). This species plays an important role in benthic communities and has a great commercial value. The literature affords the data concerning the normal physiology, histology and biochemistry of *M. yessoensis* (see: Motavkin, 1986), including the data on the reproductive cycle (Dzyuba, 1972, 1986), embryonic and larval development (Malakhov & Medvedeva, 1991). In connection with pollution of coastal waters of Peter the Great Bay (Tkalin, 1992; Tkalin et al., 1993; Khristoforova et al., 1993), it is important to find out whether this factor really influences the health of the scallops. This paper presents the results of studies of gonad morphology and offspring development of the scallops *M. yessoensis* collected during their reproductive season from polluted and clean areas in Peter the Great Bay (Sea of Japan). The second objective of this study was to check our previous suggestion that an assessment of the reproductive function state of common species of marine bottom invertebrates may be used as a sensitive index of adverse environmental conditions (Vaschenko & Zhadan, 1993).

Materials and methods

The scallops were sampled from six stations in Peter the Great Bay (Sea of Japan). For histological study, the animals were taken from stations A,B, D, F in April 24–26, 1991 at the pre-spawning stage of the reproductive cycle; and for embryological assays scallops were sampled from stations A, B, F in May 13–16, 1991 and C, D, E, F in May 26–30, 1992 at the spawning stage of the reproductive cycle (Figure 1). Four stations (A, B, C and D) were located along the eastern coast of Amursky Bay, the most polluted part of Peter the Great Bay (Tkalin, 1992; Tkalin et al., 1993, 1996; Khristoforova et al., 1993; Shul'kin & Kavun, 1995). Municipal waste of Vladivostok and contaminated water of the Razdolnaya River are the main pollution sources in Amursky Bay. High loads of suspended solids, nutrients and pollutants (trace metals, petroleum and chlorinated hydrocarbons, anionic detergents) enter via waste waters (Tkalin et al., 1993). Two other stations (E and F) were located near Popov Island, far from the main sources of pollution.

About 25 animals from each station were sampled both for histological and embryological assays.

Histology

Scallop gonads were fixed in Bouin's fluid and embedded in paraffin. The 5–6 μm thick sections were stained by hematoxylin and eosin (H&E) and studied using light microscopy.

Embryological assays

Scallops were placed into glass vessels with sea water (12–13 °C). Spawning was induced by injecting the adductor with 1–2 ml of 0.03% serotonin solution in sea water. Just after the beginning of spawning (in 1–1.5 h after injection), females were carefully rinsed with sea water and placed in individual dishes. All eggs produced by each female over a 30 min period were filtered through a 100 μm-pore mesh and washed two to three times in filtered (0.45 μm) sea water (FSW). Eggs were kept separately in 100 ml beakers at 4 °C until used (not longer than 2 h).

Sperm was collected from individual males by pipetting from cut gonads and kept at 4 °C. After the activation by dilution in sea water, three samples with highest sperm motility were chosen for further fertilisation.

In the experiment of 1991, eggs of five females and sperm of the best male of three selected (within the scallop sample from each station) were used. 4 ml aliquots of the egg suspension (10 000 eggs ml^{-1}) from each female were put into glass beakers. Then 100 μl aliquots of the sperm suspension (final dilution 40 000) were added. After 20 min, the eggs were washed two times by careful pipetting of overlying sea water and adding FSW up to original volume (4 ml). Then 0.5 ml aliquots (about 5000 fertilised eggs) were transferred into three sterile plastic beakers ($n = 15$ for each station) containing 10 ml of FSW. Eggs of *M. yessoensis*, like most bivalve species, have a thin inconspicuous fertilisation envelope, so the first cleavage (4 h after insemination) was used as an indicator of successful fertilisation. At this stage, as well as at the stage of D-veliger (86 h after insemination), part of the embryos was fixed with 2.5% solution of glutaraldehyde in sea water. The percentage of normal embryos was counted and the length of D-veliger shells (20 for each female, $n = 100$) was measured with an ocular micrometer calibrated against a micrometer slide.

In 1992, eggs of six females and pooled sperm of three males (within the scallop sample from each station) were used. Development of each female eggs separately was followed from fertilisation to veliger formation. 10 ml aliquots of the egg suspension (10 000 eggs ml^{-1}) were placed into 1 litre vessels containing 200 ml of FSW. Then aliquots of the sperm suspension (sperm concentration was previously determined using a hemocytometer) were added to the vessels up to final sperm/egg ratios 10, 25, 50 and 100, two replicate vessels per each sperm/egg ratio were prepared. After 20 min, FSW was added to each vessel up to final volume 1 litre. In 4 h, at the stage of first cleavage, part of the embryos was fixed with glutaraldehyde and counted to assess fertilisation success. Divided eggs were scored as fertilised.

After 36 h, the overlying waters (about 500 ml from each vessel) with swimming trochophores were transferred to clean vessels and FSW was added up to 1

233

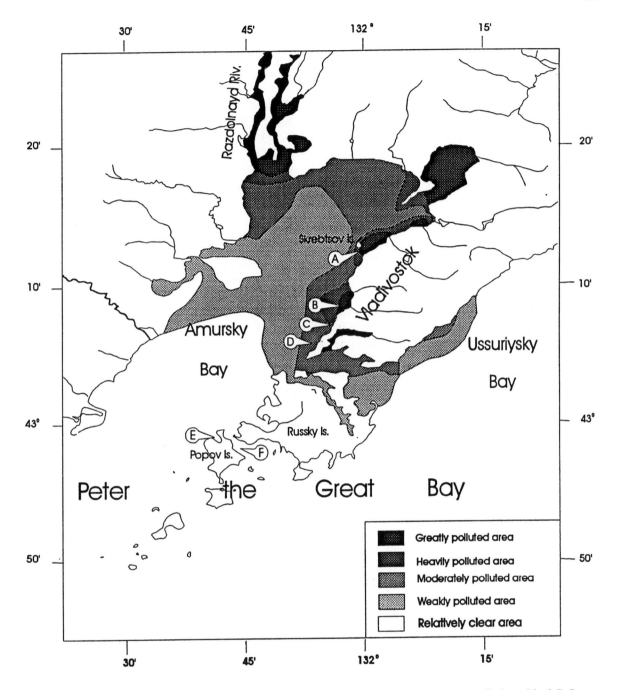

Figure 1. The scheme of pollution and location of sampling stations in Peter the Great Bay (Sea of Japan). A, Skrebtsov Island; B, Pervaya Rechka; C, Sport Harbour; D, Tokarevsky Cape; E, Alekseev Bight and F, Stark Strait. Data on pollution are taken from: Elyakov, G.V. (1992).

litre. The number of normal trochophores was counted both in overlying and bottom water and expressed as percent of both all eggs and fertilised eggs.

After 60 h, part of the larvae from each vessel was fixed to determine the percent of normal veligers.

Then the overlying waters from the vessels were pooled within each station and transferred to two large (5 l) vessels. During further development the veligers fed on natural microalgae present in natural sea water which was changed in the cultures daily. Sea water was filtered through gauze and cotton wool to remove large zooplankton organisms. Samples of larvae for abnormalities counts and growth measurements were taken on days 5, 12 and 19 of the experiment and fixed with glutaraldehyde. The length of larvae (from 85 to 230 for each station) was measured as described above.

In both experiments, the Student *t*-test was used to test for significant differences between station F and other stations.

Results and discussion

Histology of gonads

In the scallop samples from stations A, B and D, about 6% of hermaphroditic individuals (4 of 67 animals) was found while no hermaphrodites occurred in the scallop sample from station F. Three structural patterns of hermaphroditic gonads were distinguished. In two animals, female acini predominated over male ones. In one scallop, male acini prevailed over rare female ones, and another scallop had mainly male acini as well as several miscellaneous acini in which both sperm and oocytes were present. It is known that the scallop *M. yessoensis* is a dioecious species. Hermaphrodites in natural populations of this species are rare, their number does not exceed 0.3–0.4% (Bregman, 1979). Populations of some other bivalves also exhibit a low percentage of hermaphrodites: $2.5 \pm 0.5\%$ in *Crenomytilus grayanus* (Kutischev & Drozdov, 1974), 1–2% in *Mytilus edulis* (Sunila, 1981; Sprung, 1983). Nevertheless, Pekkarinen (1991) showed the number of hermaphrodites in bivalve populations may sharply increase under unfavourable ecological conditions, including pollution. The high percentage of hermaphrodites in population of *M. yessoensis* from Amursky Bay may be related to pollution of this area.

In the annual reproductive cycle of *M. yessoensis*, several stages have been determined (Dzyuba, 1986; Motavkin et al., 1990): recovery, sex inertness, beginning of development, a halt in development, active gametogenesis, pre-spawning stage and spawning. The same sampling dates (April 24–25, 1991) and depths (4.5–6 m), the lack of significant differences in water salinity and temperature in the area of study in winter and spring (Podorvanova et al., 1989), are factors that would seem to ensure the synchronisation of the reproductive cycles in different scallop populations. Nevertheless, the state of gonads in molluscs among stations markedly differed. The majority of the scallops (75%)

sampled from station F was at the pre-spawning stage of the reproductive cycle, while 25% of the molluscs had spawned, and the state of their gonads corresponded to the recovery stage. The acini were filled with hemocytes and a large number of 'brown cells', large globular cells containing one brown granule, lodged between them (Figure 2A). At the stations A, B and D, only 30, 75 and 88% of scallops respectively were at the pre-spawning stage of the reproductive cycle, while the remaining molluscs were at the earlier stage, the active gametogenesis stage. So some retardation of gametogenesis in the scallops from these stations was revealed.

Prominent histopathological changes were observed in the scallop gonads from station A: mass degeneration of the oocytes and accumulation of their debris in gonoducts of female gonads (Figure 2B), pycnosis of spermatocytes (Figure 2C) and phagocytosis of sperm by hemocytes (Figure 2D) in male gonads. Normally degeneration of sex cells and their resorption in the gonads of *M. yessoensis* occurred just after spawning (resorption of residual gametes) and during the most cold period in winter, whereas at the active gametogenesis and pre-spawning stages these processes were hardly observed (Dzyuba, 1972; Kasjanov et al., 1980). There were no histopathological alterations in the gonads of the scallops from other stations. Since station A is located in the heavily polluted zone of Amursky Bay (Figure 1), we concluded that mass resorption of gametes before spawning as well as severe retardation of gametogenesis may be caused by pollution. Previously, we have reported on the mass resorption of sex cells in the sea urchin *S. intermedius* collected at the pre-spawning stage of the reproductive cycle (August 1984, 1985, 1989) from stations A, C and D (Syasina et al., 1991; Vaschenko et al., 1992, 1993). Degeneration and resorption of sex cells as a result of exposure of marine invertebrates to different toxicants were also reported by other authors (Myint & Tyler, 1982; Sunila, 1984; Lowe & Pipe, 1985; Lipina et al., 1987; Lowe, 1988; Syasina et al., 1991) and seem to be one of the general organism's responses to polluted environment at the tissue/organ level.

Development of offspring

The results of the experiment of 1991 are presented in Table 1. The percent of fertilised eggs was significantly less in the offspring of scallops from station B compared to station F, whereas the difference between stations A and F was not reliable. The number of nor-

235

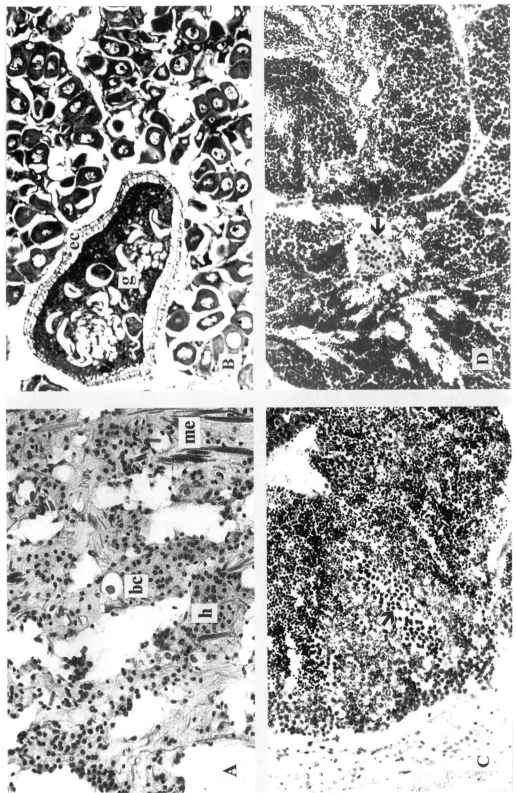

Figure 2. Morphology of the gonads of the scallops *Mizuhopecten yessoensis* sampled from different stations in Peter the Great Bay (Sea of Japan). A, hemocyte invasion in post-spawning gonads of scallops from station F (Stark Strait); B, debris of degenerating oocytes in the gonoduct of the scallop ovary from station A (Skrebtsov Island); C, pycnotic alterations in male gametes (marked by arrow) in the scallop testis from station A (Skrebtsov Island); D, phagocytosis of sperm (marked by arrow) in scallop testis from station A (Skrebtsov Island). bc, 'brown cell'; ec, epithelial cells of gonoduct; g, gonoduct; h, hemocytes; me, muscle elements. H&E; A, C, D × 312, B, × 200.

Table 1. Characteristics (mean ± standard error) of the offspring development of the scallops *Mizuhopecten yessoensis* sampled from different stations in Peter the Great Bay (Sea of Japan) in May 1991.

Station	Fertilized eggs,[a] %	Normal D-veligers,[a] %	Length of veligers,[b] μm
A	61.2±4.2	54.2±5.1*	101.0±1.2
B	17.6±2.4*	47.3±4.0*	108.7±1.7*
F	75.2±4.6	84.4±3.4	99.3±0.6

[a] Means are from five experiments (with three replicates), [b] means are from 100 larvae mesured. *Differences are significant at $p < 0.01$ (compared to station F).

mal D-veligers was sharply reduced (more than 30%, $p < 0.01$) in the scallop offspring from both station A and B compared to station F. There was no significant difference in the shell lengths of D-veligers from stations A and F, but the mean size of larvae in the scallop offspring from station B was significantly higher compared to stations A and F.

In the experiment of 1991, sperm/egg ratio was not determined, therefore it was difficult to make a correct conclusion on the gametes' capability of fertilisation in the scallops from polluted stations. In the experiment of 1992, a broad range of sperm/egg ratios (from 10 to 100) was used to assess fertilisation success more accurately.

The percentage of fertilised eggs depended on sperm/egg ratio and varied among stations (Figure 3). There were no significant differences between stations E and F, while fertilisation in the scallops from stations C and D was severely depressed especially at the lower sperm/egg ratios. The sperm/egg ratio of 50/1 was recognised as the optimal one for the scallops from stations E and F while for the scallops from station D the optimal sperm/egg ratio seemed to be higher. Thus, decrease in fertilisation capability of gametes in scallops from stations C and D was detected.

The number of normal trochophores in the offspring of scallops from station F scored as percent of fertilised eggs (Figure 4) tended to increase with increasing sperm/egg ratio. This tendency was not so clear in embryo cultures from other stations, and the number of normal trochophores scored both as percent of all eggs (Figure 5) and of fertilised eggs only (Figure 4) was significantly less than at station F.

Percent of normal D-veligers was the highest (about 80%) in the scallop offspring from station F at the lowest sperm/egg ratio and decreased to 52–64% at higher sperm/egg ratios (Figure 6) but this decrease was not statistically reliable. Significant diminution in percents of normal D-veligers was detected in the

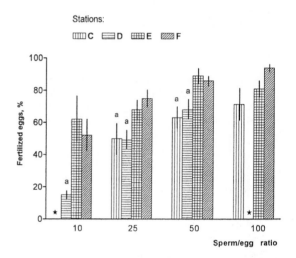

Figure 3. Effect of sperm/egg ratio on fertilization success in the scallops *Mizuhopecten yessoensis* sampled from different stations in Peter the Great Bay (Sea of Japan) in May 1992. Asterisks: no data. Error bars = SE. a, means are significantly different from station F at $P < 0.05$ ($n = 12$).

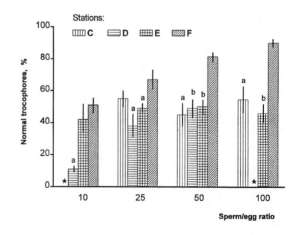

Figure 4. Number of normal trochophores (as percent of fertilized eggs) in offspring of the scallops *Mizuhopecten yessoensis* sampled from different stations in Peter the Great Bay (Sea of Japan) in May 1992. Asterisks: no data. Error bars = SE. Means are significantly different from station F at: a, $P < 0.05$; b, $P < 0.01$ ($n = 12$).

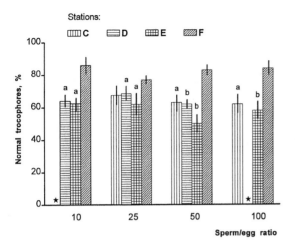

Figure 5. Number of normal trochophores (as percent of both fertilized and unfertilized eggs) in offspring of the scallops *Mizuhopecten yessoensis* sampled from different stations in Peter the Great Bay (Sea of Japan) in May 1992. Asterisks: no data. Error bars = SE. Means are significantly different from station F at: a, $P<0.05$; b, $P<0.01$ ($n=12$).

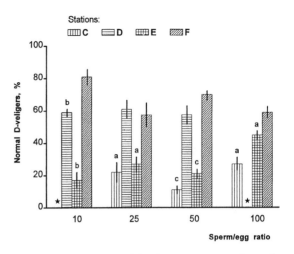

Figure 6. Number of normal D-veligers (as percent of eggs fertilized at different sperm/egg ratios) in offspring of the scallops *Mizuhopecten yessoensis* sampled from different stations in Peter the Great Bay (Sea of Japan) in May 1992. Asterisks: no data. Error bars = SE. Means are significantly different from station F at: a, $P<0.05$; b, $P<0.01$; c, $P<0.001$ ($n=12$).

offspring of the scallops from stations C, D (at the lowest sperm/egg ratio) and E (at all sperm/egg ratios with the exception of the highest one).

Further development of the veligers was accompanied by a high mortality occurring in all cultures, so the mean percent of normal veligers did not exceed 62% (Figure 7). There were no significant differences in the number of normal veligers on the 5th and 12th days of development in the scallop offspring from different stations, except the larvae culture from station C which died just after 5 days of development. In addition to this, considerable mortality/abnormalities occurred in the cultures of the scallops larvae from stations D and E by the 19th day of development.

Growth analysis revealed the most rapid growth of the larvae in the scallops offspring from station F (Figure 8) where about 15% of pediveligers with shell length >230 μm had appeared by day 19. On the contrary, the growth rate of veligers in the scallops offspring from stations C, D and E was much slower and the pediveliger stage was not attained in these cultures during the experiment.

Thus, the results of this study indicate severe disturbances in the development of offspring (decrease in fertilisation success, diminution in percents of normal trochophores, D-veligers, veligers and retardation of larval growth) of the scallops inhabiting stations A, B, C, D and E in Peter the Great Bay. These data, along with the data described above on the retarda-

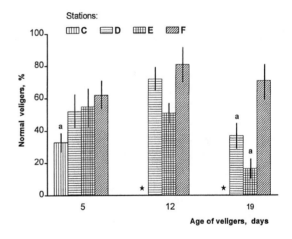

Figure 7. Number of normal veligers in offspring of the scallops *Mizuhopecten yessoensis* sampled from different stations in Peter the Great Bay (Sea of Japan) in May 1992. Asterisks: no data because of larvae death. Error bars = SE. a, means are significantly different from station F at $P<0.05$ ($n=6$).

tion of gametogenesis and an increase in the number of hermaphrodites (stations A, B, D), as well as the presence of histopathological changes in the scallop gonads (station A) are indicative of the disturbance of reproduction in the scallop *M. yessoensis* from different populations in Amursky Bay. Only molluscs from station F showed normal reproductive capacity.

Since stations A, B, C and D are located in the coastal zone adjoining to Vladivostok and affected by

238

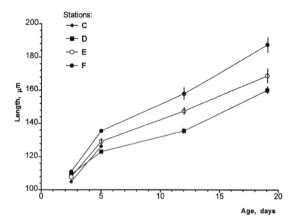

Figure 8. Growth of the veligers of the scallops *Mizuhopecten yessoensis* sampled from different stations in Peter the Great Bay (Sea of Japan) in May 1992. Error bars = 2SE. All means are significantly different ($P < 0.001$, $n > 60$) compared to station F.

industrial and domestic effluents (Figure 1), the most probable reason for the disturbance of scallop reproduction is the toxic effect of pollutants on the mollusc gametogenesis and subsequent gamete performance. As to station E (Alekseev Bight), it was attributed to 'relatively clean waters' on the basis of chemical analysis of water and sediments carried out in 1985–1989 and summarised in Elyakov (1992). Nevertheless, the results of the present study as well as the data of our previous investigations of the reproductive function state in the sea urchin *S. intermedius* in Peter the Great Bay (Vaschenko et al., 1993, 1994) suggest the presence in this area of conditions unfavourable for reproduction of bottom invertebrates. Long-term investigations of the mezoplankton structure in this bight have revealed a sharp decrease in both species diversity and abundance of the larvae of most bottom invertebrates including echinoderms and bivalves between 1986 and 1990 (Maslennikov et al., 1994). These findings also provide an evidence in favour of the conclusion made above. At least two reasons responsible for the adverse ecological conditions in Alekseev Bight may be listed: heavy organic pollution of the bight caused by a commercial farm of *M. yessoensis* that was in operation for 10 years (from 1978 to 1988) (Maslennikov et al., 1994), and increased concentrations of a number of heavy metals including mercury which were found in the water, bottom sediments, and soft tissues of several species of bivalves (Khristoforova et al., 1993; Vaschenko et al., 1993; Luchsheva, 1995).

It is known that physiology of marine molluscs and reproduction in particular greatly depend on various ecological factors, i.e. temperature, salinity, oxygen and food availability, etc. The scallop *M. yessoensis* is very sensitive to salinity (optimal range is from 26 to 32‰, according to Yaroslavtseva et al., 1988), and oxygen concentration (the lower limit of the optimal range is about 6 ml l^{-1}) (Motavkin, 1986). Gametogenesis in *M. yessoensis* (from the stage of beginning of development to the pre-spawning stage) occurred during October–April (Dzyuba, 1986). The long-term study of hydrochemistry of Peter the Great Bay (Podorvanova et al., 1989) showed that in this period salinity and oxygen concentrations in the area investigated at a depth of 5 m, varied from 31.45 to 34.02‰ and from 6.6 to 9.9 ml l^{-1} (101–151% of saturation) respectively. Variations in water temperature (1.2–2.85 °C in April and 6.3-7.4 °C in May) and in pH values (8.11–8.34) were small (The annual of the sea water quality..., 1992). Thus, it seems unlikely that natural fluctuations in the important ecological factors affect gametogenesis and offspring development in the scallop *M. yessoensis* living in Amursky Bay.

It was shown earlier, in the study of reproductive function state of the sea urchin *S. intermedius* in Peter the Great Bay, that the offspring development was the more sensitive indicator of gametogenesis disturbances compared with gonad index or gonad histology (Vaschenko & Zhadan, 1993). Significant deviations from normal development were found irrespective of whether or not there were changes in the gonad index values and in the morphology of the gonads. Krause (1994) in the field experiment on the effects of oil production effluent on gametogenesis and gamete performance in the sea urchin *S. purpuratus* has shown that the animals living closer to the outfall produced significantly larger gonads but fertilisability of eggs exhibited a positive relationship with distance away from the outfall. In the present study, the most prominent histopathological changes were found in the gonads of the scallops from one station only (A). The gonads of molluscs from station D exhibited only a slight retardation of gametogenesis but the development of their offspring was poor. This fact repeatedly demonstrates that the development of offspring is the most important and sensitive index of the reproductive function state of marine invertebrates (Vaschenko & Zhadan, 1993; Vaschenko et al., 1994).

A decline in egg viability may represent one of the general responses of an organism to environmental stress (Bayne, 1985). Thus, Bayne et al. (1978) recorded a reduction in the organic matter per egg of *M. edulis* under stress in the laboratory. The result was a slow rate of growth of the mussel offspring. A decline in gametes' capability of fertilisation and further normal development may affect the chances of offspring survival and so it represents a negative ecological consequence to the population (Bayne et al., 1985). Retardation of larval growth is also of ecological relevance: a prolongation of larval pelagic period decreases their chance of survival against predation, disease and dispersion, and may result in a reduced recruitment into the population (Calabrese et al., 1977; Bayne, 1985). Silina & Ovsyannikova (1995) investigated the community structure of *M. yessoensis* and showed that from 1982 through 1993 the mean age of scallops inhabiting station B in Amursky Bay had increased in consequence of the lack of the population recruitment. It seems to be probable that one of the reasons responsible for the changes in the scallop population structure may be disturbance of gametogenesis in the molluscs.

Conclusion

The present study demonstrates a retardation of gametogenesis, histopathological changes in the gonads, an increase in the number of hermaphrodites and severe disturbances in the offspring development in the scallop *M. yessoensis* inhabiting polluted sites in Peter the Great Bay (Sea of Japan). Development of offspring was a more sensitive index of disturbance of the reproductive function than morphology of scallop gonads. We believe that an analysis of offspring development of common species of marine invertebrates living in polluted coastal waters may be used as a sensitive index of adverse environmental conditions. Along with high sensitivity, the most important advantage of this bioassay is its prognostic feature because there is a direct relationship between the pollution effects on the reproduction of marine invertebrates and the well-being their populations. The main limitation in its use is that it is laborious and time consuming. Nevertheless, chronic larval development and growth assays should be included in a monitoring program as a very sensitive tool for the identification of environmental stress.

References

Bayne, B. L., 1985. Ecological consequences of stress. In Bayne, B. L., D. A. Brown, K. Burns, D. R. Dixon, A. Ivanovici, D. R. Livingstone, D. M. Lowe, M. N. Moore, A. R. D. Stebbing & J. Widdows (eds), The effects of stress and pollution on marine animals. Praeger, New York: 141–155.

Bayne, B. L., D. A. Brown, K. Burns, D. R. Dixon, A. Ivanovici, D. R. Livingstone, D. M. Lowe, M. N. Moore, A. R. D. Stebbing & J. Widdows, 1985. The effects of stress and pollution on marine animals. Praeger, New York, 384 pp.

Bayne, B. L., K. R. Clarke & M. N. Moore, 1981. Some practical considerations in the measurement of pollution effects on bivalve molluscs, and some possible ecological consequences. Aquat. Toxicol. 1: 159–174.

Bayne, B. L., D. L. Holland, M. N. Moore, D. M. Lowe & J. Widdows, 1978. Further studies on the effects of stress in the adults on the eggs of *Mytilus edulis*. J. mar. biol. Ass. U.K. 58: 825–841.

Bregman, Yu. E., 1979. Genetic and population structure of bivalve mollusc *Patinopecten yessoensis* Jay. Izvestiya TINRO 103: 66–78 [in Rus.].

Calabrese, A., J. R. MacInnes, D. A. Nelson & J. E. Miller, 1977. Survival and growth of bivalve larvae under heavy metal stress. Mar. Biol. 41: 179–184.

Dixon, D. R. & D. Pollard, 1985. Embryo abnormalities in the periwinkle, *Littorina 'saxatilis'*, as indicator of stress in polluted marine environments. Mar. Pollut. Bull. 16: 29–33.

Dzyuba, S. M., 1972. Morphological and cytochemical characteristics of oogenesis and reproductive cycles in Japanese scallop and Far East giant mussel. PhD Thesis, Academy of Sciences of the USSR, Institute of Marine Biology, Vladivostok, 220 pp. [in Rus.].

Dzyuba, S. M., 1986. Reproductive system and gametogenesis. In Motavkin, P. A. (ed.). Japanese scallop *Mizuhopecten yessoensis* (Jay). Academy of Sciences of the USSR, Institute of Marine Biology, Vladivostok: 117–130 [in Rus.]

Elyakov, G. V. (ed.), 1992. Long-term program of nature preservation and rational use of nature resources of Primorsky Krai till 2005, 2. Dal'nauka, Vladivostok, 276 pp. [in Rus.].

Kasjanov, V. L., L. A. Medvedeva, S. N. Yakovlev & Yu. M. Yakovlev, 1980. Reproduction of echinoderms and bivalve molluscs. Nauka, Moscow, 207 pp. [in Rus.].

Khristoforova, N. K., V. M. Shul'kin, V. Ya. Kavun & E. N. Chernova, 1993. Heavy metals in fished and cultivated species of marine molluscs in Peter the Great Bay. Dal'nauka, Vladivostok, 296 pp. [in Rus.].

Koster, A. S. & J. A. M. Van den Biggelaar, 1980. Abnormal development of *Dentalium* due to Amoco Cadiz oil spill. Mar. Pollut. Bull. 11: 166–169.

Krause, P., 1994. Effects of an oil production effluent on gametogenesis and gamete performance in the purple sea urchin (*Strongylocentrotus purpuratus* Stimpson). Envir. Toxicol. Chem. 13: 1153–1161.

Kutischev, A. A. & A. L. Drozdov, 1974. Hermaphroditism and sex structure of the population of *Crenomytilus grayanus* Dunker. Vestnik MGU, 6: 11–13 [in Rus.].

Lee, S. Y., 1986. Growth and reproduction of the green mussel *Perna viridis* (L.) (Bivalvia: Mytilacea) in contrasting environments in Hong Kong. Asian mar. Biol. 3: 111–128.

Lipina, I. G., Z. S. Evtushenko & S. M. Gnesdilova, 1987. Morphofunctional changes in the ovaries of the sea urchin *Strongylocentrotus intermedius* exposed to cadmium. Ontogenez 18: 269–276 [in Rus.].

Lowe, D. M., 1988. Alterations in cellular structure of *Mytilus edulis* resulting from exposure to environmental contaminants under field and experimental conditions. Mar. Ecol. Prog. Ser. 46: 91–100.

Lowe, D. M. & R. K. Pipe, 1985. Hydrocarbon exposure in mussels: a quantitative study of the responses in the reproductive and nutrient storage cell systems. Aquat. Toxicol. 8: 265–272.

Luchsheva, L. N., 1995. Contents of mercury in the ecosystem of Alekseev Bight (Peter the Great Bay, Sea of Japan). Russian J. mar. Biol. 21: 365–369.

Malakhov, V. V. & L. A. Medvedeva, 1991. Embryonic development of bivalve molluscs: normal and under heavy metal effects. Nauka, Moscow, 132 pp. [in Rus.].

Maslennikov, S. I., O. M. Korn, I. A. Kashin & Yu. N. Martynchenko, 1994. Long-term changes in larval plankton numbers in Alekseeva Inlet (Popov Island, Sea of Japan). Russian J. mar. Biol. 20: 107–114.

Motavkin, P. A. (ed.), 1986. Japanese scallop *Mizuhopecten yessoensis* (Jay). Academy of Sciences of USSR, Institute of Marine Biology, Vladivostok, 240 pp. [in Rus.].

Motavkin, P. A., Yu. S. Khotimchenko & I. I. Deridovich, 1990. Regulation of reproduction and biotechnology of sex cells obtaining in bivalve molluscs. Nauka, Moscow, 216 pp. [in Rus.].

Myint, U. & P. A. Tyler, 1982. Effects of temperature, nutritive and metal stressors on the reproductive biology of *Mytilus edulis*. Mar. Biol. 67: 209–223.

Pekkarinen, M., 1991. Notes on the general condition of *Mytilus edulis* L. of the southwestern coast of Finland. Bivalve Studies in Finland 1: 20–40.

Podorvanova, N. F., T. S. Ivashinnikova, V. S. Petrenko & L. S. Khomichuk, 1989. Basic characteristics of the hydrochemistry of Peter the Great Bay (Sea of Japan). Far East Branch of Academy of Sciences of USSR, Vladivostok, 201 pp. [in Rus.].

Silina, A. V. & I. I. Ovsyannikova, 1995. Long-term changes in the community of Japanese scallop and its epibionts in the polluted area of Amursky Bay, Sea of Japan. Russian J. mar. Biol. 21: 54–60.

Sprung, M., 1983. Reproduction and fecundity of the mussel *Mytilus edulis* at Helgoland (North Sea). Helgoländer. wiss. Meeresunters. 36: 243–255.

Sunila, I., 1981. Reproduction of *Mytilus edulis* L. (Bivalvia) in a brackish water area, the Gulf of Finland. Ann. zool. fenn. 18: 121–128.

Sunila, I., 1984. Copper- and cadmium-induced histological changes in the mantle of *Mytilus edulis* L. (Bivalvia). Limnologica, Berlin 15: 523–527.

Sunila, I., 1986. Histopathological changes in the mussels *Mytilus edulis* L. from a titanium dioxide plant in Northern Baltic. Ann. zool. fenn. 23: 61–70.

Shul'kin, V. M. & V. Ia. Kavun, 1995. The use of marine bivalves in heavy metal monitoring near Vladivostok, Russia. Mar. Pollut. Bull. 31: 330–333.

Syasina, I. G., M. A. Vaschenko & V. B. Durkina, 1991. Histopathological changes in the gonads of sea urchins exposed to heavy metals. Soviet J. mar. Biol. 17: 244–251.

The annual of the sea water quality in Far East Seas by the hydrochemical parameters in 1991. 1992. Far Eastern Regional Hydrometeorological Research Institute, Vladivostok, 192 pp. [in Rus.].

Tkalin, A. V., 1992. Bottom sediment pollution on some coastal areas of the Japan Sea. Ocean Res. (Rep. Korea) 14: 71–75.

Tkalin, A. V., T. A. Belan & E. N. Shapovalov, 1993. The state of the marine environment near Vladivostok, Russia. Mar. Pollut. Bull. 26: 418–422.

Tkalin, A. V., B. J. Presley & P. N. Boothe, 1996. Spatial and temporal variations of trace metals in bottom sediments of Peter the Great Bay, the Sea of Japan. Envir. Pollut. 92: 73–78.

Vaschenko, M. A. & P. M. Zhadan, 1993. Ecological assessment of marine environment using two sea urchin tests: disturbance of reproduction and sediment embryotoxicity. Sci. total Envir. (Suppl.) 2: 1235–1245.

Vaschenko, M. A., P. M. Zhadan, A. L. Kovaleva & N. M. Chekmasova, 1992. Morphological and morphometrical characteristics of the gonads of the sea urchin Strongylocentrotus intermedius inhabiting under conditions of anthropogenic pollution. Ecologiya 1: 46–54 [in Rus.].

Vaschenko, M. A., P. M. Zhadan, E. M. Karaseva & O. N. Lukyanova, 1993. Disturbance of reproductive function in the sea urchin *Strongylocentrotus intermedius* in the polluted areas of Peter the Great Bay (Sea of Japan). Russian J. mar. Biol. 19: 57–66.

Vaschenko, M. A., P. M. Zhadan & L. A. Medvedeva, 1994. Disturbances of larval development of the sea urchin *Strongylocentrotus intermedius* from polluted areas of Peter the Great Bay (Sea of Japan). Russian J. mar. Biol. 20: 104–111.

Widdows, J., 1985. Physiological measurements. In Bayne, B. L., D. A. Brown, K. Burns, D. R. Dixon, A. Ivanovici, D. R. Livingstone, D. M. Lowe, M. N. Moore, A. R. D. Stebbing, J. Widdows (eds), The effects of stress and pollution on marine animals. Praeger, New York: 3–45.

Yaroslavtseva, L. M., T. Kh. Naidenko, E. P. Sergeeva & P. V. Yaroslavtsev, 1988. Effect of decreased salinity on different ontogenetic stages of the scallop Mizuhopecten yessoensis. Russian J. mar. Biol. 14: 293–297.

Hydrobiologia **352**: 241–250, 1997.
Y.-S. Wong & N. F.-Y. Tam (eds), Asia–Pacific Conference on Science and Management of Coastal Environment.
©1997 *Kluwer Academic Publishers. Printed in Belgium.*

The use of malformations in pelagic fish embryos for pollution assessment

H. von Westernhagen[1] & V. Dethlefsen[2]
[1]*Biologische Anstalt Helgoland, Notkestrasse 31, 22607 Hamburg, Germany*
[2]*Institut für Fischereiökologie, Deichstraße 12, 27427 Cuxhaven, Germany*

Key words: fish embryos, malformations, monitoring, biological effects, temperature effects

Abstract

Malformations in fish embryos have been monitored for several years in the southern part of the North Sea. Their occurrence was thought to be related to pollution because malformation rates were highest in coastal waters, known to receive high pollution loads. For the embryos of all species synchronous trends for the fluctuation of the occurrence of malformations over time were registered in the areas covered, with intermediate prevalences at the beginning of the studies in 1984 and maxima in 1987. Thereafter, malformations in all species decreased significantly, followed by an abrupt increase in 1996. It was found that in addition to a close correlation between the concentrations of environmental pollutants in the water and in the fish a significant negative correlation existed between surface water temperature and the occurrence of malformed embryos of dab (*Limanda limanda*). These same correlations were also observed for other fish species over time and space. This temperature-related correlation became increasingly visible with decreasing concentrations of organochlorines in fish. From these findings it is concluded that aside from environmental pollutants, natural factors such as temperature may predispose developing fish embryos to the impact of pollutants.

Introduction

During the life cycle of a fish the embryonic phase is usually considered the most sensitive (Westernhagen, 1988). During this phase man-made pollutants as well as adverse natural factors may exert a strong effect on the survival of fish eggs and larvae, thus influencing recruitment. Johnson & Landahl (1994) in their treatise on the English sole (*Parophrys vetulus*) are of the opinion that the chronic effects of environmental pollutants affect primarily embryos, larvae and juvenile fish, but less so adults. Toxicants may exert their negative effects either through accumulation in the various organs of fish such as liver, kidney or gonads or directly through their presence in the water. Both mechanisms may also act in combination (Dethlefsen, 1995).

Investigations on a possible correlation between the pollution of surface waters and the occurrence of cytological and cytogenetic alterations in fish embryos have first been conducted in 1975 in the New York Bight by Longwell & Hughes (1981). The authors could prove reduced survival of fish embryos from areas which were heavily contaminated.

In similar investigations on the regional occurrence and distribution of malformations in pelagic fish embryos from the southern North Sea Cameron & Berg (1993) showed that malformation incidences were higher in river estuaries and in coastal areas off the Dutch, British and German coast than in offshore areas. Usually the percent malformations in the embryo population decreased with the distance from shore. The fact that regional differences in the occurrence of malformations in pelagic fish eggs followed the patterns and concentrations of pollutants in the respective areas (Cameron et al., 1992; Cameron & Berg, 1993), which generally displayed a gradient with declining concentrations towards the open sea, gave rise to the initiation of a biological effects monitoring programme that today has been in operation for more than a decade. The fascination of the method is that it employs a relatively simple bioindicator (normal or abnormal fish embryos) with an acknowledged societal value (fish)

242

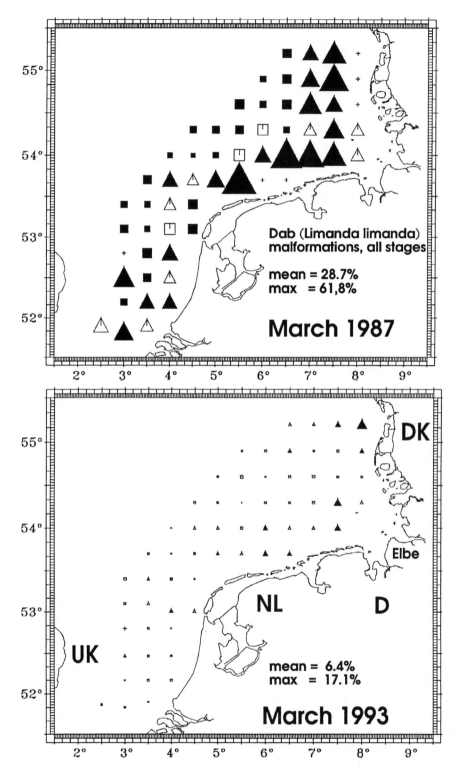

Figure 1. Dab (*Limanda limanda*). Malformations in embryos of all developmental stages in 1987 and 1993 in the southern part of the North Sea. Data for 1987 from Cameron et al., 1990. +: not enough eggs for evaluation (threshhold $n = 20$); X^2: evaluation: black triangles – significantly above the mean; black squares – significantly below the mean; white triangles and squares – not significantly above or below the mean. DK: Denmark; UK: United Kingdom; D: Germany, NL: The Netherlands.

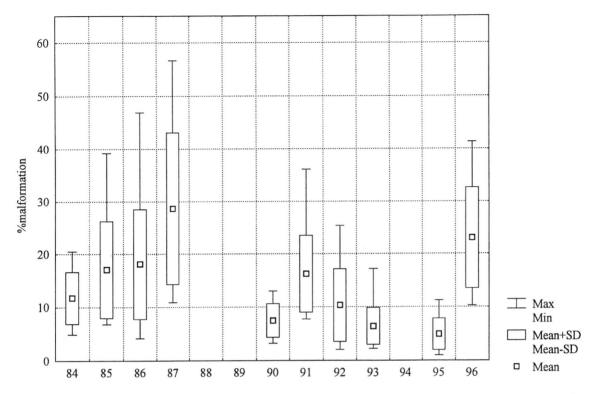

Figure 2. Dab (*Limanda limanda*). Malformations in embryos of all developmental stages during the years 1984–1996. 1990: Cameron & Berg, 1993; 1991–1992: Cameron & Westernhagen (1996).

and that it provides a good coverage even for larger marine areas (see Cameron et al., 1996).

After more than a decade of monitoring the North Sea a particular marine environment that had the 'reputation' of belonging to one of the most polluted marine areas of the world, we are facing a situation in which, due to a multitude of pollution abatement activities, the situation seems to be changing slowly for the better. If this is really so, then a potent biological effects indicator should be able to pick up these signals. Thus, in a recent review of the ten years of monitoring malformations of pelagic fish eggs in the southern part of the North Sea, Dethlefsen & Westernhagen (1996) compared the time-dependent expression of prevalences of malformations during the period of 1984–1995 in view of potential factors that may have had influence on malformation rates. Those parameters that had the potential to influence the occurrence of malformation rates were heavy metals, the residues of chlorinated hydrocarbons in water, sediment and livers of fish, including some regional hydrographical factors such as temperature and salinity. Dab (*L. limanda*) were used for monitoring, because dab is the most common

fish in the southern North Sea and the available information on xenobiotics in fish tissue is best documented for dab. A significant positive correlation between the concentrations of PCBs and DDE in the liver of dab and prevalences of embryonic malformations was demonstrated. In addition, a negative correlation between the temperature of the surface waters and the occurrence of malformed embryos of dab and some other species was found.

Thus it appears that there are at least two potential factors which are interacting in the expression of fish embryo malformations in the sea. One man-made factor (pollution) and one natural factor (the sea water temperature). The present paper tries to demonstrate the potential effects of both factors and tries to elucidate how, on the basis of an incomplete hypothesis, pollution management can be lead into a wrong direction assuming too simple a situation in an ecosystem.

244

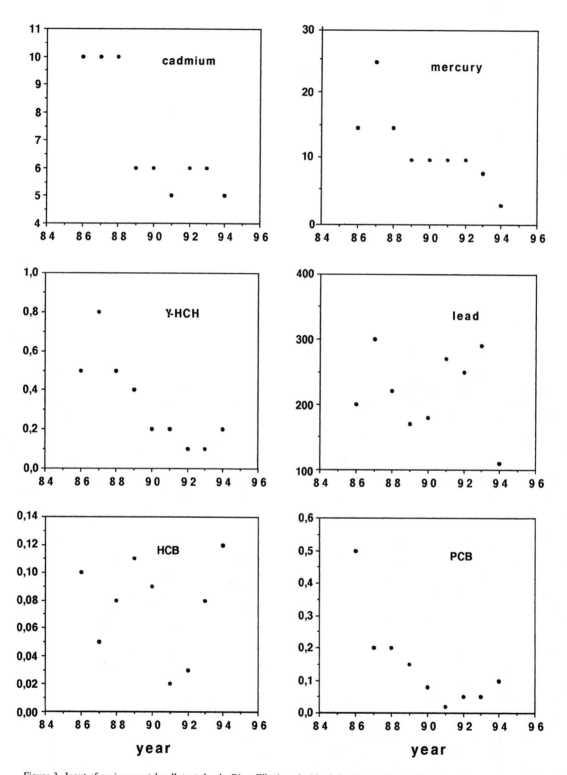

Figure 3. Input of environmental pollutants by the River Elbe into the North Sea from 1986 to 1994 (in metric tons per year).

Material and methods

Investigations on fish embryo health were initiated in the southern part of the North Sea during the spring spawning season (February/March) in 1984 and from then on these investigations were conducted until 1987, and at intervals in 1990, 1993, 1995 and 1996 (for logistics no sampling took place in 1988, 1989 and 1994). The general station grid employed is depicted in Figure 1, although, depending on sea conditions, slight deviations from this scheme had to be accepted from year to year. Intervals between stations were 18 nautical miles. All stations were located in the coastal (territorial) waters of either Denmark, The Netherlands or Germany. Pelagic fish eggs were caught with a circular plankton net (1-m in diameter; mesh size 300–500 μm) which was towed horizontally below the surface (2–5 m depth) for 5–15 min. Towing speed did not exceed 0.5 m s^{-1} in order to avoid embryo mortalities or injuries due to high turbulance in the cod end of the net. Immediately after the catch, embryos were examined for morphological alterations. During each cruise, between 20 000 to 40 000 embryos were examined. A description and photographic documentation of malformations found in the embryos of pelagic fish eggs are provided by Cameron et al. (1992). Data storage and retrieval were accomplished with a programme especially designed for this purpose (EGGMOND, Cameron et al., 1996). Data for temperature and salinity have been taken from the Helgoland Roads data base of the Biologische Anstalt Helgoland, which is part of the German monitoring activities in the North Sea, providing, among other oceanographic data, surface water temperatures on all working days of the year. In the case of the years 1993, 1995 and 1996, additional temperature data have been derived from direct recordings during the respective cruises.

Results

Since dab (*Limanda limanda*) eggs are the most common ones in the southern North Sea during the months of February to March, the following analysis refers mainly to data collected from dab embryos. The behaviour of other species towards pollutants or oceanographic parameters is generally comparable. A general species-specific sensitivity exists in which dab embryos assume a medium position, i.e. they are neither hypo- nor hypersensitive.

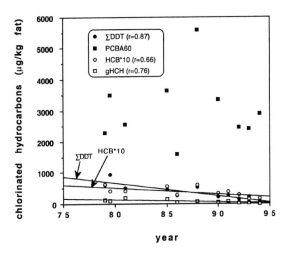

Figure 4. Concentrations of organochlorines in liver of dab (*Limanda limanda*) from the German Bight (mg kg^{-1} lipid base). PCBA60: PCB content calculated as $(CB138 + CB153 + CB180)0.3^{-1}$; gHCH = γHCH.

Generally, the results from surveys are depicted on maps with symbols of different type and size indicating the findings. Results from two cruises with high (1987) and low (1993) malformation incidence are depicted in Figure 1. The frequent overall occurrence of malformations in dab eggs dropped from 21.6% in 1987 to 6.4% in 1993. During the respective cruises the highest values were recorded near river estuaries and in areas with intensive ship traffic, but also in certain 'hot spots' which cannot immediately be accounted for. Highest prevalences for dab embryo deformities at particular stations reached 62% in 1987 while maximum values for 1993 were only 17.1%.

The time dependent fluctuations of the occurrence of malformations in dab embryos for all cruises conducted so far are depicted in Figure 2. Data are collected from all stations south of 55°20′N and east of 05°40′E. Starting in 1984 until 1987, malformations increased from mean values around 11% to values around 28%. Lower values (below 10%) were found in 1990, 1992, 1993 and 1995. In 1996 the value (mean) had drastically increased, reaching about 24%, but was still lower than in 1987; so far the highest value recorded.

These fluctuations and the decrease of malformations in time is very striking and we try to explain these phenomena by relating them to environmental factors. Thus we have compared the development of the input of pollutants into the North Sea through the river Elbe (one of the main contributors of xenobiotics

246

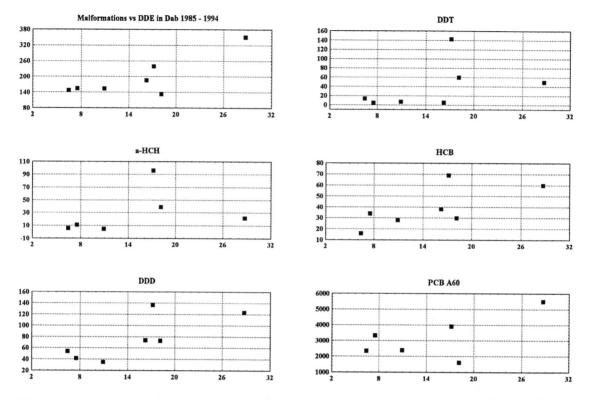

Figure 5. Correlations between organochlorine residues in liver of dab from the German Bight, southern part of the North Sea, during the period 1985–1993, and malformation rates in dab embryos from the same area. DDE versus malformations significant at the 5% level. a-HCH = α-HCH.

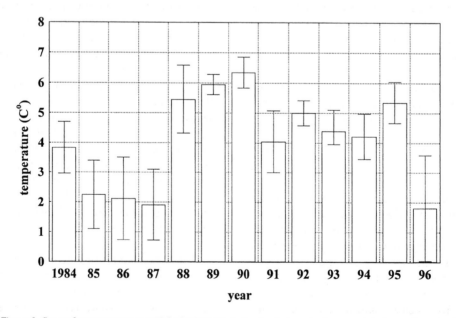

Figure 6. Sea surface temperatures at Helgoland Roads during the first quarter (measured during working days).

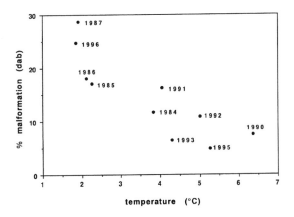

Figure 7. Relation between malformations and mean first quarter sea surface temperature at Helgoland Roads for dab embryos from the German Bight from 1984–1996.

to the southern North Sea) with time. The quantities of the major components of pollutants in the River Elbe water for the period 1986 to 1994 are depicted in Figure 3. In particular, the amounts of γ-HCH (Lindane), PCBs and the metals Cd and Hg showed a decline in concentration from 1986 to 1994. The reduced river input of the substances was correlated with a reduction in organochlorines accumulated in the biota. In Figure 4 the development of concentrations of chlorinated hydrocarbons in dab liver is depicted. With the exception of PCBs, concentrations of xenobiotics in dab liver fat decreased with time. A positive correlation exists between organochlorine residues DDE, DDD, HCB and PCBs (calculated as PCB A60 in Figure 5) in liver of dab and detected malformations (Figure 5). But at present the data base is too small to demonstrate a significant relationship other than for DDE, as seen from Table 1. Both DDE and PCBs are known to reduce viable hatch in marine fish eggs when present in the gonadal or liver tissue in sufficiently high concentrations (Westernhagen et al., 1981; Hansen et al., 1985).

Apart from xenobiotics, also oceanographic parameters such as temperature, salinity, silicate, phosphate and nitrite taken from the Helgoland Roads Monitoring data set of the Biologische Anstalt Helgoland have been tested for potential correlations with malformations. For this purpose, mean water temperatures at Helgoland Roads during the first quarter from 1984–1996 have been calculated (Figure 6). From 1984 to 1987 there is a clear trend towards relatively low water temperatures from a mean value around 4 °C in 1984 to only 2 °C in 1987 (lowest value recorded in 1987 was −0.6 °C). From 1988 to 1990 mean first quar-

ter surface water temperatures increased to between 5.5 °C and 6.4 °C. From 1991 to 1995 Helgoland Roads mean temperatures during the first quarter were between 4.0 °C and 5.4 °C. In 1996 the mean first quarter temperature was below 2 °C with extreme values at −1.0 °C. The large standard deviation in the 1996 first quarter water temperature reflects a change in oceanographic conditions in mid January. Until mid January 1996 the surface water temperature was around 4 °C, when prevailing easterly winds completely changed the oceanographic conditions in the German Bight and minimum surface water temperatures dropped as low as −1.0 °C.

High rates of malformations correspond with low water temperatures (Figure 7). The temperature sensitivity of developing fish eggs is further demonstrated by the temperature-dependent range of malformation rates in dab and plaice (*Pleuronectes platessa*) from the 1996 survey (Figure 8), which was displayed by embryos caught at different sites.

In view of the varying malformation rates, the natural (temperature-dependent) effects had to be subtracted from the anthropogenic effects in order to get a true picture of the situation. We have tried to calculate the mere temperature effects by using data from the 1996 survey and calculating the temperature-dependent change in malformation rates for one year in dab. This resulted in a non linear temperature effect with increasing malformations towards low temperatures conforming to the equation $y = 31.205*10^{-0.15796x}$ (R^2=0.694). On the basis of this equation the fluctuation of malformation rates over time, assuming a 4% naturally occurring malformation rate (see Klumpp & Westernhagen, 1995), were calculated. The results demonstrate (Figure 9) that in 1993 and 1995, the observed malformations ceased to be pollution-related, but could be contributed entirely to the naturally occurring background and the temperature effect. This situation is well in line with the reduced pollution load of the rivers Rhine and Elbe and the recent development of the organic contaminants in the southern North Sea (Haarich, 1996; Theobald pers. comm.). The high value for 1996 appears to derive from the extreme situation during this year. The extremely low temperatures during that winter had a dominating effect both on the occurrence of malformations, but also on the effects contributable to pollutants (Figure 9). The particularly strong pollution-effect in 1996 can probably be explained as a synergistic effect of low temperature and pollutants. A similar mechanism has been described by Weis et al. (1981) for the occurrence

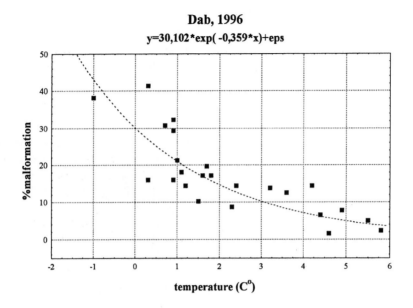

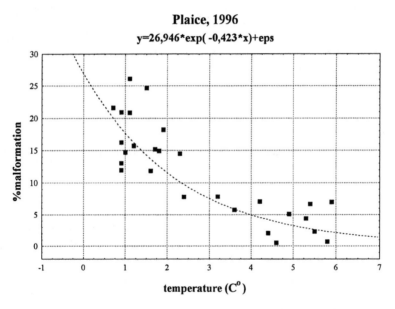

Figure 8. Malformations in dab (*Limanda limanda*) and plaice (*Pleuronectes platessa*) embryos in the German Bight, North Sea, in February, 1996.

of congenital anomalies in embryos of the killifish *Fundulus heteroclitus* exposed to methylmercury.

Discussion

Experimental evidence from previous laboratory and *in situ* investigations on developmental defects in fish embryos shows that a suite of natural and man-made parameters are capable of causing abnormal development in fish embryos. Among natural influences, temperature, salinity and oxygen (Alderdice & Forrester, 1971; Braum, 1973; Westernhagen, 1970) are the major factors. Anthropogenic influence is exerted by both inorganic and organic pollutants. Of the latter, DDE and PCBs at low doses have been shown experi-

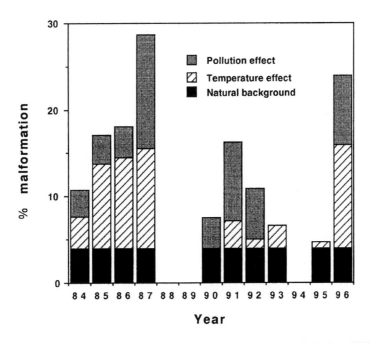

Figure 9. Embryo malformations (%) in dab (*Limanda limanda*) eggs from the German Bight, North Sea from 1984–1996. Temperature effects calculated on the basis of actual *in situ* measurements.

Table 1. Correlation matrix of malformations in dab embryos from 1985–1993 and liver residues in dab from the southern part of the North Sea.

Variable	Malf.	DDE	DDD	DDT	HCB	α-HCH	γ-HCH	PCBA60
Malf.	1.00	0.83*	0.77	0.48	0.71	0.35	0.45	0.65
DDE	0.83*	1.00	0.79	0.40	0.80	0.28	0.52	0.95*
DDD	0.77	0.79	1.00	0.86*	0.90*	0.77	0.85*	0.68
DDT	0.48	0.40	0.86*	1.00	0.79	0.98*	0.94*	0.30
HCB	0.71	0.80	0.90*	0.79	1.00	0.74	0.91*	0.78
α-HCH	0.35	0.28	0.77	0.98*	0.74	1.00	0.94*	0.22
γ-HCH	0.45	0.52	0.85*	0.94*	0.91*	0.94*	1.00	0.51
PCBA60	0.65	0.95*	0.68	0.30	0.78	0.22	0.51	1.00

*: significant positive correlation $P < 0.05$.

mentally to cause embryo mortality and developmental aberrations (Westernhagen et al., 1981; Hansen et al., 1985). The present data on the occurrence of malformations in dab embryos show that for reliable pollution assessment, natural effects have to be subtracted from observed phenomena to give a true picture of the situation. In some cases (1993–1995), observed effects may have been entirely natural, or, as was the case in 1990 and 1992, are predominantly caused by pollution.

The obvious interference of the physical environment in the expression of a bioindicator is a phenomenon that has to be accounted for in the interpretation of data. Temperature sensitivity, for instance, is par-

ticularly apparent in enzymatic processes, responsible for the expression of EROD activity, a biomarker for the determination of organic pollutants (Stegemann, 1979; Krüner et al., 1996). There, fish kept at colder temperatures are less responsive to inducers and, given similar xenobiotic exposure, induction depends strongly on the ambient temperature. The increased sensitivity towards pollutants of fish embryos developing at sub-optimum temperatures (Weis et al., 1981) has already been mentioned above.

In summing up, we may say that, in order to be able to separate man-made signals from natural ones, a profound knowledge of the organism or system used as the bioindicator is needed in order to form a balanced scientific judgement, and to unravel possible cause and effect relationships, due to the inherent complexity of the natural ecosytem. In spite of this complexity, the present example has demonstrated that pollution abatement measures are reflected in the limnic and marine environment by decreasing concentrations of contaminants in water and biota, thus underlining the usefulness of biological effects monitoring exercises.

References

Alderdice, D. F. & C. R. Forrester, 1971. Effects of salinity, temperature, and dissolved oxygen on early development of the Pacific cod (*Gadus macrocephalus*). J. Fish. Res. Bd Can. 28: 883–902.

Braum, E., 1973. Einflüsse chronischen exogenen Sauerstoffmangels auf die Embryogenese des Herings (*Clupea harengus*) (Influence of chronic exogenous oxygen deficiency on the embryogenesis of herring (*Clupea harengus*)). Neth. J. Sea Res. 7: 363–375.

Cameron, P. & J. Berg, 1993. Fortpflanzungsfähigkeit der Fische (Reproductive potential of fish). In Schutzgemeinschaft Deutsche Nordseeküste (ed.), Geht es der Nordsee besser? (Is the North Sea recovering?). Clausen & Bosse, Leck 1: 120–129.

Cameron, P., J. Berg, V. Dethlefsen & H. von Westernhagen, 1992. Developmental defects in pelagic embryos of several flatfish species in the southern North Sea. Neth. J. Sea Res. 29: 239–256.

Cameron, P., J. Berg & H. von Westernhagen, 1996. Biological effects monitoring of the North Sea employing fish embryological data. Envir. Monit. Assess. 40: 107–124.

Dethlefsen, V., 1995. Biological changes in the German Bight of the North Sea as indicators of ecosystem health. In Rapport, J., C. L. Gaudet & P. Calow (eds), Evaluating and Monitoring the Health of Large-Scale Ecosystems. Springer, Berlin, Heidelberg, 153–177.

Dethlefsen, V. & H. von Westernhagen, 1996. Malformations in North Sea pelagic fish embryos during the period 1984–1995. ICES J. mar. Sci. 53: 1024–1035.

Haarich, M., 1996. Schadstoff-Frachten durch die Flüsse (Pollutant load of the rivers). In Lozan, J. L. & H. Kausch (eds), Warnsignale aus den Flüssen (Warning signals from the rivers). Parey, Berlin, 144–148.

Johnson, L. L. & J. T. Landahl, 1994. Chemical contaminants, liver disease, and mortality rates in English sole (*Pleuronectes vetulus*). Ecol. Appl. 4: 59–68.

Krüner, G., D. Janssen & H. von Westernhagen, 1996. Wissenschaftliche Grundlagen sowie Beschaffung und Bewertung von Daten für das Biologische Monitoring der Nordsee (Scientific basis and accquisition and evaluation of data for the biological effects monitoring of the North Sea). Ber. Biol. Anst. Helgoland 11: 1–82.

Stegemann, J. J., 1979. Temperature influence on basal activity and induction of mixed function oxygenase activity in *Fundulus heteroclitus*. J. Fish. Res. Bd Can. 36: 1400–1405.

Von Westernhagen, H., 1970. Erbrütung der Eier von Dorsch (*Gadus morhua*), Flunder (*Pleuronectes flesus*) und Scholle (*Pleuronectes platessa*) unter kombinierten Temperatur- und Salzgehaltsbedingungen (Incubation of the eggs of cod, flounder and plaice under combined temperature and salinity conditions). Helgoländer wiss. Meeresunters. 21: 21–102.

Von Westernhagen, H., 1988. Sublethal effects of pollutants in fish eggs and larvae. In Hoar, W. & D. Randall (eds), Fish Physiology, 11-A. Academic Press, London, 253–346.

Weis, J. S., P. Weis & J. L. Ricci, 1981. Effects of cadmium, zinc, salinity, and temperature on the teratogenicity of methylmercury to the killifish (*Fundulus heteroclitus*). Rapp. P.-v. Réun. Cons. int. Explor. Mer 178: 64–70.

Hydrobiologia **352**: 251–262, 1997.
Y.-S. Wong & N. F.-Y. Tam (eds), Asia–Pacific Conference on Science and Management of Coastal Environment.
©1997 *Kluwer Academic Publishers. Printed in Belgium.*

Polynuclear Aromatic Hydrocarbons (PAHs) in fish from the Red Sea Coast of Yemen

Ali A-Z DouAbul*, Hassan M. A. Heba & Khalid H. Fareed
Department of Oceanography, Sana'a University, P.O. Box 11315, Yemen
*(*Author for correspondence; fax +9671 234267)*

Key words: Red Sea, PAHs, oil pollution, HPLC, GC/MS, fish, origin, toxicity

Abstract

A detailed analytical study using combined normal phase high pressure liquid chromatography (HPLC), gas chromatography (GC) and gas chromatography/mass spectrometry (GC/MS) of Polynuclear Aromatic Hydrocarbons (PAHs) in fish from the Red Sea was undertaken. This investigation involves a preliminary assessment of the sixteen parent compounds issued by the U.S. Environmental Protection Agency (EPA).

The study revealed measurable levels of Σ PAHs (the sum of three to five or six ring parent compounds) (49.2 ng g^{-1} dry weight) and total PAHs (all PAH detected) (422.1 ng g^{-1} dry weight) in edible muscle of fishes collected from the Red Sea. These concentrations are within the range of values reported for other comparable regions of the world. Mean concentrations for individual parent PAH in fish muscles were; naphthalene 19.5, biphenyl 4.6, acenaphthylene 1.0, acenaphthene 1.2, fluorene 5.5, phenanthrene 14.0, anthracene 0.8, fluoranthene 1.5, pyrene 1.8, benz(a)anthracene 0.4, chrysene 1.9, benzo(b)fluoranthene 0.5, benzo(k)fluoranthene 0.5, benzo(e)pyrene 0.9, benzo(a)pyrene 0.5, perylene 0.2, and indenol(1,2,3-*cd*)pyrene 0.1 ng g^{-1} dry weight respectively. The Red Sea fish extracts exhibit the low molecular weight aromatics as well as the discernible alkyl-substituted species of naphthalene, fluorene, phenanthrene and dibenzothiophene. Thus, it was suggested that the most probable source of PAHs is oil contamination originating from spillages and/or heavy ship traffic.

It is concluded that the presence of PAHs in the fish muscles is not responsible for the reported fish kill phenomenon. However, the high concentrations of carcinogenic chrysene encountered in these fishes should be considered seriously as it is hazardous to human health. Based on fish consumption by Yemeni's population it was calculated that the daily intake of total carcinogens were 0.15 μg/person/day.

Introduction

The need to identify organic and inorganic pollutants in the Red Sea has become a major concern for all countries in the region within the past few years. There are good reasons for this concern, among which is the need for baseline data (or background levels), the chronic pollution from industrial and other anthropogenic sources, and the acute oil pollution of the area which is a fishing ground. Because of the potential impact on marine life and fisheries, it is therefore important to know the extent of the pollution, how much it has affected marine life, and how long that effect will last, particularly from oil pollution. Due to the variety of

organic and inorganic compounds that can be present in the marine environment and the complexity involved in analyzing all of them, the present study was limited to investigating the presence of certain Polynuclear Aromatic Hydrocarbons (PAHs).

PAHs are among the most ubiquitous organic pollutants in the marine environment. The study of PAHs in coastal marine environments is of great importance since these areas are biologically active and receive considerable pollutant inputs from land-based sources via coastal discharge. The carcinogenic properties of some compounds coupled with the stability of PAHs during their atmospheric and aquatic transport and their widespread occurrence have, in recent years, generated interest in studying their sources, distribution, transport mechanisms, environmental impact, and fate (Bouloubassi & Saliot, 1993). The proposed primary sources of the environmental complex assemblages of PAHs include petroleum-related sources (ship traffic, oil seepage, or spillage) as well as combustion of various fossil fuels, natural fires, and road runoff/street dust. Other sources of more localized significance but worthy of note include domestic and industrial waste waters and sewage (Bouloubassi & Saliot, 1991).

Carcinogenic PAHs are suspected toxicants to marine organisms (Malins et al., 1984) and can be transported over long distances adsorbed onto airborne and waterborne particles. Being hydrophobic, they tend to accumulate in sediment and biota (Govers, 1990). Adsorption onto suspended particles and accumulation on bottom sediments may remove PAHs from the water column and reduce the chance of their photo-decomposition (Literathy et al., 1991).

There is evidence linking haemosiderosis, and also internal and external lesions, in bottom dwelling/feeding fish with PAHs in sediments (Malins et al., 1984; Gibbs et al., 1986). It is, therefore, essential to assess temporal influences on the accumulation of PAHs in fish.

In Yemen the marine and coastal areas are of a major economic significance. Marine resources are exploited for local consumption as well as export. The report of Haskoning (1991) suggested that the main impacts presently affecting Yemen marine environment are pollution and over-exploitation of certain natural resources. Qualitative and quantitative data are still lacking on the expected source/s of pollution, and their impacts on the coastal environment. Thus, we present here, for the first time to the best of our knowledge, a detailed analytical study using gas chromatography (GC) and gas chromatography/mass spectrometry (GC/MS) of anthropogenic PAH in fish samples from the Southern Red Sea along the coast of Yemen. This investigation involves screening of PAHs in several fish species from the Red Sea to determine if these animals show evidence of oil contamination. Furthermore, coastal sediments as well as mussels were also collected to assess damage to there resources.

Description of the region

The Red Sea is a long, narrow body of water, separating north-east Africa from Arabian Peninsula. Its nearly 2000 km of navigable waters connects at the

south with the Indian Ocean via Bab el-Mandeb. The average width of the Red Sea is 280 km, however, the width is only 28 km at the strait of Bab el-Mandeb. The maximum depth is 2246 m with an average of 700 m. The mean surface temperature increase southward, maximum surface water temperature is observed from June to September and attain 30 to 32 °C in the south (Edwards & Head, 1987). The shallow coastal waters may reach a temperature of 38 °C. The average salinity is about 35‰, but it is higher in shallow coastal areas as a result of evaporation.

The tides are semi-diurnal and spring tide ranges vary from 0.6 m in the north to 0.9 m in the south. The sea level is strongly influenced by the rate of evaporation and the balance between the inflow and outflow of the water from and to the Gulf of Aden. Surface water transport in summer is directed south by the prevailing northerly winds for about 4 months, at a velocity of 12–50 cm s^{-1}, while in winter the flow is reversed, pushing water into the Red Sea from the Gulf of Aden, the net value of the latter movement is greater than the summer outflow.

The Red sea is unique amongst deep bodies of water for having an extremely stable warm temperature throughout its deeper water. Below about 250–300 m, the water maintains a constant temperature of about 21.5 °C, which extends down to the sea floor in all areas except where heated brine pools exist (Edwards & Head, 1987).

Materials & methods

Materials

All solvents were redistilled in an all-glass distillation apparatus equipped with a 150 cm vacuum jacketed fractionation column filled with 3 mm diameter glass helices. Blanks of 1000-fold concentrates were determined by gas chromatography with flame ionization detection. Water used for cleaning the adsorption resin and sample work-up was purified with a Millipore Milli-Q system. Sodium chloride and sodium sulfate were kiln fired at 450 °C overnight and cooled in a greaseless desiccator. Silicagel used for column chromatography was solvent extracted with n-hexane in a glass cartridge inserted into an extraction apparatus, as described by Ehrhardt (1987). After extraction, the silica gel was first dried in the same cartridge by passing ultrapure nitrogen through it and was then activated by heating the cartridge in an electric tube oven to 200 °C for 6 h with the stream of nitrogen reduced to a few ml per minute.

The high performance liquid chromatograph (HPLC) was a Perkin-Elmer series 4 equipped with a microprocessor controlled solvent delivery system and fitted with a Reodyne 7125 injection valve. The detection system was composed of a Perkin-Elmer 560S scanning fluorescence spectrophotometer and an LC 75 variable wave length spectrophotometric detector with auto-control and equipped with a Perkin-Elmer analytical LC-PAH 0258 column (250 mm × 5 mm i.d.) with acetonitrile-distilled water gradient elution at a flow rate of 1 ml min^{-1}. Quantification of peaks and identification of PAHs in chromatograms was achieved by an LCI-100 laboratory computing integrator (Perkin-Elmer). The gas chromatograph was a Hewlet Packard HP5890-GC with split/splitless injector furnished with a 25 m × 0.32 mm fused silica capillary with a chemically bonded gum phase SE54 (J & W Scientific, Inc.).

Sampling

The study was carried out along the Yemen coast in the Red Sea during January–March 1995. The sites were chosen according to their importance as hot spots and the suitability of obtaining samples (Figure 1). Fish samples were also taken from fishermen fishing off the Yemen coast. After collection, the fish samples were wrapped in aluminum foil, stored in a cool box and frozen upon return to the city center (2 hours on average). A composite sample of fish, having similar size (length and weight) were chosen for each species. Sub-sample of each of the following species *Solea solea, Scombermorus malculatus, Rhochycentron canadum, Chorinemus lysan* and *Variola louti.*

Surface sediments from the sub-tidal coastal areas were collected by scuba divers, wrapped in aluminum foil, stored in a cool box and then frozen. The mussel *Thais sarignyi* (Reshayes, 1844) was found and collected from three locations only. The whole samples were used for analysis after removing the shells, and the body was washed with distilled water to remove any traces of sand, and then taking a known weight of the composite samples for chemical analysis.

Methods

The bulk sediment and tissue extraction procedure used was adapted from a method developed by Mcleod et al. (1985). Approximately 15 g of wet tissue were used for PAH analysis. After the addition of internal standards

(surrogates) and 50 g of anhydrous Na_2SO_4, the tissue was extracted three times with dichloromethane using a tissuemizer. A 20 ml sample was removed from the total solvent volume and concentrated to one ml for lipid percentage determination. The 380 ml of remaining solvent was concentrated to approximately 20 ml in a flat-bottom flask equipped with a three-ball Synder column condenser. The tissue extract was then transferred to a Kuderna-Danish tube heated in a water bath (60 °C) to concentrate the extract to a final volume of 2 ml. During concentration, the dichloromethane was exchanged for hexane.

The tissue extracts were fractionated by alumina:silica (80–100 mesh) open column chromatography. The silica gel was activated at 170 °C for 12 h and partially deactivated with 3% distilled water (v/w). Twenty grams of silica gel were slurry-packed in dichloromethane over 10 g of alumina. Alumina was activated at 400 °C for 4 h and partially deactivated with 1% distilled water (v/w). The dichloromethane was replaced with pentane by elution. The extract was then applied to the top of the column. The extract was sequentially eluted from the column with 50 ml of pentane (aliphatic fraction) and 200 ml of 1:1 pentane:dichloromethane (aromatic fraction). The aromatic fraction was further purified by HPLC to remove lipids. The lipids were removed by size exclusion using dichloromethane as isocratic phase (7 ml min^{-1}) and two 22.5 × 250 mm Phenogel 100 columns (Krahn et al., 1988). The purified aromatic fraction was collected from 1.5 min prior to the elution of 4,4′-dibromofluoro-biphenyl to 2 min after the elution of perylene. The retention times of the two marker peaks were checked prior to the beginning and at the end of a set of 10 samples. The purified aromatic fraction was concentrated to 1 ml using Kuderna-Danish tube heated in a water bath at 60 °C.

An initial screening of the 16 EPA-listed PAHs [naphthalene, acenaphthylene, acenaphthene, fluorene, phenanthrene, anthracene, fluoranthene, pyrene, benz(*a*)anthracene, chrysene, benzo(*b*)fluoranthene, benzo(*k*)fluoranthene, benzo(*a*)pyrene, dibenz(*ah*)anthracene, indeno(1,2,3-*cd*)pyrene, and benzo(*ghi*)perylene] was carried out (see Table 1). Results of HPLC analysis were complemented by a detail GC and GC/MS analysis at Texas A&M University, for the above EPA-listed PAHs as well as alkyl-substituted PAHs. PAHs were separated and quantified by GC-MS (HP5890-GC interfaced to a HP5970-MSD). The samples were injected in the splitless mode on to a 30 × 0.32 mm (0.32 μm film thick-

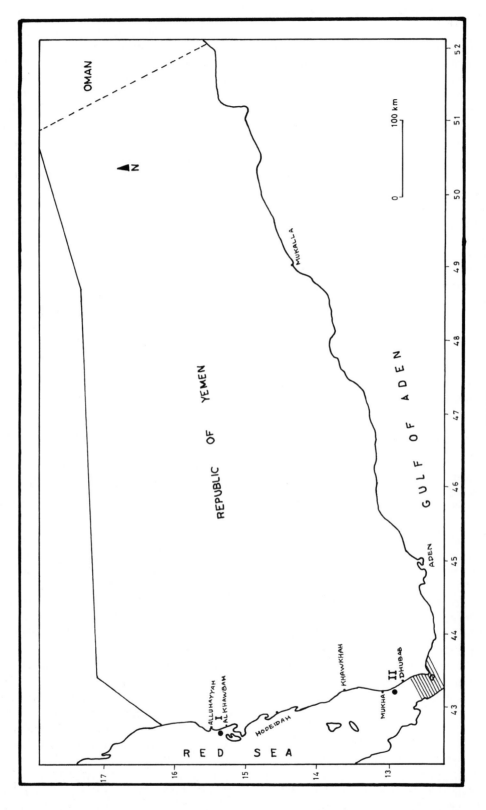

Figure 1.

Table 1. Mean * concentrations (μg g^{-1} wet weight) of 16 EPA-PAH in fish, mussels and coastal sediments from the Red Sea as determined by HPLC.

PAH Analyte Site	Fish		Mussel (*T. sarignyi*)				Sediment			
	Solea solea		*S. malculatus*		*R. canadum*		*C. lysan*		*V. louti*	
	I	II	I	II	I	II	I	II	I	II
Napthalene	<0.05	<0.05	<0.05	<0.05	<0.05	<0.05	<0.05	<0.05	<0.05	<0.05
Acenaphthylene	<0.05	<0.05	<0.05	<0.05	<0.05	<0.05	<0.05	<0.05	<0.05	<0.05
Acenaphthene	<0.05	<0.05	<0.05	<0.05	<0.05	<0.05	<0.05	<0.05	<0.05	<0.05
Fluorene	<0.05	<0.05	<0.05	<0.05	<0.05	<0.05	<0.05	<0.05	<0.05	<0.05
Phenathrene	0.01	0.02	0.01	0.01	0.01	0.02	0.02	0.03	<0.01	<0.01
Anthracene	<0.01	<0.01	<0.01	<0.01	<0.01	<0.01	<0.01	<0.01	<0.01	<0.01
Fluoranthene	<0.01	<0.01	<0.01	<0.01	<0.01	<0.01	<0.01	<0.01	<0.01	<0.01
Pyrene	<0.01	<0.01	<0.01	<0.01	<0.01	<0.01	<0.01	<0.01	<0.01	<0.01
Benz(*a*) anthracene	<0.01	<0.01	<0.01	<0.01	<0.01	<0.01	<0.01	<0.01	<0.01	<0.01
Chrysene	<0.01	<0.01	<0.01	<0.01	<0.01	<0.01	<0.01	<0.01	<0.01	<0.01
Benzo(*b*) fluoranthene	<0.01	<0.01	<0.01	<0.01	<0.01	<0.01	<0.02	<0.02	<0.02	<0.02
Benzo(*k*) fluoranthene	<0.01	<0.01	<0.01	<0.01	<0.01	<0.01	<0.01	<0.01	<0.01	<0.01
Benzo(*a*) pyrene	<0.01	<0.01	<0.01	<0.01	<0.01	<0.01	<0.01	<0.01	<0.01	<0.01
Dibenz(*ah*) anthracene	<0.01	<0.01	<0.01	<0.01	<0.01	<0.01	<0.01	<0.01	<0.01	<0.01
Benzo(*ghi*) perylene	<0.01	<0.01	<0.01	<0.01	<0.01	<0.01	<0.02	<0.02	<0.02	<0.02
Indeno(1,2,3-*cd*) pyrene	<0.01	<0.01	<0.01	<0.01	<0.01	<0.01	<0.02	<0.02	<0.02	<0.02
% Fat	3.4	2.7	2.7	3.04	0.78	14.91	0.41	0.50		

* = Mean of at least 3 determinations.

ness) DB-5 fused silica capillary column at an initial temperature of 60 °C and temperature programmed at 12 °C min^{-1} to 300 °C and held at the final temperature for 6 min. The mass spectral data were acquired using selected ions for each of the PAH analytes.

The GC/MS was calibrated and its linearity determined by injection of a standard containing 11 analytes at five concentrations ranging from 0.01 ng μl^{-1}. Sample component concentrations were calculated from the average response factor for each analyte. Analytical identifications were based on correct retention time of the quantitation ion (molecular ion) for the specific analyte and confirmed by the ratio of quantitation ion to confirmation ion. Calibration check samples were run with each set of samples (beginning, middle, and end), with no more than 6 h between calibration checks. The calibration check must maintain an average response factor within 10% for all analytes, with no one analyte greater than +25% of the known concentration. A laboratory reference sample (oil spiked solution) was also analyzed with each set of samples to confirm GC/MS system performance and calibration.

Quality assurance for each set of ten samples included a procedural blank, matrix spike, and tissue standard reference material (NIST-SRM 1974) which were carried through the entire analytical scheme. Internal standards (surrogates) were added to the sample prior to extraction and were used for quantitation. The surrogates were d-naphthalene, d-acenaphthene, d-phenethrene, d-chrysene, and d-pyrylene. Surrogates were added at a concentration similar to that expected for the analytes of interest. To monitor the recovery of the surrogates, chromatography internal standards d-fluorene and d-benzo(*a*)pyrene were added just prior to GC/MS analysis.

Recovery studies with fortified samples indicated that the recovery efficiency for naphthalene, acenaphene, phenantherene, chrysene, and perylene were 76.8%, 87.9%, 72.4%, 85.0% and 72.1% respectively. Results were not adjusted for percent recovery.

Results

The use of HPLC for the initial screening of the 16 EPA-listed PAHs *viz.* naphthalene, acenaphthylene, acenaphthene, fluorene, phenanthrene, anthracene, fluoranthene, pyrene, benz(*a*)anthracene, chrysene,

Table 2. Mean * concentrations (ng g^{-1} dry weight) of 16 EPA-PAH in fish, mussels and coastal sediments from the Red Sea as determined by GC.

PAH Analyte	Fish (*Solea solea*)		Mussel (*T. sarignyi*)			
Site	I	II	I	II	I	II
Napthalene	19.6	20.4	21.0	21.3	12.2	19.0
Acenaphthylene	1.0	1.0	0.8	0.9	0.6	0.9
Acenaphthene	1.2	1.1	1.8	1.4	0.2	1.5
Fluorene	5.5	5.0	2.7	2.1	1.2	1.4
Phenathrene	14.0	13.0	8.4	7.6	2.0	1.6
Anthracene	0.8	0.7	0.7	0.7	0.3	0.3
Fluoranthene	1.5	1.5	2.2	2.0	0.7	1.0
Pyrene	1.8	1.7	2.4	2.0	0.5	0.5
Benz(*a*) anthracene	0.4	0.3	0.3	0.3	0.2	0.3
Chrysene	1.9	1.6	0.8	0.7	0.7	0.2
Benzo(*b*) fluoranthene	0.5	0.3	0.3	0.3	0.2	0.2
Benzo(*k*) fluoranthene	0.5	0.3	0.3	0.3	0.2	0.2
Benzo(*a*) pyrene	0.9	0.7	0.3	0.3	0.3	0.2
Dibenz(*ah*) anthracene	0.1	0.1	0.2	0.1	0.2	0.1
Benzo(*ghi*) perylene	0.5	0.2	0.2	0.2	0.3	0.1
·Indeno(1,2,3-*cd*) pyrene	0.1	0.1	0.1	0.1	0.2	0.3
ΣPAH	50.3	48.0	42.5	40.3	20.0	27.8
% Fat	2.92	2.89	0.42	0.47		

* = Mean of at least 3 determinations.

benzo(*b*)fluoranthene, benzo(*k*)fluoranthene, benzo(*a*)pyrene, dibenz(*ah*)anthracene, indeno(1,2,3-*cd*)pyrene, and benzo(*ghi*)perylene in fish from the Red Sea has revealed that these pollutants were below the detection limits of 0.05, 0.05, 0.05, 0.05, 0.01, 0.01, 0.01, 0.01, 0.01, 0.01, 0.01, 0.01, 0.01, 0.01, 0.01, and 0.01, respectively (Table 1). However, the combined use of normal phase HPLC, GC and GC/MS analysis in the present study provided a powerful tool to isolate and further resolve the complex fish muscle PAH assemblages into simpler fractions. Thus, more than 40 compounds were identified and quantified. They comprise unsubstituted two- to six-ring PAHs (i.e. parent PAH), several alkyl-substituted homologs as well as sulfur-heterocyclics (benzothiophenes, benzonaphthiophenes and their alkylated homologs). The concentrations of individual parent PAH in fish muscles collected from the Red Sea ranged from less than 1 to tens of nanograms per gram dry weight (Table 2). In many environmental studies dealing with PAH, the concentrations have been reported as the sum of three- to five- (or six-) ring parent compounds, i.e. PAH with molecular weight 178 (phenanthrene, anthracene), 202 (fluoranthene, pyrene), 228 (chrysene + triphenylene, benzo(*a*)anthracene), 252 (benzofluoranthenes, benzo(*a*)pyrene, benzo(*e*)pyrene) and 276 (indeno [1,2,3-*cd*]pyrene, benzo[*ghi*]perylene). Their sum is referred to here as ΣPAH (mean concentration of ΣPAH in the Red Sea fish was 49.2 ng g^{-1} dry weight), whereas the sum of all PAHs detected is noted as totPAH (mean concentration of totPAH in the Red Sea fish was 422.1 ng g^{-1} dry weight). While the widespread use of ΣPAH for assessing pollution levels can facilitate comparison between various studies, this parameter does not however, represent the bulk amount of PAH occurring in environmental samples especially in a region like the Red Sea and Gulf of Aden. Our data showed that the concentration of total PAH (tot PAH) were higher by a factor of more than 8. This underlines the quantitative importance of the compounds not-included in the ΣPAH parameter. Although this can facilitate comparisons, it may also underestimate, sometimes severely, the bulk amount of PAH occurring in investigated samples. Moreover, from a qualitative point of view, this parameter does not take into account PAH derived mainly from fossil sources, since the latter are characterized by high abundances of alkylated homologes and sulfur-heterocyclics (Readman et al., 1991) as it will be discussed bellow. The latter source is particularly important in a region like the Red Sea and Gulf of Aden where crude oil contribution of PAH exceed that of pyrolytic/urban source. In terms of toxic effects, some alkylated homologes have mutagenic or carcinogenic properties (e.g. 1-methyl-phenethrene whose mean concentration in fish muscle was 5.0 ng g^{-1} dry weight). For this reason we preferred to use in this study the sum of all PAH compounds identified and quantified as summarized in Table 3. These constituents derived from anthropogenic sources with the exception of two compounds, tetrahydrochrysenes (αTHC and βTHC) which have natural terrestrial precursors.

Discussion

Unsubstituted (parent) PAH

Unsubstituted compounds were the minor fraction of PAH components in all fish samples. Among them, two-ring (naphthalene MW 128, biphenyl MW 154) and three-ring PAH (phenanthrene MW 178 and fluorene MW 166) dominated the distribution. Such patterns are characteristic of PAH mixtures generated by petrogenic pollution (Sauer et al., 1993). Naphthalene was the most prevalent parent compound (mean concentration was 19.5 ng g^{-1} dry weight) because it is more water soluble and has lower particulate affinity than the larger molecular weight aromatic hydrocarbons. In common with our findings, studies of PAHs indicated that naphthalenes are the compounds accumulated in highest concentrations by marine organisms (DouAbul et al., 1987). The second more abundant parent compound was phenathrene (mean concentration was 14.0 ng g^{-1} dry weight) which is a principal PAH component of crude oil. PAHs generated during high temperature combustion are mainly higher molecular weight (>4 aromatic ring) non-alkylated compounds, many of which are carcinogenic. One such PAH, chrysene, is normally produced through combustion (Readman et al., 1986) and was present in fish muscle extracts at a mean concentration of 1.9 ng g^{-1} dry weight.

The rather low concentrations of individual PAH observed in the Red Sea fishes may be attributed to rapid metabolism of PAHs by fish coupled with the rather limited source of pyrolytic/urban PAHs in the

Table 3. Mean *concentrations (ng/g dry weight) of PAHs(parent + alkyl substituted) in fish, mussel and coastal sediment from the Red Sea

PAH Analyte	Fish (*Solea solea*)	Mussel (*T. sarignyi*)	Sediment
Naphthalene	19.5	21.0	12.2
C1-Naphthalenes	20.8	27.6	10.3
C2-Naphthalenes	28.6	90.6	8.9
C3-Naphthalenes	50.4	92.3	10.9
C4-Naphthalenes	25.4	32.5	8.0
BiPhenyl	4.6	3.7	1.2
Acenaphthylene	1.0	0.8	0.6
Acenaphthene	1.2	1.8	0.2
Fluorene	5.5	2.7	1.2
C1-Fluorenes	9.8	6.6	ND
C2-Fluorenes	20.2	9.8	ND
C3-Fluorenes	21.7	ND	ND
Phenanthrene	14.0	8.4	2.0
Anthracene	0.8	0.7	0.3
C1-Phen-Anthr	19.7	9.8	ND
C2-Phen-Anthr	18.3	8.1	ND
C3-Phen-Anthr	14.6	13.3	ND
C4-Phen-Anthetr	ND	ND	ND
DiBenzoThio	10.3	2.5	0.8
C1-DiBen	22.1	8.5	1.8
C2-DiBen	24.2	20.6	3.4
C3-DiBen	12.1	16.8	3.6
Fluoranthene	1.5	2.2	0.7
Pyrene	1.8	2.4	0.5
C1-Fluoran-Pyr	3.7	3.1	ND
Ben(a)Anthracene	0.4	0.3	0.2
Chrysene	1.9	0.8	0.7
C1-Chrysene	4.6	3.0	ND
C2-Chrysene	7.7	2.6	ND
C3-Chrysene	ND	1.4	ND
C4-Chrysene	ND	ND	ND
Ben(b)Fluoran	0.5	0.3	0.2
bEN(k)Fluoran	0.5	0.3	0.2
Ben(e)Pyrene	0.9	0.3	0.3
Ben(a)Pyrene	0.5	0.3	0.3
Perylene	0.2	0.2	0.4
I123cdPyrene	0.1	0.1	0.2
DBahAnthra	0.1	0.2	0.2
BghiPerylene	0.5	0.2	0.3
2-MethylNaph	10.5	14.9	5.6
1-MethylNaph	10.2	12.7	4.7
2,6-DiMethNaph	11.3	32.9	2.6
1,6,7-TriMethNaph	15.4	22.0	1.7
1-Methyl Phen	5.0	2.3	1.5
Total PAHs	421.5	480.3	85.1

* Mean of at least 3 determinations. ND = Below the detection limit of 0.1 ng/g dry weight.

Red Sea. This led to steady state tissue levels of these compounds, and accounts for the failure in the present study to demonstrate appreciable levels of most parent PAHs in the examined samples. Accumulation and depuration of PAHs in fish can be influenced by various factors including route and length of exposure, lipid content of tissues, environmental factors (e.g. salinity, temperature... etc.), differences in species, age and sex and exposure to other xenobiotics (Varanasi et al., 1989). It is also indicated in the literature that fish efficiently metabolize PAH (Ahokas & Pelkonen, 1984). However, the metabolism of benzo(a)pyrene can lead to the formation of products which are more toxic than B(a)P itself (Gmur & Varanasi, 1982). During the metabolism of petroleum-derived aromatic hydrocarbons, the ability of an organism to process PAHs may be altered by the presence of polar components (Varanasi & Stein, 1991), including the concentrations of produced PAH metabolites (Schmeltz et al., 1978). Furthermore, the pattern of distribution of carcinogenic PAHs (including the levels in edible flesh) could be potentially affected by exposure to other xenobiotics which induce or inhibit xenobiotic-metabolizing enzyme systems (Gooch et al., 1989) or alter excretory pathways (Pritchard & Renfro, 1984).

The significance of our data (mean concentration of ΣPAH was 49.2 ng g^{-1} dry weight) is best appreciated, however, by comparing them with values of ΣPAH found in fish collected from other part of the world. For example it is lower than the maximum PAHs concentration in fish from Puget Sound, Washington USA (160 ng g^{-1} dry weight) reported by Landlot et al. (1987). It is comparable to the maximum PAHs concentration in sand flathead fillet collected from Port Phillip Bay, Victoria (55.7 ng g^{-1} dry weight) (Nicholson et al., 1994). DouAbul et al. (1987) have reported a maximum PAHs value in fish from the Arabian Gulf of 118 ng g^{-1} dry weight.

For the purpose of comparison between our results and those reported in the literature in areas of variable PAH contamination, it is necessary to convert the PAH concentration from ng g^{-1} dry weight to ng g^{-1} wet weight. In this investigation, an average reduction of weight of 75% during drying was used in the estimation. The mean ΣPAH concentration 49.2 ng g^{-1} dry weight $\times$ 0.25 dry weight/(wet weight) = 12.3 ng g^{-1} wet weight indicates that the level of PAH encountered in edible muscles of fish from the Red Sea lies within the range of values reported for zones defined as unpolluted (<0.5–148.0 ng g^{-1} wet weight) (Pancirov & Brown, 1977; Losifidou et al., 1982; Takatsu-

ki et al., 1985). Rainio et al. (1986) found few PAH components and at low levels in fish muscle collected from the Finnish archipelago sea. Neff & Anderson (1981) state that the work conducted by a number of researchers indicates that the majority of marine fin fish contain very low or undetectable levels of PAHs. Strikingly enough, we found that the mean concentrations of ΣPAHs in our samples were only 50% of the average concentration of ΣPAH (105.3 ng g^{-1} dry weight) in edible tissue of fish collected from the Arabian Gulf after the 1991 oil spill (Al-Yakoob et al., 1993). Despite the fact that between 6 to 10 million barrels of crude oil were released to the Gulf during the conflict (Thorhaug, 1992).

Alkyl-substituted PAH

The average concentration of totPAHs in the edible parts of fish from the Red Sea was 422.1 ng g^{-1} dry weight. However, the alkyl-substituted species comprise the bulk of this total (mean concentration was 372.9 ng g^{-1} dry weight), which indicates that the major source of PAHs in the Red Sea and the Gulf of Aden is petrogenic. It is unfortunate that to the best of our knowledge, there is no available relevant data to compare our results with. Most of the published work dealt with the parent PAH compounds. This is reasonable in the light of the fact that the bulk of these studies were carried out in industrialist nations whose major source of PAH originate from pyrolytic/urban source rather than from crude oil spillage's. Mean concentrations of individual alkyl-substituted PAH in the fish muscles from the Red Sea ranged from ND (below the detection limit of 0.1 ng g^{-1}) to tens of nanograms per gram dry weight (Table 3). Prominent among these are; methyl-naphthalenes, alkyl-fluorenes, Dibenzothiophenes, and methyl-phenantheren.

The low molecular weight compounds (<3 aromatic ring) and their alkylated homologues are principal constituents of crude oil (Gundlach et al., 1983; Readman et al., 1986), and our analyses demonstrated that the total resolved aromatic components generally covaried with the total petroleum hydrocarbons in fish samples. The concentration of alkyl-substituted PAHs exceed the concentrations of parent PAHs in all classes, as is typical for petroleum residue (Ehardht & Burns, 1993).

High concentrations of petroleum-related PAHs e.g. alkylated phenantherens and dibenzothiophenes were evident in all samples of fish muscle by GC/MS. In particular, the alkylated dibenzothiophenes were present in higher proportions than were other PAHs (Table 3). These compounds are known to comprise a high proportion of Marib light oil a crude that is produced from Yemen and exported via a terminal in the Red Sea (Geochem Group Limited, 1990). However, pyrogenic PAHs were present in low or non-detectable concentrations. Analogous results i.e. high concentrations of petroleum-related PAHs characteristic of Kuwait crude oil were found by HPLC and GC/MS analyses of fish following the Gulf War (Krahn et al., 1993a). Similarly high concentrations of petroleum-related PAHs characteristic of Prudhoe Bay crude oil were found by HPLC and GC/MS analyses of sediments following the *Exxon Valdez* oil spill in Alaska (Krahn et al., 1993b).

Among the sulfur-heterocyclics detected in the edible tissues of fish from the Red Sea, dibenzothiophene (DBT) and its alkylated homologs (C_1-, C_2- and C_3-DBT) were the most abundant. The sum of their concentration (ΣDBT) was 68.7 ng g^{-1} dry weight (Table 3). The compounds dominance of the dibenzothiophenes in Marib light oil samples which is a distinct chemical feature of crude oil in this area of the world. In the Red Sea fishes this distribution was similar in almost all samples, characterized by the predominance of the mono- and di-methyl DBT.

Origin of PAH

Assessment of the origin of PAHs in environmental samples requires detailed analysis of individual components in order to compare their composition with those of known pollutant emissions. This approach is also necessary for evaluating the fate and impact of PAHs since their environmental behavior and hazardous properties depend on various components (Bouloubassi & Saliot, 1991). In many cases fossil imprints are less readily recognizable in environmental PAH mixtures than pyrolytic ones, unless important petroleum inputs have occurred. Low molecular weight, mono-, bi- and tri-cyclic aromatics are generally the most abundant constituents of unburned fossil fuels with only minor relative amounts of tetracyclic and larger PAH (Neff, 1979). In contrast, combustion PAH mixtures are dominated by compounds of three or more condensed rings. The low molecular weight PAH of primarily fossil origin is known to degrade more severely than the larger PAH through physical-chemical and microbial processes. Jones et al. (1986) reported preferential biodegradation of oil-derived PAH with respect to pyrolytic ones. They suggested that the latter show specific association with particles (sequestration, occlusion) which render them relatively inaccessible to bacterial action, while oil-derived PAH are mainly introduced in the aquatic environment as emulsions presenting a large surface area to the degrading organisms. Hence, the commonly observed apparent predominance of pyrolytic PAH in environmental samples may results from both high contributions of related sources and better preservation of these imprints.

In order to gain some information on the probable source of PAHs (Hites et al., 1980), fish extracts were subjected to capillary gas chromatography/mass spectrometry. To investigate oil contamination (Neff, 1979), the following compound specifications were selected; m/z 128, 142, 156, 170-naphthalene and alkyl-naphthalenes: 166, 180, 194-fluorene and alkyl-fluorenes: 178, 192, 206, 220-phenanthrene and alkyl-phenanthrenes: 184, 198, 212, 226-dibenzothiophene and alkyl dibenzothiophenes. In the Red Sea fishes the discernible parent and alkyl-substituted species are for and in particular dibenzothiophenes which represent evidence for petrogenic contamination (Hites et al., 1977; DouAbul et al., 1987; Bouloubassi & Saliot, 1993).

To investigate combustion/urban runoff the molecular ions for the typical 'parent' (unsubstituted) PAHs (Blumer & Youngblood, 1975; Herrmann, 1981) were selected: m/z 178-phenanthrene/anthracene; 202-fluoranthene/pyrene; 228 benzoanthracene/chrysene; 252-benz-fluoranthenes/benz-pyrenes; 276-benz(ghi) perylene/indeno (1,2,3,-cd)pyrene. It is apparent that these compounds are present in the Red Sea fishes thus indicating combustion/urban runoff.

Molluscs have been used for monitoring contaminants in the environment (Farrington et al., 1983). These are sentinel organisms which concentrate pollutants from the marine environment, yet do not readily metabolize contaminants such as petroleum hydrocarbons and heavy metals (Farrington & Quinn, 1973). The concentration of a contaminant in a mussel is the difference between uptake and excretion of that contaminant. Thus, the contaminants found in mussels reflect the current contaminant burden of an ecosystem (Jackson et al., 1994). Furthermore, mussels were also collected to assess damage to these resources. It is well established that aquatic sediments are the final accumulation site of water-borne constitutes derived from natural sources (living organisms and their detritus) *in situ* and surroundings, and artificial (domestic, urban-industrial and agricultural wastes) sources (DouAbul et al., 1984). The aquatic sediments can thus provide not only a historic record of sedimentary environment, but also reserve the features of average sedimentary environmental constituents. Besides they are *vice versa*, also a possible source of chemicals in water. Based upon the foregoing facts mussels and sediments were collected from the Red Sea and analyzed for their PAHs contents in order to correlate the distribution patterns of the pollutants found in them with that of the fish, consequently it will be possible to pin-point the origin of contamination/s i.e. whether it is local or transported. Furthermore, mussels were also collected to assess damage to these resources.

Mussel and sediment extracts were also subjected to capillary gas chromatography/mass spectrometry in order to investigate the probable source of PAHs (Table 3). Similarly, in both mussels and sediments naphthalene was the most prevalent compound which gives some evidence that fish, mussels and sediments were subjected to the same source of PAHs contamination in the Red Sea namely crude oil. Again, in the Red Sea mussels and sediments the discernible parent and alkyl-substituted species are for and in particular dibenzothiophenes which represent further evidence for petrogenic contamination (Hites et al., 1977; DouAbul et al., 1987).

In the light of the above reasoning we may thus conclude that the dead fishes were subjected to the same source of oil contamination as the life sentinel mussels and sediments. This pollution is a consequence of localized oil operations (Rushdi et al., 1991) and/or heavy ship traffic crossing Bab el-Mendeb, currently 100 million tons of oil transit the Red Sea annually (PERSGA,1995).

Toxicity

Based on qualitative classification of PAH carcinogencity (IARC, 1983) detected PAHs classified as having sufficient or limited evidence for carcinogenicty [benzo(a)pyrene, benzo(a)anthracene and chrysene] were lower in concentration and frequency of detection than those classified as having insufficient or no evidence for carcinogenicity [naphthalene, biphenyl, acenaphthylene, acenaphthylene, Fluorene, anthracene, phenanthrene, fluoranthene, and pyrene].

Among PAH, benzo(a)pyrene, a well known PAH because of its carcinogenic properties, shows a detectable level (0.5 ng g^{-1} dry weight). Chrysene (1.9 ng g^{-1} dry weight) is also classified as having sufficient evidence for carcinogenicty (IARC, 1983),

thus the presence of this PAH in the edible muscles of the Red Sea fishes should be treated with absolute care (Futoma et al., 1981).

To assess health implications associated with PAHs, benzo(a)pyrene is often used as a toxicologic surrogate for all carcinogenic PAHs (Collins et al., 1991). Based on fish consumption by Yemeni's population (in 1994 the total fish consumption was 84 150 tones divided over 15 400 000 persons = 0.546 kg wet wt/person/month) estimated by Ministry of Fish Wealth, and considering the likely preference of consumption of highly contaminated fish, the daily intake of carcinogenic PAHs concentration as a potential extreme in exposure:

$$[50 \ \mu g/kg \text{dry wt} \times 0.5\text{Kg wet wt /mo} \times$$

$$\text{mo} \ (30 \ \text{d})^{-1} \times 0.2\text{dry wt (wet wt)} =$$

$$0.15 \ \mu g/\text{person/day}]$$

This estimate accounts for 6%, 18.8%, (5.17–12.5%) and 15% of the total intake of benzo(a)pyrene from food reported by Dennis et al. (1983), Vaessen et al., (1988), DeVos et al. (1990) and Menzie et al. (1992) respectively. Evidently that the total intake of chrysene is a much higher 0.0063 μg/person/day.

The contribution of fish to food [the major source of exposure of man to PAHs (Santodonato et al., 1981)] is influenced mainly by the way fish is prepared (Fazio & Howard, 1983). Charboiled or smoked fish are major additional sources of carcinogenic PAHs (Menzie et al., 1992).

Evaluation of possible risks imposed on different population segments in Yemen and/or the rest of the Red Sea region due to prolonged seafood consumption requires further investigations involving seasonal analysis of edible tissues of different species/size/age categories of locally consumed fish in addition to conducting seafood consumption surveys.

Conclusions

In the light of the above reasoning we may thus conclude:

(1) the presence of carcinogenic PAHs in fish muscles are hazardous to human health and should be considered seriously. Based on fish consumption by Yemeni population it was calculated that the daily intake of carcinogenic PAHs was 0.15 μg/person/day.

(2) The fish extracts exhibit the discernible parent and alkyl-substituted species of naphthalene, fluorene,

phenanthrene and dibenzothiophene. Hence the most probable source of PAHs is oil contamination originating from spillage's and/or heavy ship traffic.

(3) The Red Sea fishes are subjected to the same source of oil contamination as the life sentinel mussels. This pollution is a consequence of oil operations and heavy ship traffic crossing Bab el-Mendeb.

Acknowledgments

The authors would like to extend their sincere gratitude to Professor James M. Brook, Director, Geochemical & Environmental Research Group, Texas A&M University for conducting the gas chromatographic and gas chromatographic/mass spectrometric analysis in his laboratories, without such collaboration the present project could not be executed.

References

Ahokas, J. T. & O. Pelkonen, 1984. Metabolic activation of polycyclic aromatic hydrocarbons by fish liver cytochromeP-450. Mar. envir. Res. 14: 59–69.

Al-Yakoob, S., T. Saeed & H. Al-Hashash, 1993. Polycyclic aromatic hydrocarbons in edible tissue of fish from the Gulf after the 1991 oil spill. Mar. Pollut. Bull. 27: 297-301.

Blumer, M. & W. W. Youngblood, 1975. Polycyclic aromatic hydrocarbons in soil and recent sediments. Science 188: 53–55.

Bouloubassi, J. & A. Saliot, 1991. Composition and sources of dissolved and particulate PAH in surface waters from the Rhone Delta (NW Mediterranean). Mar. Pollut. Bull. 22: 588–594.

Bouloubassi, J. & A. Saliot, 1993. Investigation of anthropogenic and natural organic inputs in estuarine sediments using hydrocarbon markers (NAH, LAB, PAH). Oceanol. Acta 16: 145–161.

Collins, J. P., J. P. Brown, S. V. Dawson & M. A. Marty, 1991. Risk assessment for benzo(a)pyrene. Regulat. toxicol. Pharmacol. 13: 170–184.

Dennis, M. J., R. C. Massey, D. J. McWeeney & M. E. Knowles, 1983. Analysis of polycyclic aromatic hydrocarbons in U.K. total diets. Fd. Chem. Toxic. 21: 569–574.

DeVos, R. H., V. Dokkum, A. Schouten & P. De Jong-Berkhoot, 1990. Polycyclic aromatic hydrocarbons in Dutch total diet samples (1984–1986). Fd. Chem. Toxic. 28: 263–268.

DouAbul, A. A-Z., H. T. Al-Saad & S. A. Darmoian, 1984. Distribution of petroleum residues in surficial sediments from the Shatt al-Arab River and the NW region of the Arabian Gulf. Mar. Pollut. Bull. 15: 198.

DouAbul, A. A-Z., J. K. Abaychi, T. E. Al-Edanee, A. A. Ghani & H. T. Al-Saad, 1987. Polynuclear aromatic hydrocarbons (PAHs) in fish from the Arabian Gulf. Bull. envir. Contam. Toxicol. 38: 546–552.

Edwards, F. J. & S. M. Head (eds), 1987. The Red Sea Key Environment. Pergamon Press, Oxford, 441 pp.

Ehrhardt, M., 1987. Lipophilic organic material: an apparatus for extracting solids used for their concentration from seawater. ICES Techn. envir. Sci. 4: 1–14.

Ehrhardt, M. & K. A. Burns, 1993. Hydrocarbons and related photooxidation products dissolved in Saudi Arabia coastal waters and hydrocarbons in underlying sediments and bioindicator bivalves. Mar. Pollut. Bull. 27: 187–197.

Farrington, J. W. & J. G. Quinn, 1973. Petroleum hydrocarbons in Narragansett Bay. I. Survey of hydrocarbons in sediments and clams (Mercenaria mercenaria). Estuar. coast. mar. Sci. 1: 71–79.

Farrington, J. W., E. D. Goldberg, R. W. Risebrough, J. H. Martin & V. T. Bowen, 1983. US mussel watch 1976–1978: An overview of the trace metal, DDE, PCB, hydrocarbon and artificial radionuclide data. Envir. Sci. Technol. 17: 490–496.

Fazio, T. & J. W. Howard, 1983. Polycyclic aromatic hydrocarbons in foods. In Bjorseth (ed.), Handbook of Polycyclic Aromatic Hydrocarbons, New York: 461–505.

Futoma, D. J., S. R. Smith, T. E. Smith & J. Tanaka, 1981. Polycyclic aromatic hydrocarbons in water system. CRC Press, Boca Raton.

Geochemical Group Limited, 1990. A geochemical evaluation of origin of Red Sea oil slick 'Safer' area, Yemen. Prepared for Yemen Hunt Oil Company.

Gibbs, C. F., J. W. J. Wankowski, J. S. Langdon, J. D. Humphery, J. S. Andrews, P. V. Hodson, J. G. Fabris & A. P. Murray, 1986. Investigations following a fish kill in Port Phillip Bay, Victoria, during February 1984. Tech. Report No. 63. Marine Science Laboratories, Department of Conservation, Forests and Lands, Victoria, Australia.

Gmur, D. & U. Varanasi, 1982. Characterization of benzo(a)pyrene metabolites isolated from muscle, liver, and bile of juvenile flatfish. Carcinogenesis 3: 1397–1403.

Gooch, J. W., A. A. Elskus, P. J. Klopper-Sama, M. E. Hahn & J. J. Stegeman, 1989. Effect of ortho- and non-ortho- substituted polychlorinated biphenyls congeners on the hepatic monooxygenase system in scup (Stenotomus chrysops). Toxicol. appl. Pharmacol. 98: 422–433.

Govers, H. A. J., 1990. Predication of environmental behaviour of polycyclic aromatic hydrocarbons by PAR an QSAR. In W. Karcher & J. Devillers (eds), Practical Applications of Quantitative Structure-Activity Relationship (QSAR) in Environmental Chemistry and Toxicology. Kluwer Academic Publishers, The Netherlands: 411–432.

Gundlach, E. R., P. D. Boehm, M. Marchand, R. M. Atlas, D. M. Ward & D. A. Wolfe, 1983. The fate of Amoco Cadiz oil. Science 2201: 122–129.

Haskoning, 1991. Proposal for developing a coastal management plan in the Republic of Yemen. Report by Haskoning (Royal Dutch Consulting Engineers and Architects). Support to the Secretariat of the Environmental Protection Council, Yemen: 53 pp.

Herrmann, R., 1981. Transport of polycyclic aromatic hydrocarbons through a partly urbanised river basin. Wat. Air Soil Pollut. 16: 455–453.

Hites, R. A., R. E. Laflamme & J. W. Farrington, 1977. Sedimentary polycyclic aromatic hydrocarbons: the historical record. Science 198: 829–831.

Hites, R. A., R. E. Laflamme & J. G. Windsor Jr., 1980. Polycyclic aromatic hydrocarbons in an anoxic sediments core from the Pettaquamscutt River, USA. Geochim Cosmochim Acta 44: 873–878.

International Agency for Research on Cancer (IARC) (1983) IARC Monogr. Eval. Carcinog. Risk Chem. Hum., Suppl. 7.

Jackson, T. J., T. L. Wade, T. J. McDonald, D. L. Wilkinson & J. M. Brooks, 1994. Polynuclear aromatic hydrocarbon contaminants in oysters from the Gulf of Mexico (1986–1990). Envir. Pollut. 83: 291–298.

Jones, D. M., S. J. Rowland & A. G. Douglas, 1986. An examination of the fate of Nigerian crude oil in surface sediments of the

Humber estuary by gas chromatography/mass spectrometry. Int. J. envir. analyt. Chem. 24: 227–247.

Krahn, M. M., G. M. Ylitalo, J. Buzitis, S.-L. Chan & U. Varanasi, 1993a. Rapid high-performance liquid chromatographic methods that screen for aromatic compounds in environmental samples. J. Chromatogr. 642: 15–32.

Krahn, M. M., G. M. Ylitalo, J. Buzitis, S.-L. Chan, U. Varanasi, T. L. Wade, T. J. Jackson, J. M. Brooks, D. A. Wolfe & C.-A. Manen, 1993b. Comparison of HPLC/fluorecence screening and GC/MS analysis for aromatic compounds in sediment sampled after the *Exxon Valdez* oil spill. Envir. Sci. Technol. 27: 699–708.

Landlot, M., D. Kalman, A. Nevissi, K. Van Ness & F. Hafer, 1987. Potential toxicant exposure among consumers of recreationally caught fish from urban embayments of Puget Sound. NOAA Technical Memorandum, NOS OMA-33.

Literathy, P., G. Morel & A. Al-Bloushi, 1991. Environmental transformation, photolysis of fluoresing petroleum compounds in marine waters. Wat. Sci. Technol. 23: 507–516.

Losidifdou, H. G., S. D. Kilikidis & A. P. Kamarianos, 1982. Analysis for polycyclic aromatic hydrocarbons in mussels (*Mytilus galloprovincialis*) from the Thermaikos Gulf, Grees. Bull. envir. Contam. Toxicol. 28: 535–541.

Malins, D. C., B. B. McCain, J. T. Landahl, M. S. Meyrs, M. M. Kahn, D. W. Brown & W. T. Roubal, 1984. Chemical pollutants in sediments and diseases of bottom dwelling fish in Puget Sound, Washington. Envir. Sci. Technol. 18: 705–713.

Mcleod, W. D., D. W. Brown, A. J. Friedman, D. G. Burrows, O. Maynes, R. W. Pearce, C. A. Wigren & R. W. Boger, 1985. Standard analytical procedures of the NOAA National Analytical Facility 1985–1986. Extractable Toxic organic compounds (2nd ed.) US Department of Commerce, NOAA/NMFS. NOAA Tech. Memo NMFS F/NWC-92.

Menzie, C. A., B. P. Bonnie & J. Santodonato, 1992. Exposure to carcinogenic PAHs in the environment. Envir. Sci. Technol. 25: 1278–1284.

Neff, J. M., 1979. Polycyclic aromatic in the aquatic environment – source, fate and biological effects. Appl. Sci. Publ. London.

Neff, J. M. & J. W. Anderson, 1981. Response of Marine Animals to Petroleum and Specific Petroleum Hydrocarbons. Appl. Sci. Publ. Ltd., London.

Nicholson, G. J., T. Theodoropoulos & G. J. Fabris, 1994. Hydrocarbons, pesticides, PCB and PAH in Port Phillip Bay (Victoria) Sand Flathead. Mar. Pollut. Bull. 28: 115–120.

Pancirov, R. & R. Brown, 1977. Polycyclic aromatic hydrocarbons in marine tissues. Envir. Sci. Technol. 11: 989–992.

Parker, G. J., P. Howgate, P. R. Mackie & A. S. McGill, 1990. Flavour and hydrocarbon assessment of fish from gas fields in the southern North Sea. Oil Chem. Pollut. 6: 263–277.

PERSGA, 1995. Regional programme of action for protection of the marine environment of the Red Sea and Gulf of Aden from land-base sources pollution, 30 pp.

Pritchard, J. B. & J. L. Renfro, 1984. Interactions of xenobiotics with teleost renal function. In L. J. Webber (ed.), Aquatic Toxicology, Vol. 2. Raven Press, New York: 51–106.

Rainio, K., R. R. Linko & L. Ruotsila, 1986. Polycyclic aromatic hydrocarbons in mussels and fish from the Finnish archipelago sea. Bull. envir. Contam. Toxicol. 37: 337–343.

Readman, J. G., S. W. Fowler, J.-P. Villeneuve, C. Cattini, B. Oregioni & L. D. Mee, 1992. Oil and combustion-product contamination of the Gulf marine environment following the war. Nature 358: 662–664.

Readman, J. W., M. R. Preston & R. C. F. Mantoura, 1986. An itegral technique to quantify sewage oil and PAH pollution in estuarine and coastal environments. Mar. Pollut. Bull. 17: 298–308.

Rushdi, A. I., A. A. Ba'Issa & A. Ba'Bagi, 1991. Preliminary investigations of oil pollution along the Red Sea coast of Yemen. In Proceedings of the Seminar on the Status of the Environment in the Republic of Yemen: 175–186.

Santodonato, J., P. Howard & D. Basu, 1981. Health and ecological assessment of polyaromatic hydrocarbons. J. envir. Path. Toxicol., 1–376.

Sauer, T. C., J. S. Brown, P. D. Boehm, D. V. Aurand, J. Michel & M. O. Hayes, 1993. Hydrocarbon characterization of itertidal and subtidal sediment of Saudi Arabia from the Gulf War oil spill. Volumme II. Marine Spill Response Corporation, Washington DC, MSRC Technical Report Series.

Schmeltz, I., J. Tosk, J. Hilfrich, N. Hirota, D. Hoffman & E. L. Wynder, 1978. Bioassay of naphthalene and alkylnephthalene for carcinogenic activity: relation to tobacco carcinogenesis. In Jones, P. W. & R. I. Freundenthal (eds), Carcinogenesis, Polynuclear Aromatic Hydrocarbons, Vol. 3, Raven Press, New York: 47 pp.

Takatsuki, K., S. Suzuki, N. Sato & I. Ushizawa, 1985. Liquid chromatographic determination of polycyclic aromatic hydrocarbons in fish and shellfish. J. Ass. Off. Analyt. Chem. 68: 945–949.

Thorhaug, A., 1992. The environmental future of Kuwait. In Al-Shatti, A. K. & J. M. Harrington (eds), Proceedings of The International Symposium on the Environmental and Health Impacts of the Kuwaiti Oil Fires. University of Birmingham, Edgbaston: 69–72.

Vaessen, H. A., A. A. Jekel & A. A. M. M. Wibers, 1988. Dietary intake of polycyclic aromatic hydrocarbons. Toxicol. envir. Chem. 16: 281–294.

Varanasi, U. & J. E. Stein, 1991. Disposition of xenobiotic chemicals and metabolites in marine organisms. Envir. Health Perspec. 90: 93–100.

Varanasi, U., J. E. Stein & M. Nishimoto, 1989. Biotransformation and disposition of polycyclic aromatic hydrocarbons (PAH) in fish. In Varanasi, U. (ed.), Metabolism of Polycyclic Aromatic Hydrocarbons in the Aquatic Environment, CRC Press, Boca Raton: 93–149.

Hydrobiologia **352**: 263–278, 1997.
Y.-S. Wong & N. F.-Y. Tam (eds), Asia–Pacific Conference on Science and Management of Coastal Environment.
©1997 *Kluwer Academic Publishers. Printed in Belgium.*

The application of gene transfer techniques to marine resource management: recent advances, problems and future directions

F. Y. T. Sin[1]*, U. K. Mukherjee[1], L. Walker[1] & I. L. Sin[2]

[1]*Department of Zoology, University of Canterbury, Christchurch, New Zealand*
[2]*Department of Obstetrics and Gynaecology, Christchurch Womens Hospital, Christchurch, New Zealand*
(Author for correspondence; fax: 03-364 2024; e-mail: f.sin@csc.canterbury.ac.nz)

Key words: transgenic fish, gene transfer, growth enhancement, lopifection, particle bombardment, electroporation, fish sperm, fish embryos

Abstract

Recent advantages in transgenic fish research are reviewed, with special reference to the methods for gene transfer. These include microinjection, electroporation, particle bombardment, and lipofection. The success and problems associated with each of these methods, and the possible applications of transgenic fish research to aquaculture are discussed.

Introduction

Advances in our understanding of gene structure and function have tempted molecular biologists to manipulate the expression of genes and to develop artificial genes with the aim of producing new phenotypes, and to study how these artificial genes function in the host organisms. The transfer of "man-made" genes into an organism will allow one to learn about the control of complex biological phenomena such as growth and reproduction. The first such experiment was reported by Palmiter and co-worker in 1982. In this experiment a growth hormone gene (cDNA) of a rat driven by a mouse metallothionein-I promoter, a novel man-made gene, was injected into the pronuclei of mouse embryos and the embryos were implanted into a foster mother. After birth, the new born mice carrying the novel gene showed enhanced growth rate as compared with their siblings. This "giant" mouse carrying the foreign gene is called a transgenic mouse and the gene is called a transgene. This report has had a profound influence on horticulture and agriculture. Using transgenesis new plant varieties were produced (Gasser & Fraley, 1989). The first example of a transgenic cultivar is a tomato which is commercially available (Leemans, 1993 for review). This cultivar has a gene which when expressed

inhibits the enzyme polygalacturonase, hence retarding the rotting process (Sheehy et al., 1988). The method is also being tested in farm animals for stock development (Pursel et al., 1989; Pursel & Rexroad, 1993). Transgenic research may also play a role in aquaculture and marine resource management. For instance, the development of a freeze resistant salmon will allow commercial sea-cage farming in the cold environment such as the Atlantic coast of Canada (Fletcher & Davies, 1991). The potential of transgenic research in aquaculture has been discussed previously (Fletcher & Davies, 1991). Transgenic technology can improve the economics of fish, shellfish and crustacean cultures. Potential advances include enhanced growth rates, increased overall size, improved freeze resistance, increased disease resistance, improved feed conversion efficiency, increased brood stock fecundity, and possibly larval survival. In this review advances in transgenic fish research will be reviewed, including its problems and their possible solutions, and future research directions will be discussed.

Table 1. Teleost, mollusc and crustacean species used in transgenic research

Common name	Scientific name
Teleost	
Model fish	
Gold fish	*Carassius auratus*
Zebrafish	*Danio rerio*
Medaka	*Oryzias latipes*
Cultivated fish	
Channel catfish	*Ictalurus punctatus*
Tilapia	*Oreochromis niloticus*
Loach	*Misgurnus anguillicandatus*
Red crucian carp	*Carassius auratus auratus*
Common carp	*Cyprinus carpio*
Black porgy	*Acanthopagrus schlegeli*
Northern pike	*Esox lucius*
Brown trout	*Salmo trutta*
Atlantic salmon	*Salmo salar*
Rainbow trout	*Oncorhynchus mykiss*
Cutthroat trout	*Oncorhynchus clarkii*
Coho salmon	*Oncorhynchus kisutch*
Chinook salmon	*Oncorhynchus tshawytscha*
Mollusc	
Abalone	*Haliotis rufescens*
Crustacean	
Artemia	*Artemia franciscan*

Table 2. Methods used for gene transfer into fish embryos

Method	Mechanism	Site of delivery
Microinjection of embryos	Membrane disruption	nuclei or cytoplasm
Electroporation of embryos/ sperm	Multiple pore formation	cytoplasm
Particle bombardment of embryos	Multiple membrane puncture	cytoplasm
Lipofection of embryo/ sperm	Charge interaction between DNA and lipid, and membrane	cytoplasm
Virus-mediated	Virus-host membrane interaction	nucleus

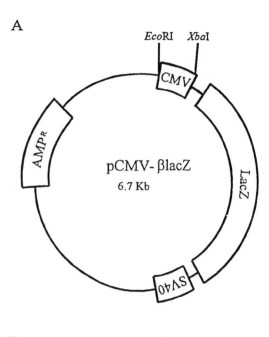

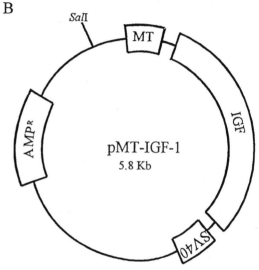

Figure 1. Plasmids carrying fusion genes. (A) Reporter fusion gene containing a human cytomegalovirus promoter (CMV), a *E. coli* β-galactosidase, and the Simian virus 40 polyA signal sequence. (B) Growth factor fusion gene containing a chinook salmon metallothionein promoter, a coho salmon IGF-1 cDNA, and the Simian virus 40 polyA signal sequence.

Marine and freshwater teleosts, mollusc and crustacean in transgenic research

Over the last decade a number of fish with short reproductive cycles and life cycles have been used as models for the testing of promoter function, transgene expression analysis, and transmission genetics (Table 1). Marine and freshwater fish species have also been used in transgenic research with the aim of developing broodstock with fast growth rate or freeze resistance (Table 1). Commercially valuable shellfish species such as abalone and oysters, and crustacean such as prawns and lobsters will become more prominent in transgenic research.

Gene constructs and detection

Various fusion genes have been used in transgenic fish research (Table 4). These can be classed into three broad groups: reporter genes such as those conferring the activities of chloramphenicol acetyltransferase (CAT), *E. coli* β-galactosidase (LacZ), firefly luciferase (luc), and the green fluorescent protein (GFP) from the bioluminescent jellyfish (*Aequorea victoria*); growth factors such as growth hormone (GH) and insulin-like growth factor (IGF); and structural and functional proteins including α-crystalline and antifreeze protein (Table 4). These artificial fusion genes consist of three parts: a regulatory sequence containing the promoter and enhancer(s); the gene proper, usually made of cDNA (DNA copied from the messenger RNA of a specific gene); and a transcription termination signal sequence (Figure 1). The promoters may be of viral origin, e.g. Simian virus 40 (SV40), cytomegalovirus (CMV), or from structural genes such the metallothionein gene and the actin gene.

For the development of a fusion gene for transfer, it is vital to determine the ability of the promoter sequence to regulate expression of the gene (e.g. Gong et al., 1991). Reporter genes are useful for this purpose (Alam & Cook, 1990). To test the promoter, it is ligated to a reporter gene such as the CAT gene (Figure 2). Figure 2 shows the expression of the CAT gene driven by the metallothionein promoter-1 of chinook salmon. The level of expression of the CAT gene will indicate the presence of certain sequences in the promoter which is essential for optimal expression.

Reporter genes are also useful for testing gene transfer techniques, may it be by microinjection or mass transfer techniques such as electroporation and particle bombardment (e.g. Muller et al., 1993; Sin et al., 1993; Gendreau et al., 1995). Since the reporter genes are usually controlled by a viral promoter, such as the long terminal repeat (LTR) of SV40, the expression of the reporter gene is constitutively expressed after transfer into the fish and the level of expression is usually sufficiently high to be detected by histochemical staining in tissue sections, or whole embryos and fry. The LacZ reporter gene is one that has been used extensively (e.g. Takagi et al., 1994; Ueno et al., 1994). It can be visualized in whole embryos *in situ*. CAT

can be detected through enzymatic reaction carried out using tissue extracts (Gorman et al., 1982; Neumann et al., 1987). Luciferase activity can be detected and quantified using a scintillation counter (Nguyen et al., 1988; Tamiya et al., 1990; Alestrom et al., 1992; Gendreau et al., 1995), or by screening with photographic film (Gibbs et al., 1994). The green fluorescent protein gene will become more prominent in transgenic research as the protein can be induced to fluoresce and it can be detected in live organisms (Chalfie et al., 1994; Pines, 1995; Hodgkinson, 1995; Haseldoff & Amos, 1995).

The detection of transgenes after transfer into the host fish can be carried out by a number of techniques. These include searching for the transgene co-extracted with the host fish DNA by Southern blotting (Southern, 1975) followed by molecular hybridization using either radiolabelled or non-radiolabelled DNA as a probe, or by the polymerase chain reaction (PCR) using primers specific for the transgene. Southern blotting is a less sensitive method than PCR for the screening of the P1 fish because the copy number of the transgene may be too low to be detected as a result of its mosaic distribution in the host. PCR is a more sensitive method for the screening of transgenic fish. It overcomes the problem of mosaicism.

Expression of growth hormone can be established by the detection of mRMA by northern blotting, followed by molecular hybridization using gene specific DNA sequences as probes (e.g. Chen et al., 1993) or by the direct detection of the growth hormone by radioimmuno assay (Devlin et al., 1994).

Methods for gene transfer

The most common methods used in gene transfer involve egg membrane disruption by microinjection, pore formation in the cell membrane by exposure to high voltage (electroporation), multiple membrane puncture by microprojectiles or charge interactions between DNA/lipid and membrane (lipofection) (Table 2). The most common method employed is by microinjection of DNA solution directly into the nuclei or cytoplasm of fertilized eggs (Table 3).

Microinjection

The basic items of equipment for microinjection are a stereoscopic dissecting microscope, a micromanipulator and a device for the dispensing of solutions. Glass

A

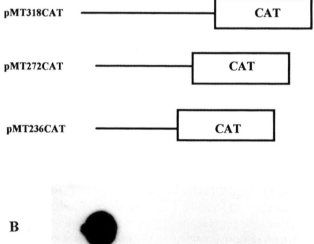

```
  1  CTGATTAAGT TTTGTATAGT TAAATAAATA TAGGTGTAGC CTTAATTAAT CGATGATCAA
 61  CGTGGTAATC AGGTTTATGT AACAGGCTAT GGAATTTGGA AACAATAGGA AACTCTTCCT
121  TGATTATTTT CGCGCAGTAT AATGAAATAA CCCGGGTGCA AACCCTGATC GTCTGAACGC
181  GAGACTGTTT TGCACACGGC ACCCGTCTGT CCCTGACGC AIAAAAACGG TCTTCGCCAA
241  AGAGAAATTT AAAGCTTACA ACTCACAAGT GAAATTGAGC TGAAATACTT CATTTGACTA
301  AAGAAGCGCG ATCGAAAA
```

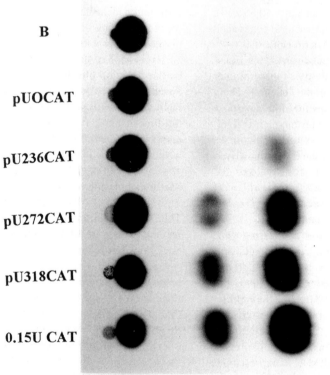

B

pUOCAT

pU236CAT

pU272CAT

pU318CAT

0.15U CAT

Figure 2. Nucleotide sequence of the chinook salmon metallothionein promoter. (A) Full length, 318 bp. The underlined section is 272 bp and the triangles indicate the 236 bp sequence. (B) CAT assay showing promoterless CAT (pUOCAT), truncated promoters (pU236CAT) and pU272CAT), full length promoter (pU318CAT), and enzyme control.

Table 3. Needle size, volume and amount of DNA transferred by microinjection

Fish species	Needle size	Volume, copies of DNA	Ref
Gold fish	3 μm	1–2 nl, 7×10^6	Zhu et al., 1985
Medaka	3–4 μm	20–30 pL	Ozato et al., 1986
Rainbow trout	10 μm	20 nl, 200 pg	Chourrout et al., 1986
Channel catfish	2–10 μm	20 nl, 10^6	Dunham et al., 1987
Tilapia	5 μm	10^6	Brem et al., 1988
Atlantic salmon		2–3 nl, 10^6	Fletcher et al., 1988
Atlantic salmon	NA	20 nl, 10^7	McEvoy et al., 1988
Zebrafish		10 pl, 30 pg	Stuart et al., 1988
Rainbow trout	10 μm	10 nl, 10^9	Rokkones et al., 1989
Goldfish	2 μm	2 nl, 25 ng/μl	Yoon et al., 1990
Common carp	2–10 μm	20 nl, 10^6	Zhang et al., 1990
Zebrafish	NA	10 pl, 30 pg	Culp et al., 1991
Channel catfish	2–10 μm	20 nl, 10^6	Hayat et al., 1991
Common carp	2–10 μm	20 nl, 10^6	Hayat et al., 1991
Northern pike	2 μm	50 pl, 25 ng/μl	Gross et al., 1992
Medaka	3–5 μm	200 pl, 10^5-10^6	Lu et al., 1992
Atlantic salmon	ND	2–3 nl, 10^6	Du et al., 1992
Goldfish	1–5 μm	4–20 nl, 50–250 pg	Moav et al., 1993
Coho salmon	5–10 μm	1–2 nl, 100 μg/ml	Devlin et al., 1994
Rainbow trout	5–10 μm	1–2 nL, 100 μg/ml	Devlin et al., 1995

needles are drawn with a needle puller. Bore sizes, ranging from 2 μm to 10 μm in diameter can be used, depending on the size of the fish eggs (Table 3). The volume of the injection is between 200 pl to 20 nl according to the size of the eggs, and the amount of DA injected is about 10^6 copies (Table 3).

For mouse embryos the DNA solution is delivered to the pronuclei which are easily visible (Palmiter et al., 1982). However, the pronuclei of the fertilized fish eggs are invisible and the eggs are opaque in most fish species. Thus, the DNA is delivered to the cytoplasm (e.g. Chen & Powers, 1990; Penman et al., 1990). The second problem associated with microinjection is the hard chorion of some fish species which renders microinjection difficult. This has led to the adoption of various approaches to overcome this problem. For example, the DNA can be delivered through the micropyle of the eggs (e.g. Brem et al., 1988; Fletcher et al., 1988; Devlin et al., 1993); or a two-step method involving puncturing the chorion with a metal needle for the insertion of the glass needle for delivery (Chourrout et al., 1986; Penman et al., 1990). Others tried manual dechorionation of the eggs (Ozato et al., 1986), or using pronase to digest the chorion before microinjection (Zhu et al., 1985).

In general, fusion genes are usually injected into the cytoplasm of embryos at the 1-, 2- or 4-cell stage. Survival of the injected embryos ranges from 16% in zebrafish embryos (Stuart et al., 1988) to 85% in salmon (Devlin et al., 1995). The number of transgenic founder fish (P1) detected ranged from none in common carp (Hayat et al., 1991), 5% in zebrafish (Stuart et al., 1988), 6% in salmon (Devlin et al., 1994), 13% in tilapia (Martinez et al., 1996) to over 70% in rainbow trout (Chourrout et al., 1986; Penman et al., 1990). However, the presence of the transgene in the founder fish does not mean that these genes will be transmitted to the progeny, or that the transgene is integrated into the host DNA.

Other factors affecting gene transfer efficiency include buffer, DNA concentration and whether linear or circular DNA is used (Penman et al., 1990; Fletcher & Davies, 1991). Penman et al. showed that when rainbow trout was injected with EDTA (0.125 m) the survival rate of fish embryos decreased by 10%, but retention of the DNA was increased by about 10 to 13%. When 10^8 copies of DNA injected as compared with 10^6 copies, the survival rate was reduced by 10%. Linear DNA was 15-fold more efficient in integration than circular DNA.

Table 4. Efficiency of integration of foreign gene transfer by microinjection

Fish species	Gene construct	% transgenic founder (N)	Expression	Ref
Gold fish	BPVMG.	50 (6)	ND	Zhu et al., 1985
Medaka	pδC-1B	33 (30)	protein	Ozato et al., 1986
Rainbow trout	pSV507(hGH)		ND	Chourrout et al., 1986
	linear	75 (37)		
	circular	40 (30)		
Channel catfish	mthGH	20 (10)	ND	Dunham et al., 1987
Tilapia	MThGH	3/18;1/14	ND	Brem et al., 1988
Atlantic salmon	AFP	7 (30)	ND	Fletcher et al., 1988
Atlantic salmon	pMTLacZ	NA	protein	McEvoy et al., 1988
Zebrafish		5 (NA)	ND	Stuart et al., 1988
Rainbow trout	MThGH	NA	RNA, GH	Rokkones et al., 1989
Goldfish	pRSVneo	ND	RNA	Yoon et al., 1990
Common carp	pRSVrtGH1	5.5 (365)	GH, growth	Zhang et al., 1990
Zebrafish	pRSV-βGal	0–25 (19-54)	ND	Culp et al., 1991
Channel catfish	pRSVrtGH	1.5–12.5	ND	Hayat et al., 1991
Common carp	pRSVrtGH	0–11.5	ND	
Northern pike	sGH	6 (1398)	GH, growth	Gross et al., 1992
Rainbow trout	pCMVCat	MT	protein	Tewari et al., 1992
Medaka	hGH	22 (22)		Lu et al., 1992
	cBA-hGH	27 (100)	RNA, growth	
	mMT-hGH	17 (126)	RNA, growth	
	vTK-hGH	24 (36)		
	rtGH2	22 (NA)		
Atlantic salmon	opAFPGH	12 (50)	GH, Growth	Du et al., 1992
Zebrafish	pLuc	T-F1	protein	Kavumpurath et al., 1993
Zebrafish, goldfish	pactCAT	NA	protein	Moav et al., 1993
Common carp	pRSVLTR-rt	GH1T-f1	growth	Chen et al., 1993
Coho salmon	pOnMTGH1	6.2 (1073)	GH, growth	Devlin et al., 1994
Coho salmon	opAFPGHc	4 (427)	growth	Devlin et al., 1995
Rainbow trout	opAFPGHc	14 (131)	growth	
Cutthroat trout	opAFPGHc	ND	growth	

N, total number of fish analyzed. NA, not available. MT, Mendelian transmission to F2. T-F1, transmitted to F1.

Despite the technical problems associated with microinjection this method of gene transfer has produced transgenic fish with enhanced growth rates (Table 4). Du et al. (1992) demonstrated enhancement in growth of Atlantic salmon after microinjection of a growth hormone gene fused with an antifreeze protein gene promoter. The transgenic fish had an average increase of 2 to 6-fold in growth rate. Devlin et al. (1994) demonstrated remarkable growth in coho salmon after microinjecting embryos with a fusion gene consisting of sockeye salmon growth hormone gene driven by a sockeye salmon MT promoter. Analysis of the yearlings showed that 6.2% of the fish contained the fusion gene, and the transgenic fish

had an average growth rate of over 11-fold greater than the non-transgenic control fish. Growth enhancement was also observed in coho salmon and rainbow trout containing a chinook salmon growth hormone fusion gene (Devlin et al., 1995). Chen et al. (1993) microinjected rainbow trout growth hormone cDNA driven by the avian Rous sarcoma virus promoter into embryos of the common carp and obtained a 20% to 40% increase in growth rate. Recently, Martinez et al. (1996) microinjected a tilapia growth hormone cDNA fused to the human cytomegalovirus enhancer-promoter. This fusion gene was stably integrated and transmitted to F1 and F2 progeny. F1 progeny showed an increased growth rate of about 80% over that of the control fish. Recently, Scotland and New Zealand have commenced field trials using the antifreeze protein gene promoter-growth hormone gene construct.

Fletcher et al. (1988) transferred the winter flounder (*Pseudopleuronectes americanus*) antifreeze protein (AFP) gene into Atlantic salmon embryos. The transgene was found to integrate in 7% of the founder fish. Detectable levels of proAFP, an AFP precursor, were found in the transgenic founder salmon and in the F1 progeny (Shears et al., 1991). However, the level of AFP was about 100-fold lower than that in winter founder. Thus, in order to produce salmon with low temperature tolerance, a different gene construct with a stronger promoter would be required.

Electroporation

The induction of changes in the permeability of a cell by an electric field is called electroporation. Electroporation is a commonly used technique for the transfer of macromolecules such as DNA and protein into plant and animal cells, and bacterial cells (Shigekawa & Dower, 1988; O'Hara et al., 1989; Sukharev et al., 1992; Weaver, 1993). When cells are exposed to an electric field, the components of the cell membrane become polarized and a voltage potential develops across the membrane. When the potential differences exceed a threshold level between the inside and the outside of the cell membrane, the membrane breaks down in localized areas and pores are formed, and the cells become permeable to exogenous molecules (Knight, 1981; Knight & Scrutton, 1986). The change in the membrane permeability is reversible provided that the magnitude or duration of the electric pulse does not exceed a critical limit (Tsong, 1983; Serpersu et al., 1985; Sowers & Lieber, 1986). The pore size can be controlled by varying the pulse width, field strength, and ionic strength of the medium (Tsong, 1983). A molecule such as dextran, 156 kDa, with a radius of greater than 8.4 nm was able to escape ghost cells after short pulses of electric shock (Sowers & Lieber, 1986).

In order to transfer DNA molecules into cells, an electric pulse is passed through a cell suspension containing the DNA to be transferred. Two different types of electric pulses, square wave and exponential wave, have been used for gene transfer into fish embryos (e.g. Inoue et al., 1990; Powers et al., 1992, 1995) and fish sperm (e.g. Sin et al., 1990, 1993; Muller et al., 1992; Tsai et al., 1995; Patil & Khoo, 1996). The voltage of the square wave is raised to the required amplitude which is kept for a set length of time (pulse width),

then is returned to zero. For the exponential wave, the voltage is raised to a desired amplitude, then is allowed to decay exponentially.

The most important electrical parameters in an electroporation experiment are field strength and pulse length. Field strength, expressed in volts per centimetre (V/cm), is the voltage applied by the power supply to the electrodes. Pulse length is determined by the combination of resistance (ohms) and capacitance (farads, F) in the circuit. The capacitance is dependent on the incorporation of capacitors in the circuit, and resistance is mainly determined by the sample. In the exponential wave method, the voltage of the pulse starts at a pre-determined value and decays exponentially as a function of time.

Cells become permeabilized when the voltage across the cells reaches a threshold value. This drop in voltage is a function of the field strength (E) and the size of the cell. The field strength is equal to the voltage between the electrodes divided by the distance (d) between the electrodes.

The resistance of the sample is determined by the ionic strength of the electroporation medium, the distance between the electrodes, and the cross sectional area of the electrodes. The ionic strength is inversely proportional to the resistance. Thus higher ionic strength of the medium will result in a lower resistance. The greater the distance between the electrodes the greater will be the resistance; the smaller the electrode, the greater the resistance. Discharge of a given capacitor into a low resistance medium (high ionic strength) will have a short pulse length (r) (Shigekawa & Dower, 1988).

Gene transfer by electroporation of fish embryos
Foreign DNA has been transferred by electroporation into fertilized embryos of medaka (Inoue et al., 1990; Lu et al., 1992), goldfish and crucian carp (Yamada et al., 1988), loach and red crucian carp (Xie et al., 1989, 1993), channel catfish and common carp (Powers et al., 1992), zebrafish (Buono & Linser, 1992; Zhao et al., 1993), black porgy (Tsai & Tseng, 1994), and red abalone (Powers et al., 1995) (Table 5).

The survival and hatching rates of the treated embryos are species dependent (Powers et al., 1992) and the conditions of electroporation also can affect the hatching rate.

Inoue et al. (1990) showed that 90% of the medaka embryos survived the electric shock, but only 25% of the surviving embryos hatched and approximately 4%

Table 5. Conditions for the electroporation of different fish embryos

Species	DNA µg/ml	Pulse length ms/µs	Field strength (V/cm)	Ref
Medaka	100	50 µs	750[4]	Inoue et al.,[1] 1990
Medaka	NA	0.32–0.4 ms	250,375,500[4]	Lu et al.,[1] 1992
Channel catfish, carp	50	160 µs	2.5–10 kV	Powers et al.,[6] 1992
Zebrafish	NA	NA	NA	Kavumpirath et al.,1993
Crucian carp[3], loach	100	0.97 ms	250	Xie et al.,[1] 1993
Black porgy	5	160 µs	3 kV[5]	Tsai & Tseng,[2] 1994
Red abalone	285	160 µs	10 kV	Powers et al.,[2] 1995

1, capacitor based power supply.
2, square wave: 0.4s or 0.8s burst time; 2^{11} pulses/cycle.
3, dechorionated embryos.
4, electrode distance = 2 mm; pulse interval = 1 s.
5, electrode distance = 1 mm.
6, square wave: 4 cycles, 0.8s burst time; 2^6-2^{11} pulses/cycle.
NA, not available.

of the hatchlings were found to contain the transgene. The number of the transgene in these hatchlings varied between 1 to over 100 copies. Lu et al. (1992) showed that 70% of the surviving medaka embryos hatched and 20% of the surviving fish showed integration of the transgene. Furthermore, Powers et al. (1992) demonstrated in common carp and channel catfish that the survival rate of the electroporated eggs and fertilized embryos was dependent on the stage of development at which electroporation was applied, and the rate of integration was also dependent on the age of the eggs and embryos.

In loach and crucian carp the efficiency of gene transfer and the number of copies (<0.1 to >100) of transgene per genome were found to increase with increasing pulse duration within the range of 0.48–4.8 ms and the voltage of the pulse. The highest efficiency of transfer was over 60% when 300 V and 0.97 ms was used for electroporation (Xie et al., 1993). Kavumpurath et al. (1993) demonstrated successful gene transfer by electroporation of zebrafish embryos but the transgene, pLuc, was not expressed.

Survival of black porgy embryos after electroporation varied between 60 to 70% and hatching rates were about 60% relative to the control groups. The efficiency of transfer, measured at seven days after hatching, was found to be about 15% (Tsai & Tseng, 1994). The gene transfer efficiency using a similar electropo-

ration system (square wave) was between 35 to 75% for zebrafish embryos (Powers et al., 1992).

Recently, Powers et al. (1995) have further demonstrated that electroporation can be used to transfer DNA into abalone embryos. Larval survival was about 70% to 84% of that of the untreated embryos. The efficiency of transfer ranged between 70% to 100% when measured 2 h after electroporation. The optimal time for electroporation was between 30 and 35 min after fertilization, that was about 1.5 h before the first mitotic division. Using Southern blot analysis these authors concluded that the DNA was integrated into the host genome.

In these experiments the detection of the transgene was carried out using fry of various ages, and therefore a direct comparison of the transfer efficiency may not be valid because a greater number of embryos may have initially been carrying the transgene but have later lost it due to degradation of the foreign DNA. Powers et al. (1992, 1995) have observed that the number of zebrafish and abalone embryos containing the foreign gene was lower when the fish and embryos were assayed for the transgene at an older age, suggesting that the foreign gene was not integrated. Thus, these experiments were measuring the retention of the transgene by the host up to certain age rather than the true integration rate of the transgene.

Southern blotting of fish DNA was used to detect the transgene in fish genomes. Although Southern blot is a reliable method for the detection of homologous DNA, its sensitivity may not be high enough to detect the low copy number of the transgene in some fish. It is well established that transgenes often have a mosaic distribution in the founder generation (Chen et al., 1993). Southern blot analysis may underestimate the efficiency of transfer. Polymerase chain reaction is a far more sensitive method for the detection of foreign DNA with a mosaic distribution in founder fish (e.g. Chen et al., 1993; Symonds et al., 1994b).

In order to determine whether the transferred DNA is integrated into the host genome it is necessary to determine whether the transgene is detected in the high molecular weight genomic DNA. Restriction digestion of the fish DNA and analysis by Southern blotting without analysing the undigested genomic DNA does not allow one to conclude successful integration unequivocally. Germline transmission may be tested by crossing transgenic founder fish to wild-type fish.

The above experiments demonstrated that electroporation is capable of transferring foreign DNA into fish and shellfish embryos. The efficiency of transfer will require empirical testing of various electroporation parameters, namely pulse length, field strength and pulse number, to determine the optimal conditions for a particular species. Although transfer efficiency will increase when dechorionated embryos are used, the survival rate of these embryos are likely to be lower than those without such manipulation. The age of the embryos can also affect the survival and hatching rate (Tsai & Tseng, 1994).

Electroporation of sperm cells for gene transfer
Sperm cells of a number of mammals: porcine (Horan et al., 1991, 1992); bovine (Gagne et al., 1991); mouse (Camaioni et al., 1992; Francolini et al., 1993); of chicken (Nakanishi & Iritani, 1993), salmon (Symonds et al., 1994a) and abalone (Sin et al., 1995) have been shown to be capable of binding DNA *in vitro*. Horan et al. (1991) estimated that about 3.8×10^2 DNA molecules were associated with each porcine sperm head within 15 to 20 min after mixing of sperm and DNA. They further showed that motile sperm were more capable of picking up DNA; about 30% of the motile sperm carried DNA. Using light microscopic and ultrastructural autoradiography, Francoline et al. (1993) and Camaioni et al. (1992) showed that most DNA was attached to the mouse sperm head, between

Table 6. Sperm-mediated gene transfer in mammals

Animal	Electric shock	Reference
Rabbit	no	Brackett et al., 1971
Mouse	no	Lavitrano et al., 1989
	no	Hochi et al., 1990
	no	Bachiller et al., 1990
	yes	Baranov et al., 1990
Cattle	yes	Gagne et al., 1991
	no	Perez et al., 1991
Pig	yes	Horan et al., 1992
Chicken	yes;no	Nakanishi & Iritani, 1993

15 to 22% of sperm bound DNA was associated with the mouse nuclei, and the amount of DNA internalized increased with the time of incubation. DNA molecules have also been found to be internalized in the bovine sperm head (Gagne et al., 1991). DNase digestion of sperm of mouse, salmon and abalone after incubation in a DNA solution removed most of the sperm associated DNA, further suggesting that some DNA molecules were bound to the outside of the sperm cells (porcine: Horan et al., 1991; chicken: Nakanishi & Iritani, 1993; salmon: Symonds et al., 1994; abalone: Sin et al., 1995; zebrafish: Patil & Khoo, 1996;).

Sperm cells have been shown to be capable of transferring foreign DNA into eggs of rabbits (Brackett et al., 1971). Some concern regarding the reliability of this method for gene transfer was raised when Lavitrano's experiments (Lavitrano et al., 1989) were unable to be repeated by some laboratories. It was suspected that sperm-mediated transfer is not controllable (Birnstiel & Busslinger, 1989; Gavora et al., 1991). Nevertheless, in spite of the possible problems of using sperm cells for gene transfer a number of laboratories have successfully demonstrated the ability of sperm cells to transfer genes in mammals (Table 6).

Sperm-mediated gene transfer has been demonstrated by simply soaking the sperm in a DNA solution and using the treated sperm to fertilize the eggs of the sea urchins, *Paracentrotus lividus, Strongylocentrotus purpuratus* and *Arbacia lixula* (Arezzo, 1989). The transfer of genes such as the CAT gene into eggs of *S. purpuratus* was demonstrated by the expression of the transgene in the embryos of the swimming blastula stage.

Sperm cells of zebrafish have also been shown to be capable of picking up DNA molecules, e.g. pRSVlacZ plasmid DNA, in solution and transferring the foreign DNA into eggs (Khoo et al., 1992; Patil & Khoo, 1996).

Table 7. Sperm-mediated gene transfer in fish and sea urchin

Fish	Electric shock (+ or -)	DNA	References
Sea urchin	-	pSV2CAT	Arezzo, 1989
Brown trout	+	pRSVlacZ	Sin et al., 1990
Zebrafish	-	pRSVCAT	Khoo et al., 1992
Common carp	+	pRSVlacZ pHSVtkneo	Muller et al., 1992
African catfish	+	pRSVlacZ pHSVtkneo	
Tilapia	+	pGMH4CAT	
Chinook salmon	+	pRSVlacZ	Sin et al., 1993
Goldfish	-	AFP	Yu et al., 1994
Chinook salmon	+	pRSVlacZ pRSVCAT	Symonds et al., 1994
Chinook salmon	+	pRSVlacZ	Walker et al., 1995
Loach	+	poAFPGHc	Tsai et al., 1995
Zebrafish	+	pMTL	Patil & Khoo, 1996
Zebrafish	+	pCMVlacZ pcMTcoIGF	Walker et al., (Unpublished observation)

The efficiency of transfer as determined by Southern blot analysis of fin clips of adult fish ranged between 10 to 53%. The transgene was transmitted from the parental transgenic fish to F2 progeny through crosses between transgenic and control fish. However, it appears that the CAT gene was not expressed and it might exist extrachromosomally (Khoo et al., 1992). Yu et al. (1994) have also demonstrated that goldfish sperm soaked in DNA solution can transfer the DNA to eggs during fertilization and 26% of the fish analyzed were shown to retain the transgene. However, sperm of brown trout and chinook salmon appeared to be much less efficient in picking up DNA from solution (Sin et al., 1990; Sin et al., 1993; Symonds et al., 1994b).

In view of the low efficiency of gene transfer after soaking of sperm in DNA, electroporation of sperm in the presence of DNA was tested in brown trout (Sin et al., 1990), common carp, African catfish, and tilapia (Muller et al., 1992), chinook salmon (Sin et al., 1993, Symonds et al., 1994b), loach (Tsai et al., 1995) and zebrafish (Patil & Khoo, 1996). In general the motility of sperm, a measure of sperm viability, is found to decrease with increasing field strength and pulse length, and the survival of the embryos fertilized by the electroporated sperm is similar to the control eggs when approximately 30 millions of motile sperm are used to fertilize 500 eggs.

Chinook salmon sperm respond to electric shock (exponential decay wave) in similar ways to cultured cells. The viability, measured as motility, of sperm decreases with increasing field strength and pulse length. Hatching rate of embryos fertilized by electric shocked sperm is similar to the control embryos which have been fertilized by untreated sperm (Sin et al., 1993; Symonds et al., 1994b). Using optimized electroporation conditions, two pulses, each of 27.4 ms duration and 1000 V/cm (Cell-Porator, BRL) the transfer efficiency measured by the polymerase chain reaction of alevin DNA was found to be over 85% (Symonds et al., 1994b). The transfer of a gene construct containing a chinook salmon metallothionein promoter and a coho salmon insulin-like growth factor-I cDNA into zebrafish eggs increased the growth rate of the transgenic zebrafish by 20 to 80% (Tables 9 and 10) (Mukherjee, Walker & Sin, unpubl. observ.).

Enhancement of DNA binding to sperm of African cat fish, tilapia and common carp has been carried out by electroporation (Muller et al., 1992). These authors found that the efficiency of transfer was between 2.6 to 4.2% using dot blot analysis of DNA from free-swimming fry. In zebrafish the efficiency of transfer ranged between 8 to 14.5%, dependent the electroporation conditions (Patil & Khoo, 1996).

Recently, Tsai et al., (1995) used electroporated sperm cells of loach (square wave: 8 kV; burst time,

Table 8. Advances in gene transfer methods into fish embryos

Methods	Expression in embryos/fish	Growth enhancement in P fish	Transmission to progeny
Microinjection	+	+	+
Electroporation			
sperm	+	+	+
embryo	+	NA	ND
Particle			
bombardment	+	NA	ND
Lipofection			
sperm	+	NA	ND
embryo	+	NA	ND
Virus-mediated	+	NA	+
Chromosome-mediated	+	NA	ND

NA, not applicable because reporter genes were transferred.
ND, not determined.
P, parental fish containing the transgene.

Table 9. Percentage increase in growth of transgenic zebrafish over the control fry

Fry	Topoisomerase (units/μl)	MT-IGF (PCR positive)	Length (mm)	% of control mean
1	0.1	-	8.9	101
2	0.1	-	9.8	111
3	0.1	++	10.3	117
4	0.1	++	11.0	125
5	0.1	++	12.2	139
6	1	-	8.6	98
7	1	++	10.4	118
8	1	++	10.6	120
9	1	++	11.3	128
10	1	++	11.3	128
11	5	+	10.5	119
12	5	++	12.4	141
13	5	++	13.0	147
14	5	++	13.3	151
15	5	++	14.6	166
16	5	++	16.6	188
C	0	-	8.8±1.0(28)	100

++, strong band; +, faint band.

1.6 s; pulse number, 2^8; pulse time, 120 μs; 6 cycles) to transfer a gene construct consisting of an antifreeze protein gene promoter and a salmon growth hormone cDNA into loach eggs. They found an efficiency of transfer of 50%; the transgenic loach showed a 2.5 fold increase in growth rate.

Table 10. Effect of MT-IGF fusion gene on the growth of zebrafish treated with different concentrations of topoisomerase I.

Topoisomerase (units/25 μl)	DNA (μg/25 μl)	37 days	75 days
0	0	8.8±1.0(28)	16.9±2.0(23)
0.1	2	10.4±1.1(5)*	ND
1.0	2	10.4±1.0(5)*	18.9±2.2(15)
5.0	2	13.4±1.9(6)*	24.2±2.8(6)

* Statistically significantly different from the control fry ($P<0.003$, Student's t test)

So far integration of the transgene transferred by sperm has not been demonstrated unequivocally. Expression of the transgene in the founder fish does not mean that these genes are integrated. Thus growth enhancement of fish carrying growth factor genes may be simply reflecting the expression of the transgene. Genetic crosses should be carried out to establish whether the transgene will be transmitted. If the transgene is integrated into the germline, a genetic cross between the transgenic founder fish and the wild-type fish will produce progeny with a phenotypic ratio in accordance with Mendelian genetics. When half of the progeny carries the transgene, it would suggest that the transgene has been integrated and exists in a heterozygote state. When the ratio is over 50% it would suggest multiple integration sites are present; while less than 50% would suggest that the integration of the transgene is not uniformly found in all the germ cells. However, successful transmission of the transgene to the progeny may not necessarily mean that the transgene is integrated (Khoo et al., 1992; Walker & Sin, unpubl. observ.). Further experiments, such as sequencing of fusion gene/host DNA junction, will be required to establish an integration event.

Particle bombardment

The use of microprojectile bombardment for the transfer of foreign DNA was first demonstrated in the transformation of plants (Sandford, 1988). Later the method was adopted for gene transfer into fertilized eggs of loach, rainbow trout and zebrafish (Zelenin et al., 1991; Kavumpurath et al., 1993), and brine shrimp (Gendreau et al., 1995).

The DNA to be transferred is coated onto tungsten or gold particles (about 10 μg DNA/1 to 12 mg particles per shot) which are then propelled into the eggs placed in a chamber. Microprojectile bombardment device (PDA 1000/He, BioRad) uses a gunpowder discharge to accelerate the particles. The eggs are placed in a vacuum chamber to reduce air resistance on the projectiles. The sizes of the tungsten particles used are dependent on the size of the fish eggs, for example 0.1 to 1.5 μm in diameter were used for loach and zebrafish eggs, 1.0 to 3.5 μm for rainbow trout eggs (Zelenin et al., 1991). The efficiency of transfer is dependent on the distance between the discharge barrel and the eggs: 20–25 cm for loach and zebrafish, and 10 to 12 cm for rainbow trout eggs were found to be the optimal distances (Zelenin et al., 1991). Using these conditions, Zelenin et al. (1991) showed that about 70% of the eggs survived the bombardment and some of the embryos expressed the transferred genes, pRSVlacZ and pSV3-neo. Kavumpurath et al. (1993) demonstrated successful transfer of a luciferase gene into zebrafish eggs, and expression of the luciferase gene was detected 24 h after transfer, but the embryos failed to develop. Gendreau et al. (1995) have also demonstrated successful transfer of a luciferase reporter gene (pDrLuc), coated onto gold particles of 1 μm in diameter, into artemia embryos. The success was dependent on the distance between the barrel and the embryos: 35 mm being the optimal distance and a pressure of 450-psi.

To date, particle bombardment as a technique for gene transfer is still in its infancy. It is not known whether the transferred genes will be integrated into the host genome, and whether the foreign gene will be transmitted to the progeny. However, the method may prove to be useful for certain species when other methods for gene transfer are not applicable.

Liposome-mediated gene transfer (lipofection)

Cationic lipids such as N1-(2,3-dioleyloxy) propyl-N,N,N-trimethylammonium chloride (DOTMA), dimethyl dioctadecylammonium bromide (DDAB) and 2,3 - dioleyloxy - N [2 (sperminecarboxamido) ethyl]-N , N - dimethyl - 1 - propanamonium trifluoroacetate (DOSPA) (GIBCO BRL), interact spontaneously with the negatively charged nucleic acids such as DNA and RNA, to form a complex consisting of the liposome and nucleic acid. Three models of DNA-lipid interactions have been suggested, namely: the electrostatic model, the internal model and the coated electrostatic model (reviewed by Smith et al., 1993). The first model predicts that the cationic liposomes associate with DNA through electrostatic charges. The number of liposomes associated with a particular DNA molecule is dependent on the size of the DNA molecule.

The second model suggests that the DNA is entrapped in the liposome. The DNA interacts electrostatically with the inner surface of the liposome. The third model proposes that the DNA is surrounded by a pair of lipid bilayers which form a ribbon with the nucleic acid. The first and third model appear to be consistent with electron microscopic studies of DNA-liposome complexes (Gershon et al., 1993). Since the surface of biological membranes are also negatively charged, liposomes can associate with the negatively charged plasma membrane readily and transfer nucleic acids into mammalian cells (Felgner et al., 1987; Behr et al., 1989; Gershon et al., 1993; Smith et al., 1993). The mechanism involved in the internalization of the DNA is not yet understood. It has been shown that the rate at which the liposome binds to the plasma membrane determines the overall uptake rate; and lipid specificity for a specific cell type also plays an important role for uptake (Smith et al., 1993).

Lipofection has been used to enhance sperm-DNA association for subsequent sperm-mediated gene transfer in chicken. However, the transferred DNA was not integrated into the host genome (Rottmann et al., 1992). When mouse eggs were fertilized by sperm which had been treated with liposome to enhance DNA-sperm association, no transgenic mouse was detected (Bachiller et al., 1991).

Liposomes have also been used successfully to transfer two plasmids containing a gene coding for chloramphenicol acetyltransferase and the neo^r gene coding for aminoglycoside-phosphotransferase (APT) into African catfish embryos (Szelei et al., 1994). However, the number of larvae expressing the two genes were found to decrease with the age of the larvae, falling from 80% to 30% over a period of two weeks. By four months old analysis of the DNA from tail fin clip failed to detect any transgene. This is consistent with the hypothesis that most of the foreign DNA transferred into the eggs and embryos is lost through degradation after transfer, and that the DNA is not integrated into the host genome. These experiments demonstrate one important point in transgenic research. The successful transfer of foreign DNA into embryos or eggs is not necessarily followed by integration.

Future directions

For transgenesis to take off as a method for broodstock development in aquaculture and marine resource management it is evident that more fusion genes designed for a specific purpose have to be developed. Presently, the focus of transgenic fish research is on growth enhancement. Research in the future should also be focused on the isolation and characterization of genes which confer disease resistance. This is particularly important for species which are under intensive cultivation such as some crustacean species (prawns) and molluscs (oysters). Our knowledge on marine mollusc and crustacean immunology is sparse. Research on disease resistance must be begun by identifying disease resistant animals, and the genes which confer disease resistance (Bachere et al., 1995; Hervio et al., 1995).

Parallel to the focus on identifying specific genes, tissue specific promoter sequences should be isolated and their expression evaluated. With the appropriate promoter more specific expression of the transgene in defined tissue will minimize the chances of getting deleterious side effect due to ectopic expression of the fusion gene.

The efficiency of trangene integration is generally low. Improved fusion gene construction which allows more efficient non-homologous recombination may accelerate the progress of transgenic fish research. The exploitation of retroviral sequences in fusion gene construction may enhance the integration efficiency (Ivics et al., 1993; Lin et al., 1994; Collas et al., 1996).

Although microinjection has made significant progress in transgenic fish research, more efficient mass gene transfer techniques will further accelerate progress and will also allow the application of transgenic research extended to a wider range of marine and freshwater organisms than those to which microinjection is easily applicable. Preliminary studies have demonstrated that co-transfer of growth factor with topoisomerase enhanced growth of zebrafish. More detailed analysis of topoisomerases and non-homologous recombination may be worthwhile. Mass gene transfer techniques will be particularly useful when the survival rate of the larval stage is normally low as seen in abalone.

Acknowledgments

Financial support through the Foundation for Research, Science and Technology to FYTS is gratefully acknowledged.

References

Arezzo, F., 1989. Sea urchin sperm as vector for foreign genetic transformation. Cell Biol. Int. Rep. 13: 391–404.

Alam, J. & J. L. Cook, 1990. Reporter genes: application to the study of mammalian gene transcription. Analyt. Biochem. 188: 245–254.

Alestrom, P., G. Kisen, H. Klungland & O. Andersen, 1992. Fish gonadotropin-releasing hormone gene and molecular approaches for control of sexual maturation: development of a transgenic fish model. Mol. mar. Biol. Biotech. 1: 376–379.

Bachere, E., E. Mialhe, D. Noel, V. Boulo, A. Morvan & J. Rodriguez, 1995. Knowledge and research prospects in marine mollusc and crustacean immunology. Aquaculture 132: 17–32.

Bachiller, D., K. Schellander, J. Peli & U. Ruther, 1991. Liposome-mediated DNA uptake by sperm cells. Mol. Reprod. Dev. 30: 194–200.

Baranov, V. S., I. Gapala, P. Griyach, L. Kovach & K. Bodya, 1990. Incorporating macromolecules into the sex cells of male mice using electric impulse and DMSO. Tsitologiya I Genetika 24: 3–7.

Behr, J. P., B. Demeneix, J. P. Loeffler & J. Perez-Mutul, 1989. Efficient gene transfer into mammalian primary endocrine cells with lipopolyamine-coated DNA. Proc. natl. Acad. Sci. USA 86: 6982–6986.

Birnstiel, M. L. & M. Busslinger, 1989. Dangerous liaisons: spermatozoa as natural vectors for foreign DNA. Cell 57: 701–702.

Brem, G., B. Brenig, G. Horstgen-Schwark & E.-L. Winnacker, 1988. Gene transfer in Tilapia (Oreochromis niloticus). Aquaculture 68: 209–219.

Brackett, B. G., W. Baranaska, W. Sawiicki & H. Koprosky, 1971. Uptake of heterologous genome by mammalian spermatozoa and its transfer to ova through fertilization. Proc. natl. Acad. Sci. USA 68: 353–357.

Buono, R. J. & P. J. Linser, 1992. Transient expression of RSVCAT in transgenic zebrafish made by electroporation. Mol. mar. Biol. Biotech. 1: 271–275.

Camaioni, A., M. A. Russo, T. Odorisio, F. Gandolfi, V. M. Fazio & G. Siracusa, 1992. Uptake of exogenous DNA by mammalian spermatozoa: specific localization of DNA on sperm heads. J. Reprod. Fertil. 96: 203–212.

Chalfie, M., Y. Tu, G. Euskirchen, W. W. Ward & D. C. Prasher, 1994. Green fluorescent protein as a marker for gene expression. Science 263: 802–805.

Chen, T. T. & D. A. Powers, 1990. Transgenic fish. Trends Biotech. 8: 209–215.

Chen, T. T., K. Kight, C. M. Lin, D. A. Powers, M. Hayat, N. Chatakondi, A. C. Ramboux, P. L. Duncan & R. A. Dunham, 1993. Expression and inheritance of RSVLTR-rtGH1 complementary DNA in the transgenic common carp, Cyprinus carpio. Mol. mar. Biol. Biotech. 2: 88–95.

Chourrout, D., R. Guyomard, & L-M. Houdebine, 1986. High efficiency gene transfer in rainbow trout (Salmo gairdneri Rich.) by microinjection into egg cytoplasm. Aquaculture 51: 143–150.

Collas, P., H. Husebye & P. Alestrom, 1996. The nuclear localization sequence of the SV40 T antigen promotes transgene uptake and expression in zebrafish embryo nuclei. Transgenic Res. In press.

Culp, P., C. N-V. & N. Hopkins., 1991. High-frequency germ-line transmission of plasmid DNA sequences injected into fertilized zebrafish eggs. Proc. natl. Acad. Sci. USA 88: 7953–7957.

Devlin, R. H., T. Y. Yesaki, E. M. Donaldson & C.-L. Hew, 1995. Transmission and phenotypic effects of an antifreeze/GH gene

276

construct in coho salmon (*Oncorhynchus kisutch*). Aquaculture 137: 161–169.

Devlin, R. H., T. Y. Yesaki, E. M. Donaldson, S. J. Du & C.-L. Hew, 1993. Production of germline transgenic Pacific salmonids with dramatically increased growth performance. Can. J. Fish. aquat. Sci. 52: 1376–1384.

Devlin, R. H., T. Y. Yesaki, C. A. Biagi, E. M. Donaldson, P. Swanson, W.-K. Chan, 1994. Extraordinary salmon growth. Nature 371: 209–210.

Du, S. J., Z. Gong, G. L. Fletcher, M. A. Shears, M. J. King, D. R. Idler & C. L. Hew, 1992. Growth enhancement in transgenic Atlantic salmon by the use of an 'all fish' chimeric growth hormone gene construct. Biotechniques 10: 176–180.

Dunham, R. A., J. Eash, J. Askins & T. M. Townes, 1987. Transfer of a metallothionein-human growth hormone fusion gene into channel catfish. Trans. am. Fish. Soc. 116: 87–91.

Felgner, P. L., T. R. Gadek, M. Holm, R. Roman, H. W. Chan, M. Wenz, J. P. Northrop, G. M. Ringold & M. Danielsen, 1987. Lipofection: a highly efficient, lipid-mediated DNA transfection procedure. Proc. natl. Acad. Sci. USA 84: 7413–7417.

Fletcher, G. L. & P. L. Davies, 1991. Transgenic fish for aquaculture. Genet. Engineering 13: 331–70.

Fletcher, G. L., M. A. Shears, M. J. King, D. L. Davies, & C. L. Hew, 1988. Evidence for antifreeze protein gene transfer in Atlantic salmon (*Salmo salar*). Can. J. Fish. aquat. Sci. 45: 352–357.

Francolini, M., M. Lavitrano, C. L. Lamia, D. French, L. Frati, F. Cotelli & C. Spadafora, 1993. Evidence for nuclear internalization of exogenous DNA into mammalian sperm cells. Mol. Reprod. Devel. 34: 133–139.

Gagne, M. B., F. Pothier & M. A. Sirard, 1991. Electroporation of bovine spermatozoa to carry foreign DNA in oocytes. Mol. Reprod. Devel. 29: 6–15.

Gavora, J. S., B. Benkel, H. Sasada, W. J. Cantwell, P. Fiser, R. M. Teacher, J. Nagal & M. P. Sabour, 1991. An attempt at sperm-mediated gene transfer in mice and chickens. Can. J. anim. Sci 71: 287–291.

Gendreau, S., V. Lardans, J. P. Cadoret & E. Mialhe, 1995. Transient expression of a luciferase reporter gene after ballistic introduction into *Artemia franciscana* (Crustacea) embryos. Aquaculture 133: 199–205.

Gershon, H., R. Ghirlando, S. B. Guttman & A. Minsky, 1993. Model of formatin and structural features of DNA-cationic liposome complexes used for transfection. Biochemistry 32: 7143–7151.

Gibbs, P. D. L., A. Peek & G. Thorgaard, 1994. An in vivo screen for the luciferase transgene in zebrafish. Mol. mar. Biol. Biotech. 3: 307–316.

Gong, Z., C. L. Hew & J. R. Vielkind, 1991. Functional analysis and temporal expression of promoter regions from fish antifreeze protein genes in transgenic Japanese medaka embryos. Mol. mar. Biol. Biotech. 1: 64–72.

Gorman, C. M., L. Moffat & B. Howard, 1982. Recombinant genomes which express chloramphenicol acetyl transferase in mammalian cells. Mol. Cell. Biol. 2: 1044–1051.

Grasser, C. S. & R. T. Fraley, 1989. Genetically engineering plants for crop improvement. Science 244: 1293–1299.

Gross, M. L., J. F. Schneider, N. Moav, B. Moav, C. Alvarez, S. H. Myster, Z. Liu, E. M. Hallerman, P. B. Hackett, K. S. Guise, A. J. Farae & A. R. Kapiscinski, 1992. Molecular analysis and growth evaluation of northern pike (*Esox lucius*) microinjected with growth hormone genes. Aquaculture 103: 253–273.

Haseldoff, J. & B. Amos, 1995. GFP in plants. Trends Genet. 11: 328–329.

Hayat, M., C. P. Joyce, T. M. Townes, T. T. Chen, D. A. Powers & R. A. Dunham. 1991. Survival and integration rate of channel catfish and common carp embryos microinjected with DNA at various developmental stages. Aquaculture 99: 249–255.

Hervio, D., E. Bachere, N. Boulo, V. Cochennec, V. Vuillemin, Y. Le Coguic, G. Cailletaux, J. Mazurie & E. Mialhe, 1995. Establishment of an experimental infection protocol for the flat oyster, *Ostrea edulis*, with the intrahaemocytic protozoan parasite, *Bonamia ostreae*: application in the selection of parasite-resistant oysters. Aquaculture 132: 183–194.

Hoci, S., T. Ninomiya, A. Mizuno, M. Honma & A. Yuchi, 1990. Fate of exogenous DNA carried into mouse eggs by spermatozoan. Anim. Biotech. 1: 25–30.

Hodgkinson, S., 1995. GFP in *Dictyostelium*. Trends Genet. 11: 327–328.

Horan, R., R. Powell, S. McQuaid, F. Gannon & J. A. Houghton, 1991. Association of foreign DNA with porcine spermatozoa. Arch. Andro. 26: 83–92.

Horan, R., R. Powell, J. M. Bird, F. Gannon & J. A. Houghton, 1992. Effects of electropermeabilization on the association of foreign DNA with pig sperm. Arch. Andro. 28: 105–114.

Inoue, K., S. Yamashita, J-i. Hata, S. Kabeno, S. Sada, E. Nagahisa & T. Fujita, 1990. Electroporation as a new technique for producing transgenic fish. Cell Diff. Devel. 29: 123–128.

Ivics, Z., Z. Izsvak & P. B. Hackett, 1993. Enhanced incorporation of transgenic DNA into zebrafish chromosomes by a retroviral integration protein. Mol. mar. Biol. Biotech. 2: 162–173.

Kavumpurath, S., O. Anderson, G. Kisen & P. Alestrom, 1993. Gene transfer methods and luciferase gene expression in zebrafish, *Brachydanio rerio* (Hamilton). Israeli J. aquat. Bamidgeh 45: 154–163.

Khoo, H.-W., L.-H. Ang, H.-B. Lim & K.-Y. Wong, 1992. Sperm cells as vector for introducing foreign DNA into zebrafish. Aquaculture 107: 1–9.

Knight, D. E., 1981. Rendering cells permeable by exposure to electric fields. Tech. Cell Physiol. 113: 1–20.

Knight, D. E. & M. C. Scrutton, 1986. Gaining access to the cytosol: the technique and some application of electropermeabilization. Biochem. J. 234: 497–506.

Lavitrano, M., A. Camaioni, V. M. Fazio, S. Dolci, M. G. Farace & S. Dolci, 1989. Sperm cells as vectors for introducing foreign DNA into eggs: genetic transformation of mice. Cell 57: 717–723.

Leemans, J., 1993. Ti to tomato, tomato to market: a decade of plant biotechnology. Biotechnology 11: 22–26.

Lin, S., N. Gaiano, P. Gulp, J. C. Burns, T. Friedmann, J-K. Yee & N. Hopkins, 1994. Integration and germline transmission of a pseudotyped retroviral vector in zebrafish. Science 265: 666–669.

Lu, J-K., T. T. Chen, C. L. Chrisman, O. M. Andrisani & J. E. Dixon, 1992. Integration, expression, and germ-line transmission of foreign growth hormone genes in medaka (*Oryzias latipes*). Mol. mar. Biol. Biotech. 1: 366–375.

Martinez, R., M. P. Estrada, J. Berlango, I. Guillen, O. Hernandez, E. Cabrera, R. Pimentel, R. Morales, F. Herrera, A. Morales, J. C. Pina, Z. Abad, V. Sanchez, P. Melamed, R. Lleonart & J. de la Fuente, 1996. Growth enhancement in trangenic tilapia by ectopic expression of tilapia growth hormone. Mol. mar. Biol. Biotech. 5: 62–70.

McEvoy, T., M. Stack, B. Keane, B. T. Barry, J. Sreenan & F. Gannon. 1988. The expression of a foreign gene in salmon embryos. Aquaculture 68: 27–37.

Moav, B., Z. Liu, L. D. Caldovic, M. L. Gross, A. J. Faras & P. B. Hackett, 1993. Regulation of expression of trangenes in developing fish. Transgenic Res. 2: 153–161.

Muller, F., Z. Ivics, F. Erdelyi, T. Papp, L. Varadi, L. Horvath, N. Maclean & L. Orban, 1992. Introducing foreign genes into fish eggs with electroporated sperm as carrier. Mol. mar. Biol. Biotech. 1: 276–281.

Muller, F., Z. Ivics, L. Varadi, L. Menczel & L. Orban, 1993. Efficient transient expression system based on square pulse electroporation and in vivo luciferase assay of fertilized eggs. FEBS Lett 324: 27–32.

Nakanishi, A. & A. Iritani, 1993. Gene transfer in the chicken by sperm-mediated methods. Mol. Reprod. Dev. 36: 258–261.

Neumann, J. R., C. A. Morency & K. O. Russian, 1987. A novel rapid assay for chloramphenicol acetyltransferase gene expression. Biotechniques 5: 444–447.

Nguyen, V. T., M. Morange & O. Bensaude, 1988. Firefly luciferase luminescence assays suing scintillation counters for quantitation in transfected mammalian ce;;s. Analyt. Biochem. 171: 404–408.

O'Hara, M. J., M. G. Ormerod, P. R. Imrie, J. H. Peacock & W. Asche, 1989. Electropermeabilization and electrosensitivity of different types of mammalian cells. In Neumann, E., A. E. Sowers & C. A. Jordan (eds), Electroporation and Electrofusion in Cell Biology. Plenum Press, New York: 319–341.

Ozato, K., H. Kondoh, H. Inohara, T. Iwamatsu, Y. Wakamatsu & T. S. Okada. 1986. Production of transgenic fish: introduction and expression of chicken δ-crystallin gene in medaka embryos. Cell Diff. 19: 237–244.

Palmiter, R. D., R. L. Brinster, R. E. Hammer, M. E. Trumbayer, M. G. Rosenfeld, N. C. Brinberg & R. M. Evans, 1982. Dramatic growth of mice that develop from eggs microinjected with metallothionein-growth hormone fusion genes. Nature 300: 611–615.

Patil, J. G. & H. K. Khoo, 1996. Nuclear internalization of foreign NDA by zebrafish spermatozoa and its enhancement by electroporation. J. exp. Zool. 274: 121–129

Penman, D. J., A. J. Beeching, S. Penn & N. Maclean, 1990. Factors affecting survival and integration following microinjection of novel DNA into rainbow trout eggs. Aquaculture 85: 35–50.

Perez, A. R., R. Solana, R. Castro, R. Lleonart, R. De Armas, R. Martinez, A. Aguilar, L. Herrera & J. D. L. Fuente, 1991. Sperm cells mediated gene transfer in cattle. Biotech. Apl. 8: 90–94.

Pines, J., 1995. GFP in mammalian cells. Trends Genet. 11: 326–327.

Powers, D. A., L. Hereford, T. Cole, C. Creech, T. T. Chen, C. M. Lin, K. Kight & R. Dunham, 1992. Electroporation: a method for transferring genes into the gametes of zebrafish (Brachydanio rerio), channel catfish (Ictaluruspunctatus), and common carp (Cyprinus carpio). Mol. mar. Biol. Biotech. 1: 301–308.

Powers, D. A., V. L. Kirby, T. Cole & L. Hereford, 1995. Electroporation as an effective means of introducing DNA into abalone (Haliotis rufescens) embryos. Mol. mar. Biol. Biotech. 4: 369–376.

Pursel, V. G. & C. E. Rexroad, Jr., 1993. Recent progress in the transgenic modification of swine and sheep. Mol. Reprod. Dev. 36: 251–254.

Pursel, V. G., C. A. Pinkert, K. F. Miller, D. J. Bolt, R. G. Campbell, R. D. Palmiter, R. L. Brinster & R. E. Hammer, 1989. Genetic engineering of livestock. Science 244: 1281–1288.

Rokkones, E., P. Alestrom, H. Skjervold & K. M. Gautvik, 1987. Microinjection and expression of a mouse metallothionein human growth hormone fusion gene in fertilized salmonid eggs. J. comp. Physiol. 158: 751–758.

Rottmann, O. J., R. Antes, P. Hofer & G. Maierhofer, 1992. Liposome mediated gene transfer via spermatozoa into avian egg cells. J. anim. Breed. Genet. 109: 64–70.

Sandford, J. C., 1988. The bioballistic process. Trends Biotech. 6: 299–302.

Serpersu, E. H., K. Kinosita, Jr & T. Y. Tsong, 1985. Reversible and irreversible modification of erythrocytes by electrical breakdown. Biochim. Biophys. Acta 816: 332–348.

Sheehy, R. E., M. Kramer & W. R. Hiatt, 1988. Reduction of polygalacturonase activity in tomato fruit by antisense RNA. Proc. natl. Acad. Sci. USA 85: 8805–8809.

Shears, M. A., G. L. Fletcher, C. L. Hew, S. Gauthier & P. L. Davis, 1991. Transfer, expression, and stable inheritance of antifreeze protein genes in atlantic salmon (Salmo salar). Mol. mar. Biol. Biotech. 1: 58–63.

Shigekawa, K. & W. J. Dower, 1988. Electroporation of eukaryotes and prokaryotes: a general approach to the introduction of macromolecules into cells. Biotechniques 6: 742–751.

Sin, F. Y. T., A. Bartley, S. Walker, S. Bulman, L. A. Allison, C. L. Hopkins & I. L. Sin, 1990. Electroporation of fish sperm for gene transfer. Cell Biol Int. Reports 14: 167.

Sin, F. Y. T., A. L. Bartley, S. P. Walker, I. L. Sin, J. E. Symonds, L. Hawke & C. L. Hopkins, 1993. Gene transfer in chinook salmon (Oncorhynchus tshawytscha) by electroporating sperm in the presence of pRSV-lacZ DNA. Aquaculture 117: 57–69.

Sin, F. Y. T., U. K. Mukherjee, J. C. McKenzie & I. L. Sin, 1995. Electroporation of abalone sperm enhances sperm-DNA association. J. Fish Biol. 47 (Suppl. A): 20–28.

Smith, J. G., R. L. Walzem & J. B. German, 1993. Liposomes as agents of DNA transfer. Biochim. Biophys. Acta 1154: 327–340.

Southern, E. M. 1975. Detection of specific sequences among DNA fragments separated by gel electrophoresis. J. mol. Biol. 98: 503–517.

Sowers, A. E. & M. R. Lieber, 1986. Electropore diameters, lifetimes, numbers, and locations in individual erythrocyte ghosts. FEBS Letters 205: 179–184.

Stuart, G. W., J. V. McMurray & M. Westerfield, 1988. Replication, integration and stable germ-line transmission of foreign sequences injected into early zebrafish embryos. Development 103: 403–412.

Sukharev, S. I., V. A. Klenchin, L. V. Servo, L. V. Chernomordik & Y. A. Chizmadzhev, 1992. Electroporation and electrophoretic DNA transfer into cells. The effect of DNA interaction with electropores. Biophys. J. 63: 1320–1327.

Symonds, J. E., S. P. Walker & F. Y. T. Sin, 1994a. Electroporation of salmon sperm with plasmid DNA: evidence of enhanced sperm/DNA association. Aquaculture 119: 313–327.

Symonds, J. E., S. P. Walker, F. Y. T. Sin & I. L. Sin, 1994b. Development of a mass gene transfer method in chinook salmon: optimization of gene transfer by electroporated sperm. Mol. mar. Biol. Biotech. 3: 104–111.

Szelei, J., L. Varadi, F. Muller, F. Erdelyi, L. Orban, L. Horvath & E. Dudo. 1994. Liposome-mediated gene transfer in fish embryos. Transgenic Res. 3: 116–119.

Takagi, S., T. Sasado, G. Tamiya, K. Ozato, Y. Wakamatsu, A. Takeshita & M. Kimura, 1994. An efficient expression vector for transgenic medaka construction. Mol. mar. Biol. Biotech. 3: 192–199.

Tamiya, E., T. Sugiyama, K. Masaki, A. Hirose, T. Okoshi & I. Karube, 1990. Spatial imaging of luciferase gene expression in transgenic fish. Nucleic Acids Res. 18: 1072.

Tewari, R., C. Michard-Vanhee, E. Parrot & D. Chourrout, 1992. Mendelian transmission, structure and expression of transgenes following their injection into the cytoplasm of trout eggs. Transgenic Res. 1: 250–260.

Tsai, H-J. & F. S. Tseng, 1994. Electroporation of a foreign gene into black Acanthopagrus schlegeli embryos. Fish. Sci. 60: 787–788.

278

Tsai, H. J., F. S. Tseng & I. C. Liao, 1995. Electroporation of sperm to introduce foreign DNA into the genome of loach (*Misgurnus anguillicaudatus*). Can. J. Fish. aquat. Sci. 52: 776–787.

Tsong, T. Y., 1983. Voltage modulation of membrane permeability and energy utilization in cells. Bioscience Rep. 3: 487–505.

Ueno, K., S. Hamaguchi, K. Ozato, J-h. Kang & K. Inoue, 1994. Foreign gene transfer into nigorobuna (*Carassius auratus grandoculis*). Mol. mar. Biol. Biotech. 3: 235–242.

Weaver, J. C., 1993. Electroporation: a general phenomenon for manipulating cells and tissues. J. Cell Biochem. 51: 426–435.

Winkler, C., J. R. Vielkind & M. Schartl, 1991. Transient expression of foreign DNA during embryonic and larval development of the medaka fish (*Oryzias latipes*). Mol. Gen. Genet. 226: 129–140.

Xie, Y., D. Liu, J. Zou, G. Li & Z. Zhu, 1989. Novel gene transfer in the fertilized eggs of loach via electroporation. Acta Hydrobiol. Sin. 13: 387–389.

Xie, Y., L. Dong, J. Zou, G. Li & Z. Zhu, 1993. Gene transfer via electroporation in fish. Aquaculture 111: 207–213.

Yamada, E., M. Matsuoka & F. Yamazaki, 1988. Introduction of exotic reagents into denuded eggs of goldfish and crucian carp by electroporation. Nippon Suisan Gakkaishi 54: 2043.

Yoon, S. J., E. M. Hallerman, M. L. Gross, Z. Liu, J. F. Schneider, A. J. Fareas, P. B. Hackett, A. R. Kapuscinski & K. S. Guise, 1990. Transfer of the gene for neomycin resistance into goldfish, *Carassius auratus*. Aquaculture 85: 21–33.

Yu, J. K., W. Yan, Y. L. Zhang, Y. Shen & S. Y. Yan, 1994. Sperm-mediated gene transfer and method of detection of integrated gene by PCR. Acta Zool. Sin. 40: 96–99.

Yuko, W., H. Hashimoto, M. Kinoshita, T. Iwamatsu, Y. Hyodo-Taguchi, H. Tomita, M. Sakaguchi & K. Ozato, 1993. Generation of germ-line chimerase in medaka (*Oryzias latipes*) Mol. mar. Biol. Biotech. 2: 325–332.

Zelenin, A. V., A. A. Alimov, V. A. Barmintez, A. O. Beniumov, I. A. Zelenina, A. M. Krasnov & V. A. Kolesnikov, 1991. The delivery of foreign genes into fertilised fish eggs using high-velocity microprojectiles. FEBS Lett. 287: 118–120.

Zhao, X., P. J. Zhang & T. K. Wong, 1993. Application of baekonization: a new approach to produce transgenic fish. Mol. mar. Biol. Biotech. 2: 63–69.

Zhang, P., M. Hayat, C. Joyce, L. I. Gonzalez-Villasenor, C. M. Lin, R. A. Dunham, T. T. Chen & D. A. Powers, 1990. Gene transfer, expression and inheritance of pRSV-rainbow trout-GH in the common carp, *Cyprinus carpio* (Linnaeus). Mol. Reprod. Dev. 25: 3–13.

Zhu, Z., G. Li, L. He & S. Chen, 1985. Novel gene transfer into the fertilization eggs of gold fish (*Carassius auratus* L. 1758). Z. angew. Ichthyol. 1: 31–34.

Hydrobiologia **352**: 279–285, 1997.
Y.-S. Wong & N. F.-Y. Tam (eds), Asia–Pacific Conference on Science and Management of Coastal Environment.
©1997 *Kluwer Academic Publishers. Printed in Belgium.*

Probiotic effect of lactic acid bacteria in the feed on growth and survival of fry of Atlantic cod (*Gadus morhua*)

Asbjørn Gildberg[1], Helene Mikkelsen[1], Elin Sandaker[1] & Einar Ringø[2]
[1]*Norwegian Institute of Fisheries and Aquaculture, 9005 Tromsø, Norway*
[2]*Department of Arctic Veterinary Medicine, The Norwegian College of Veterenary Medicine, 9005 Tromsø, Norway*

Key words: Probiotic feed, Atlantic cod, lactic acid bacteria, *Vibrio anguillarum*, intestinal colonisation

Abstract

A growing concern for the high consumption of antibiotics in aquaculture has initiated a search for alternative methods of disease control. Improved resistance against infectious diseases can be achieved by the use of probiotics. Probiotics are live microorganisms supplemented in food or feed which give beneficial effects on the intestinal microbial balance. In the present study a dry feed containing lactic acid bacteria (*Carnobacterium divergens*) isolated from Atlantic cod (*Gadus morhua*) intestines was given to cod fry. After three weeks of feeding the fry was exposed to a virulent strain of *Vibrio anguillarum*. The death rate was recorded during further three weeks of feeding with lactic acid bacteria supplemented feed. A certain improvement of disease resistance was obtained, and at the end of the experiment lactic acid bacteria dominated the intestinal flora in surviving fish given feed supplemented with lactic acid bacteria. No obvious growth inhibition of *V. anguillarum* was observed in an *in vitro* mixed culture of this bacterium and the *C. divergens* isolated from cod intestines.

Introduction

The substantial growth in marine aquaculture is increasingly making its impact on many coastal environments. Extensive use of antibiotics in aquaculture disease control is an ecological threat to coastal areas heavily exploited for industrial cultivation of fish and shell-fish. Hence, finding less harmful alternatives, may be of premium importance to avoid severe ecological damages. Recent developments of fish vaccines have reduced the use of antibiotics in certain regions (Newman & Deupree, 1995), and development of fish feed containing immunostimulants and probiotics will hopefully contribute to further reduction in the future.

Commercial use of fish feed supplemented with immuno stimulatory β-1,3-glucans has already been implemented (Raa, 1996). *In vitro* experiments have shown that enzymatically hydrolyzed cod muscle protein contains peptides with a potential immunostimulatory activity (Bøgwald et al., 1996). Addition of

hydrolyzed protein in feed may also improve the growth of juvenile fish (Fagbenro & Jauncey, 1994).

Probiotics are usually defined as live microbial feed supplements which beneficially affects the host animal by improving its intestinal microbial balance (Fuller, 1989). The use of probiotics has long traditions in animal husbandry (Stavric & Kornegay, 1995), but is rarely applied in aquaculture. The main strategy in the use of probiotics is to isolate intestinal bacteria with favourable properties from mature animals and include large numbers of these bacteria in the feed for immature animals of the same species. Most frequently probiotics are associated with lactic acid bacteria (Nousiainen & Setälä, 1993). This is due to the fact that these bacteria often produce bacteriocins and other chemical compounds that may inhibit the growth of pathogenic bacteria. Until recently, it has not been generally accepted that lactic acid bacteria are a part of the native microbial flora of fish intestines (Pilet et al., 1995). Thus, supplementation of such bacteria in fish feed was regarded as futile. During the last decade intestinal colonization by lactic acid bacteria has been verified in several species of wild marine fishes as well as in cultivated Atlantic salmon (Strøm, 1988; Strøm & Ringø, 1993; Pilet et al., 1995). Lactic acid bacteria isolated from fish can also inhibit the growth of fish pathogens such as *Vibrio anguillarum* and *Aeromonas salmonicida* (Strøm, 1988; Gildberg et al., 1995). However, challenge experiments with fish given probiotics are rare. Gatesoupe (1994) showed that turbot *(Scophthalmus maximus)* larvae fed with rotifers enriched with lactic acid bacteria had improved resistance against pathogenic *Vibrio*-infection, but the mortality of non-infected fish increased slightly if the level of lactic acid bacteria in the feed was too high. In a challenge experiment with salmon fry, it was shown that lactic acid bacteria supplemented in dry feed could colonize the intestines, but no protection against *A. salmonicida* infection could be detected (Gildberg et al., 1995).

In the present study Atlantic cod *(Gadus morhua)* fry given feed supplemented with lactic acid bacteria were challenged by bath exposure to *V. anguillarum*, and the cumulative mortality was compared with that of fish given two different control feeds. One control feed was supplemented with intact cod muscle protein, whereas the other control feed and the probiotic feed were supplemented with enzymatically hydrolyzed cod muscle protein. This was done to determine whether muscle protein hydrolysate containing immuno-stimulatory peptides could further enhance protection against infection.

Materials and methods

Experimental design

Approximately 105 fry of Atlantic cod *(Gadus morhua)*, with an average weight of 3.0 g (calculated from the total weigth of 31 fish), were stocked in 6 square tanks, each containing 250 l aerated sea water at 10 °C. The water exchange rate was about 4 l min^{-1}. The fish were reared *ad libitum* on three different dry pellet diets throughout the whole experiment. After 18 days of feeding, randomly selected samples of 20 fish from each tank were weighed, and specific growth rates (SGR)[1] were calculated. After further three days of feeding the fish were exposed to a strain of

Vibrio anguillarum (LFI 1243) isolated from infected Atlantic cod. The *V. anguillarum* was grown in Tryptic soy broth (Oxoid Ltd, London, England) supplemented with 0.25% glucose and 1% NaCl. The bacteria were cultured aerobically at 12 °C for 24 h. The cells were harvested in late exponential phase, washed and resuspended in 0.9% sterile NaCl solution. About 7×10^{11} colony forming units (CFU) were added to each tank after the water volume of the tanks had been reduced to 100 l, yielding an initial concentration of viable *V. anguillarum* of about 7×10^6 per ml.

During the infection period the water exchange was stopped, and oxygen was supplied to all tanks. After 1 hour the water volumes were increased to 250 l, and normal water exchange (4 l min^{-1}) was re-established. The challenge experiment lasted for three weeks, and dead fish were removed once a day. Samples from the kidney of some of the dead fish were seeded on blood-agar plates for re-isolation of bacteria. Classical vibriosis was diagnosed on the basis of the re-isolation of typical haemolytic colonies in pure culture. The diagnosis was confirmed by testing colonies with a rapid agglutination test using Mono-Va from BioNor Aqua, Skien, Norway. At the end of the experiment, intestines from 5 surviving fish from each tank were subjected to microbial analyses.

Statistical analysis

To evaluate the significance of the differences in mortality between the various groups and between parallel samples, an approximation to the standard normal distribution was used (Bhattacharyya & Johnson, 1977). This test has been found to be convenient for statistical analyses of challenge experiments with a large number of fish in each group (Bøgwald et al., 1992; Gildberg et al., 1995).

Preparation of feed

Commercial dry feed (FK Marinpellet, Felleskjøpet, Norway) served as a basis of all the diets, accounting for approximately 83% by weight of the diets. Products derived from the white muscle of Atlantic cod contributed by about 11% of the dry weight, and crude alginate (Protan, Protatek BFG-alginate, Protan, Norway), added as a binder, accounted for the rest of the diet (about 6%). Feed A was made by mixing 32% minced cod fillet, 50% powdered commercial feed, 15% water and 3% crude alginate in a homogenizer. The homogenate was squeezed through a meat

---

[1] SGR = (ln w_2 − ln w_1) ∗ 100%/t; here w_1 and w_2 are average fish weight at the beginning and the end of a period, and t is the length of the period in days

grinder fitted with a disc containing 2 mm holes. It was then frozen, lyophilized, crushed manually and pressed twice through a 2 mm metal grid using a mortar plunger. After removal of fines on a 1 mm shaking sieve, a dry pellet feed with an approximate diameter of 1.5 mm and 1–4 mm particle length was obtained. Feeds B and C were prepared in a similar manner, but the cod muscle homogenate was replaced with hydrolyzed cod muscle protein and a lactic acid bacteria culture grown on hydrolyzed cod muscle protein respectively. The alginate concentration in these mixtures was increased to 3.5% to compensate for the loss of binding capacity due to the hydrolysis of the fish muscle protein.

Preparation of cod muscle hydrolysate

Minced cod fillets were mixed with an equal weight of warm tap water (60 °C) and 0.5% (v/v) of the protease concentrate Neutrase (0.5L., Novo Industr., Denmark). The mixture was stirred and heated to about 40 °C. After 3 hours at 40 °C, the enzymes were inactivated by heating to 80–90 °C for 10 min, and the reaction mixture was cooled to about 4 °C over night. A clear supernatant was harvested and subjected to ultrafiltration according to a method described earlier (Gildberg et al., 1995). The major fraction of the hydrolysate, which was recorded as permeate from a 2000 d cut-off membrane filtration, was used in the preparation of feed B and as the basis of the growth medium for the lactic acid bacterium culture used in feed C. The chemical composition of this fraction was: Water, 95.1%; Kjeldahl protein, 4.5% and ash, 0.4%.

Chemical analyses

Dry matter was determined after drying for two days at 105 °C and ash after heating of dry samples to 580 °C for 20 hours. Crude protein was determined by the Kjeldahl procedure. Fat was determined gravimetrically by Soxhlet extraction with petroleum benzene.

Preparation of lactic acid bacteria culture

A lactic acid bacteria isolated from Atlantic cod intestines (Strøm, 1988) was classified according to species level using the criteria given by Hamnes et al. (1992), and the bacterial strain was found to be a *Carnobacterium divergens*. This bacterium was grown in shake bottles with liquid medium composed of the low molecular weight fraction of the cod muscle

hydrolysate supplemented with 0.4% glucose and standard solutions of minerals, vitamins and nucleotides (1% of each) (Ford et al., 1958). After 3–4 days at 12 °C an average viable count of about 10^9 cells per ml was achieved. The culture was then frozen at −80 °C, and part of it was lyophilized and mixed in the diet to provide 11% of feed C on dry matter basis. The viable count of lactic acid bacteria in newly prepared feed C was about 2×10^9 CFU per g.

Microbial analyses

Analysis of the intestinal microbiota was done on samples consisting of excised washed intestines pooled from five fish. The intestines, including *pyloric caeca*, were gently excised and cut open with a pair of sterile scissors. The gut content was removed by gentle scraping, and total viable counts and number of lactic acid bacteria were determined. The intestines were washed three times with saline solution to remove non-adherent bacteria. Due to its complex morphology the *pyloric caeca* could not be washed as thoroughly as the rest of the intestines. The samples were then homogenized with 10 ml sterile saline solution in Stomacher bags (Stomacher, Lab-Blender 400). Dilution series were prepared from the homogenates. Bacterial counts and the presumptive number of lactic acid bacteria were determined by the spread plate method using Tryptone soya agar (Oxoid) with 1% NaCl and 0.5% glucose as general medium. Samples from representative colonies were re-isolated and grown on new plates for classification. After one week of incubation the colonies were inspected. Water samples from the tanks were analyzed according to the same procedure as the intestine homogenates. Dilution series were prepared after taking out samples (approx. 100 ml) in sterile flasks. All incubations were performed in air at 12 °C. The classification of bacteria was done according to the Muroga scheme (Muroga et al., 1987), which is based on Gram tests, catalase tests, visual judgements of colonies on agar media and on light microscopy studies.

Bacterial growth inhibition study

Cells from freshly cultivated lactic acid bacteria and *V. anguillarum* were centrifuged (3000 × **g**, 15 min), and the pellets were resuspended in sterile 0.9% NaCl. Samples of both suspensions (30 μl) were inoculated in 30 ml fish peptone broth yielding 4×10^6 CFU per ml of *V. anguillarum* and 0.5×10^6 of *C. divergens*. The fish peptone broth had the following composition in g per litre of distilled water: NaCl, 10; glucose, 5; fish peptone (H 0100, Marine Biochemicals, Norway), 2 and standard solutions of vitamin, mineral and nucleotides, 10 of each (Ford et al., 1958). Control samples of each bacteria in monocultures were grown parallelly. Initial pH of all the cultures were 7.3, and they were grown in shake-bottles for one week at 12 °C. During the growth period samples were taken out for pH measurements and bacterial count determinations. Viable counts for each bacterium were determined by the spread plate method using Tryptone soya agar (Oxoid) with 1% NaCl and 0.5% glucose. Colonies of the two bacteria could easily be discriminated due to morphological differences. Round pale-yellowish colonies of *V. anguillarum* appeared after one day of incubation at 12 °C, whereas *C. divergens* grew slower and appeared as small white-greyish colonies after two days of incubation.

Results and discussion

The classification showed that the lactic acid bacteria used in this work was a *Carnobacterium divergens* similar to a *Carnobacterium divergens* isolated from Atlantic salmon (Ringø, unpublished results) and originally classified as *Lactobacillus plantarum* (Strøm, 1988; Gildberg et al., 1995). Both *C. divergens* and *C. piscicola* have been isolated from various fish raw materials, and it has been shown that *C. divergens* may produce bacteriocins (Pilet et al., 1995; Holck et al., 1996).

The *C. divergens* grew well on the cod muscle hydrolysate medium, and total viable counts of the bacteria in newly prepared feed C was about 2×10^9 per g. At the end of the experiment, after storage for about one month at 4 °C and thereafter for about two weeks at room temperature, the number of viable *C. divergens* in the feed was reduced to about 10^7 per g. According to Nousiainen & Setälä (1993) 10^7 lactic acid bacteria per g feed is considered to be sufficient to give good probiotic effects. The number of bacteria in feed A and B was very low ($<10^4$ per g), and no lactic acid bacteria were detected in these feeds. These results show that a dry feed with a high number of *C. divergens* isolated from cod intestines could be made, however, the survival of bacteria during storage was not as good as what was obtained previously with a similar bacterium species isolated from the intestines of Atlantic salmon (Gildberg et al., 1995). Oxygen restriction by vacuum-

Table 1. Chemical composition of the three diets in % (w/w)

	Protein	Fat	Ash	Water
Feed A	55.0	14.4	9.3	2.0
Feed B	54.3	14.2	9.7	2.6
Feed C	53.7	14.6	9.8	3.3

packaging and freezstorage would probably improve the storage stability substantially.

The three diets have the same general chemical composition (Table 1), however, feed A contains less hydrolyzed protein (peptides and free amino acids) than feed B and C. Feed A was included as a control to feed B to reveal whether peptides from the fish muscle hydrolysate had some immuno stimulatory effect leading to a reduced mortality in fish given this feed. Injection of peptides from such a fish protein hydrolysate in Atlantic salmon has shown that stimulation of leucocytes may occur (Bøgwald et al., 1996). However, since the specific growth rate before infection was substantially lower in fish given feed A than in fish given feed B and C (Table 2), the value of this feed as a control is dubious. It has been shown that supplementation of free amino acids in diets may improve the specific growth rate of juvenile fish (Espe & Lied, 1994), and supplementation of the fish protein hydrolysate in this experiment seem to have a similar effect. After infection the specific growth rate (SGR) is inversly related to the cumulative death rates in parallel tanks (Tables 2 and 3). This is as expected since a higher infection rate probably created more stress and reduced feed intake.

The cumulative mortalities (Table 3 and Figure 1) are different for fish given different diets. Fish given feed A have significantly lower mortality than fish given feed C ($p < 0.0003$) and fish given feed C significantly lower than fish given feed B ($p < 0.0002$). Hence, the cumulative mortality was lower with fish given feed supplemented with intact muscle protein than with fish given feed supplemented with protein hydrolysate and lactic acid bacteria. The reason for this is uncertain, but the significantly lower initial growth rate of the fish given feed A may have rendered this fish less susceptible to infection. Similar observations have been done with slow growing mature cod (Leifson, 1996). An alternative explanation is that the content of peptide immunostimulants in feed B and C is too high. This may have an adverse effect on leucocytes and macrophages, reducing their activity instead of improving it (Siwicki et al., 1990; Gildberg et al., 1996).

Table 2. Growth of fry of Atlantic cod (*Gadus morhua*), before, during and after infection with *Vibrio anguillarum* was performed at day 21

Feed	Tank	Average initial weight* (g)	Average weight at day 18** (g)	Average weight at day 43*** (g)	SGR day 0–18 (% day)	SGR day 18–43 (% day)
A	1	3.0	4.7	9.2	2.5	2.7
	2	"	3.7	10.6	1.2	4.2
B	3	"	5.6	11.2	3.5	2.8
	4	"	5.0	13.5	2.8	4.0
C	5	"	5.2	9.2	3.1	2.3
	6	"	5.3	10.1	3.2	2.6

* Average initial weight is calculated from the total weight of 31 fish at day 0
** Average weight calculated from the total weight of 20 fish from each tank
*** Average weight calculated from total weight of surviving fish in each tank

Table 3. Cumulative mortality of Atlantic cod (*Gadus morhua*) in each tank as a function of days after exposure to infection by *Vibrio anguillarum*

	Feed A		Feed B		Feed C	
Days	Tank 1	Tank 2	Tank 3	Tank 4	Tank 5	Tank 6
4	0	0	0	0	0	0
5	0	0	2	5	2	5
6	3	2	9	19	2	10
7	8	2	13	27	9	14
8	16	9	30	30	17	26
9	18	11	45	31	28	30
10	19	13	46	32	30	33
11	19	14	49	33	32	35
12	21	15	50	34	33	36
13	21	15	50	38	34	36
14	21	16	50	38	35	36
15	21	16	51	42	36	37
16	24	16	52	44	37	37
17	24	16	54	47	37	39
18	26	16	55	49	38	40
19	26	16	55	50	38	41
20	29	16	61	50	38	41
21	29	18	61	51	39	41
22	29	18	64	53	39	41

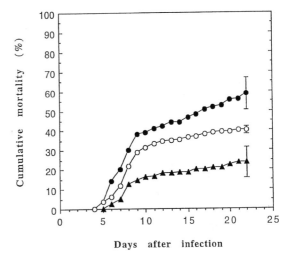

Figure 1. Average cumulative mortality as a function of time after infection for fish given feed A (▲), feed B (●) and feed C (○). Twenty two days after infection there are significant differences ($p<0.0003$) in cumulative mortality for fish given different feeds. Standard deviation bars are given for mean values of mortality in parallel tanks at the end of the experiment.

Feed B is the most relevant control to feed C in the determination of the probiotic effect of the lactic acid bacteria, since these feeds are almost identical with the exception of the content of lactic acid bacteria in feed C. Also, the average size of fish given feed B and C is very similar at the time of infection (Table 2). Twenty two days after infection the cumulative mortality of fish given feed supplemented with lactic acid bacteria has levelled off at 40%, whereas the cumulative mortality of fish given the corresponding feed without lactic acid bacteria is approaching 60% and still increasing. This clearly demonstrates a significant, although moderate, probiotic effect of the addition of *Carnobacterium divergens* isolated from the intestines of mature cod in the feed to cod fry.

There are some differences in mortality obtained in the parallel tanks where feed A and B were given. The significance of these differences between parallels ($p<0.05$ and $p<0.07$), however, are much lower than

284

the significance of the differences between the groups given different feeds. Certain differences of mortality will always occur when parallel samples are run in challenge experiments with fish. One reason for this is that small biological differences in the fish samples may result in considerable differences in secondary infection preassure.

It was confirmed by the tests described in the materials and methods section that the mortalities suffered from classical vibriosis. Microbial analyses of surviving fish at the end of the experiment showed that the total viable count of gut content was about 10^4 per g in fish given feed A and about 10^5 per g in fish given feed B and C. About 75% of the bacteria in the gut content of fish given feed C were lactic acid bacteria, whereas about 25% appeared to be *Pseodomonas*-like bacteria. No lactic acid bacteria were detected in the gut content of fish given feed A or B. The number of viable bacteria in washed intestines was very low (100–1000 bacteria per intestine). While intestines from fish given feed A and B had a mixed microbiota, the intestines from fish given feed C apparently had a monoculture of lactic acid bacteria. Since the gut content also contained *Pseodomonas*-like bacteria, this indicates a true colonisation of *C. divergens*, and that this bacteria has displaced other potential colonizers. Whether the colonization is attributed to the whole intestinal tract or solely restricted to the less thoroughly washed *pyloric caeca* remains to be clarified. The low number of colonizing bacteria indicates that the intestines of cod fry at this stage is still purely adapted for bacterial colonisation.

The *in vitro* growth experiment with a mixed culture of *V. anguillarum* and *C. divergens* did not reveal an obvious inhibition of the pathogen by the probiont (Figure 2). However, after 7 days of growth the number of viable *Carnobacterium* was slightly higher than the number of viable *Vibrio*, although this bacteria yielded almost 90% of the total viable count at the start of the experiment. In a similar experiment it was shown that lactic acid bacteria isolated from Arctic capelin (*Mallotus villosus*) strongly inhibited the growth of *V. anguillarum* (Strøm, 1988).

Studies of the use of probiotics in aquaculture are rare and often inconclusive. The present results indicate that probiotics may be useful as an ecologically favourable alternative in aquaculture disease control, but there is a great need for further research and development within this field before probiotic fish feed with defined protective effects can be promoted.

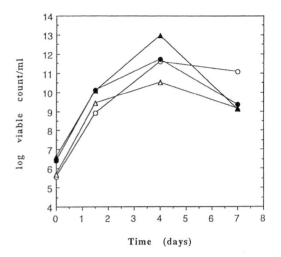

Figure 2. Growth of *Vibrio anguillarum* (▲ ●) and *Carnobacterium divergens* (Δ ○) in a mixed shake bottle culture as compared with the growth of the two bacteria in separate cultures on the same medium. Triangles are plotted for mixed- and circles for separate cultures.

Acknowledgments

Financial support from The Norwegian Research Council is acknowledged.

References

Bhattacharyya, G. K. & R. A. Johnson, 1977. Comparison of two binomial proportions. In Bradly, R. A., J. S. Hunter, D. G. Kendall & G. S. Watson (eds), Statistical Consepts and Methods. Wiley & Sons, New York: 308–333.

Bøgwald, J., R. A. Dalmo, R. M. Leifson, E. Stenberg & A. Gildberg, 1996. The stimulatory effect of a muscle protein hydrolysate from Atlantic cod, *Gadus morhua* L., head kidney leucocytes. Fish Shellfish Immunol. 6: 3–16.

Bøgwald, J., K. Stensvåg, J. Hoffmann, K. O. Holm & T. Jørgensen, 1992. Vaccination of Atlantic salmon, *Salmo salar* L., with particulate lipo-poly-saccharide antigens from *Vibrio salmonicida* and *Vibrio anguillarum*. Fish Shellfish Immunol. 2: 251–261.

Espe, M. & E. Lied, 1994. Do Atlantic salmon (*Salmo salar*) utilize mixtures of free amino acids to the same extent as intact protein sources for muscle protein synthesis? Comp. Biochem. Physiol. 107A: 249–254.

Fagbenro, O. & K. Jauncey, 1994. Growth and protein utilization by juvenile catfish (*Clarias gariepinus*) fed moist diets containing autolysed protein from stored lactic acid-fermented fish-silage. Bioresour. Technol. 48: 43–48.

Ford, J. E., K. D. Perry & C. A. E. Briggs, 1958. Nutrition of lactic acid bacteria isolated from rumen. J. Gen. Microbiol. 18: 273–284.

Fuller, R., 1989. Probiotics in man and animal. J. appl. Bacteriol. 66: 365–378.

Gatesoupe, F.-J., 1994. Lactic acid bacteria increase the resistance of turbot larvae, *Scophthalamus maximus*, against pathogenic vibrio. Aquat. Living Resour. 7: 277–282.

Gildberg, A., J. Bøgwald, A. Johansen & E. Stenberg, 1996. Isolation of acid peptide fractions from a fish protein hydrolysate with strong stimulatory effect on Atlantic salmon *(Salmo salar)* head kidney leucocytes. Comp. Biochem. Physiol. 114B: 97–101.

Gildberg, A., A. Johansen & J. Bøgwald, 1995. Growth and survival of Atlantic salmon *(Salmo salar)* fry given diets supplemented with fish protein hydrolysate and lactic acid bacteria during a challenge trial with *Aeromonas salmonicida*. Aquacult. 138: 23-34.

Hamnes, W. P., N. Weiss & W. Holzapfel, 1992. The genera *Lactobacillus* and *Carnobacterium*. In Balows, A., H. G. Truper, M. Dworkin, W. Harder & K.-H. Schleifer (eds), The prokaryotes. A Handbook on the Biology of Bacteria: Ecophysiology, Isolation, Identification, Applications. Vol. II. Springer-Verlag, New York: 1536–1594.

Holck, A., L. Axelsson & U. Schillinger, 1996. Divergicin 750, a novel bacteriocin produced by *Carnobacterium divergens* 750. FEMS Microbiol. Letters 136: 163–168.

Leifson, R. M., 1996. Oralvaksinering av Atlantisk laks *(Salmo salar* L.) ved bruk av liposom/alginat innpakket vaksine. M. Sci. Thesis. The Norwegian College of Fishery Science. 103 pp (in Norwegian).

Muroga, K., M. Higashi & H. Keitoku, 1987. The isolation of intestinal micro-flora of farmed red seabream *(Acanthopagrus schlegeli)* at larval and juvenile stages. Aquaculture 65: 79–88.

Newman, S. G. & R. Deupree, 1995. Biotechnology in aquaculture in aquaculture. Infofish Internat. 1/95: 40–46.

Nousiainen, J. & J. Setälä, 1993. Lactic acid bacteria as animal probiotics. In Salminen, S. & A. von Wright (eds), Lactic Acid Bacteria. Marcel Dekker, New York: 315–356.

Pilet, M.-F., X. Dousset, R. Barré, G. Novel, M. Desmazeaud & J.-C. Piard, 1995. Evidence for two bacteriocins produced by *Carnobacterium piscicola* and *Carnobacterium diveregens* isolated from fish and active against *Listeria monocytogenes*. J. Food Protect. 58: 256–262.

Raa, J., 1996. The use of immunostimulatory substances in fish and shellfish farming. Rev. Fish. Sci. 4: 229–288.

Siwicki, A. K., D. P. Anderson & O. W. Dixon, 1990. *In vitro* immunostimulation of rainbow trout *(Oncorhynchus mykiss)* spleen cells with levamisole. Developmental Comp. Immunol. 14: 231–237.

Stavric, S. & T. Kornegay, 1995. Microbial probiotics for pigs and poultry. In Wallace, R. J. & A. Chesson (eds), Biotechnology in Animal Feeds and Animal Feeding, Weinheim, New York: 205–231.

Strøm, E., 1988. Melkesyrebakterier i fisketarm. Isolasjon, karakterisering og egenskaper. M. Sci. Thesis. The Norwegian College of Fishery Science, 88 pp (in Norwegian).

Strøm, E. & E. Ringø, 1993. Changes in the bacterial composition of early developing cod, *Gadus morhua* (L.) larvae following inoculation of *Lactobacillus plantarum* into the water. In Walther, B. & H. J. Fyhn (eds), Physiological and Biochemical Aspects of Fish Larval Development. University of Bergen, Bergen, Norway: 226–228.

Hydrobiologia **352**: 287–293, 1997.
Y.-S. Wong & N. F.-Y. Tam (eds), Asia–Pacific Conference on Science and Management of Coastal Environment.
©1997 *Kluwer Academic Publishers. Printed in Belgium.*

Translocation of assimilates in *Undaria* and its cultivation in China

Wu Chaoyuan[1] & Meng Jianxin[2]
[1]*Institute of Oceanology, Chinese Academy of Sciences, Qingdao, China 266071*
[2]*Department of Molecular Biology, University of Wyoming, Laramie, WY 82071, USA*

Key words: Undaria pinnatifida, translocation, photoassimilates, growth, cultivation, yield

Abstract

Undaria cultivation on a commercial scale began in China only in the last decade. Today, *Undaria pinnatifida* is the main species under cultivation concentrated in two provinces, Liaoning and Shandong. The annual production in the early nineties was 8000-13 000 tons dry weight, which is two or three times the pre-1980 figures. The raft cultivation method maintaining the alga at the desired depths generally ensures the light saturated rate of photosynthesis on clear days, and enhances production. Under the cultivated condition, the calculated annual primary productivity of this alga is 160 g C m^{-2} y^{-1}.

Translocation of ^{14}C-labelled photoassimilates in rapidly growing sporophyte of *Undaria pinnatifida* was studied in the open sea. Samples from different parts of the blade with counterparts exposed to tracer (NaH^{14}CO$_3$) showed that the translocation that occurred mainly from the tip of the blade to the growing region had obvious source-sink relationship. It took 20 minutes to translocate the labelled photoassimilates from the epidermis, via cortex, to the medulla of the midrib, where rates of translocation averaging 42–48 cm h^{-1} were observed in the open sea.

Production experiments of tip-cutting of the blades showed an increased production of 9%.

Introduction

It was estimated that the total production of cultured seaweeds in the Asia-Pacific region represented more than 90% of the world's production (Csavas, 1985). In this region, five countries, China, Japan, Republic of Korea, the Philippines and Indonesia are principally engaged in the commercial cultivation of six economic species which qualify as marine crops. *Undaria pinnatifida* is one of the six. China has natural resources of *Undaria pinnatifida*, a northern type distributing in the Bohai Sea and the Yellow Sea, and a southern type distributing along the coast of Chengsi Island of Zhejiang Province. The two types are quite different in shape and living habit.

In China today, *Undaria pinnatifida* is the main species under cultivation concentrated in two provinces, Liaoning and Shandong. As the largest base of *Undaria* cultivation in China, the output of Dalian in Liaoning Province accounts for upwards of 90% of the nation's total yield. Progress has also been made in

Rongcheng City of Shandong Province in recent years. Today, *Undaria* production ranks third, next to *Laminaria* and *Porphyra* only, in the seaweed cultivation industry in China. The annual production in the early nineties was 8000–13,000 tons dry weight, two or three times the pre-1980 figures. The per hactare yield is generally around 70 tons fresh weight in Dalian City.

Seaweeds are known to be potentially very productive. *Undaria* is believed to be one of the potential food products of the subtidal zone. Under the cultivation conditions in North China, the calculated annual primary productivity of this alga is 160 g C m^{-2}. The effect of temperature on growth, translocation phenomena and the blade tip-cutting method as a means for increasing production were studied in the search to find ways to enhance production.

Materials and methods

Culture conditions

All experiments were conducted in the open sea in the Pier area, Qingdao City, where the nitrate nitrogen content of higher than 1.71×10^{-4} mol m^{-3} (Wu,1962), favours rapid growth of *U. pinnatifida*. Experimental plants were cultured in the 1 meter below sea surface layer under a floating raft.

Growth and temperature effect

Experiments were carried out in the spring of 1981. Plants of 15 cm in length were twisted around a palm rope and cultured under a floating raft in the sea every 20 days. In order to know the actual growth rate, holes were punched at a place about 3/4 of the total length from the base parts of the blade. Length measurements were made every 10 days and growth rate calculated.

Translocation study

1. Plant materials and culture conditions
Healthy sporophytes of *U. pinnatifida* 40 to 60 cm in length were collected in spring of 1982. The experiment was done on a clear day when the light intensity was 40 000 lx, well exceeding the light saturation point (ca 20 000 lx, Wu et al., unpublished results).

2. Direction and rate of translocation
The blade tip or middle part or basal part of the blades was immersed in NaH^{14}CO$_3$ seawater (specific activity 2.479×10^4 Bq ml^{-1}) separately. The labelling time was 5 min. The treated plants together with controls were immediately put back to the open sea, and were then sampled after 20 min, 30 min, 40 min, 1 h, 2 h, and 23 h respectively using previously described area sampling method (Nicholson & Briggs, 1972; Schmitz & Srivastava, 1975). Starting from the tip 1 cm $\times$ 1 cm squares were cut at 1 cm intervals along the midrib of the blade. Samples were digested using the perchloric acid-hydrogen peroxide method (Lobban, 1974). Radioactivies in digested samples were counted by using FJ-353 double channel liquid scintillation counter (302 factory, Xian, China). Quenching was compensated by channel ratio methods. Counting time for each sample was based on its radioactivity in order to let the standard error be less than 2% of total counts (Liquid Scintllation Group, Institute of Bio-

physics, Academia Sinica, 1979). Samples from control plants were used for background counts, which did not show any contamination during the experiments.

Tip-cutting experiment

On the basis of the downward translocation of photoassimilates from the tip to the base part of the blade, and the phenomenon of natural casting off of the old tissue in the distal part of the blade, the authors proposed the 'tip cutting method' to increase production. Plants of 20 cm in length were twisted in palm ropes in late November of 1984. As soon as the plants grew to about 80 cm in length in the next spring, the blades were cut at places about 3/4 and 1/2 of the total length from the base part of the blades. After harvest on March 15, 1985, the fresh weight of the plants were measured.

Results and discussion

Growth and temperature effect

Length and weight increments are given in Figure 1 and Figure 2. Figure 1 shows that the bigger the plant was, the earlier the decrease in growth occurred. Sporophytes of the first 2 sets stopped growing basically when water temperature exceeded 10 in mid-April. When the plant length exceeded 130 cm, the plants of 100 cm long still grew until water temperature exceeded 18. This means that smaller sporophytes can tolerate higher temperature. Rapid growth occurred at 5–15 regarded as favourable temperature range for growth in length. Figure 2 shows that the smaller the alga was, the later the peak of wet weight occurred. The peak of wet weight of the transplants of the first set occurred on mid-April, whereas those of the fourth set occurred on late-April. Weight increment of the sporophytes decreased rapidly after late-April when water temperature was over 12. Sporophytes transplanted after March 2 were small, but were still growing upon harvest. The results of all sets of plants showed that the peak of wet weight appeared 10 days later than the peak of the length. The wet weight of transplants of the first set reached 700 g per plant by the end of April.

Translocation study

1. The results of translocation direction experiments are described below.

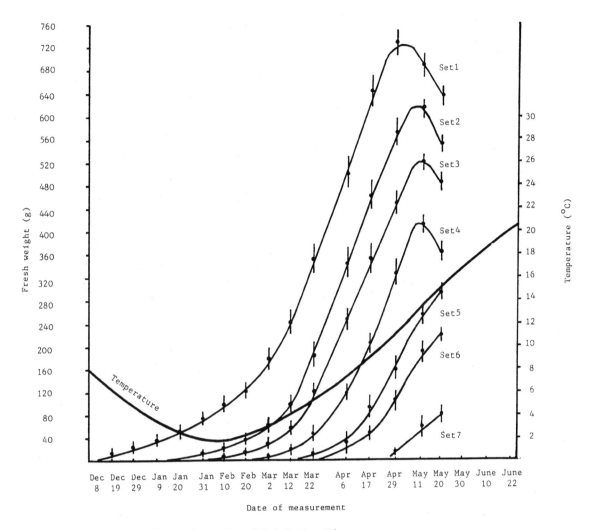

Figure 1. Effect of temperature on the growth sporophyte of *Undaria pinnatifida.*
Standard deviations of the mean (N = 30 each) measurements are represented by vertical bars.

A. Labelled tips. Plants were divided into three groups and cultured in the open sea for 1, 2, and 23 h, respectively. The results are shown in Figure 3. It is clear that after 1 h culture, photoassimilates just reached the blade base. The two hour culture showed some accumulation of photoassimilates in the growing region. After 23 h, however, a large quantity of photoassimilates was translocated and accumulated there. The distribution of photoassimilates in different parts of midrib had a distinct pattern. In the upper part, the closer it was to the labelling position, the more ^{14}C-radioactivity could be detected, indicating of the presence of more photoassimilates. The level of radioactivity gradually decreased down the midrib, until it reached the upper middle part of the blade when it was

851 cpm. From this position down, the closer it was to the growing region, the more accumulation of photoassimilates could be seen. This trend was obvious.

B. Labelled in the middle part of blade. The blade in this position had already stopped growth. Figure 4 illustrates the results after 220 min open sea culture. It clearly showed that the majority of translocated photoassimilates were transported towards the basal part of plant and accumulated in the growing region. Less radioactivity was detected above the labelled position, and the front only moved about 6 cm. This kind of photoassimilates movement might be caused by redistribution. There was some lateral movement to neighboring branched parts, but the quantity was less compared to the radioactivity detected in the growing region.

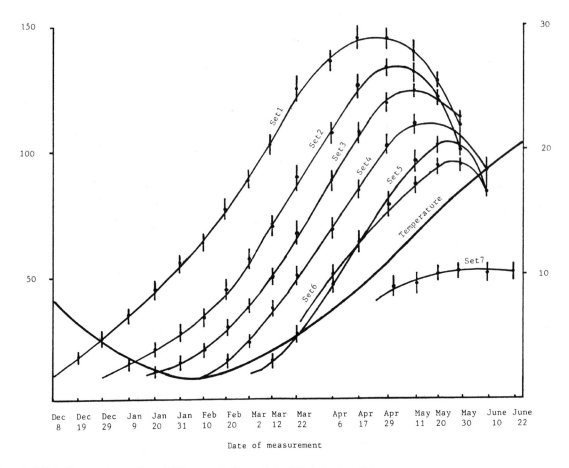

Figure 2. Effect of temperature on the weight increment of sporophyte of *Undaria pinnatifida*. Standard deviations of the mean (N = 30 each) measurements are represented by vertical bars.

C. Labelled in blade base. Photoassimilates only moved upward 12 cm in one hour. The rate and quantity were low (shown in Figure 5).

2. It was found from translocation rate experiments that after the tips of the blades were labelled, it took 20 min for radioactivity to be detected below the labelled position. This is an indication that the process of ^{14}C fixation in the epidermis and cortex, and the subsequent transport of photoassimilates to the medulla took about 20 min. After culturing for 30 min, ^{14}C could be found in the blade 8±2.3 cm down the tips, and 40 min culture showed ^{14}C at the 14±2.7 cm position. It was calculated that, under natural growth conditions, translocation of photoassimilates in the midrib had an average rate of ca 42–48 cm h^{-1}. This value was the average of three separate experiments, and 3 plants were used for each experiment.

In all these ^{14}C-labelling experiments, there was no detectable contamination during both labelling and culture periods in the control plants. When calculating translocation rate, there are a few factors that should be taken into consideration in order to ensure accurate rates. First, fixation of carbon into photoassimilates by photosynthesis and subsequent transport of ^{14}C-labelled metabolites from the epidermis and cortex to the medulla requires time. Steinbiss & Schmitz (1973) found that the whole process took about 30 min in *Laminaria hyperborea*. In *U. pinnatifida*, it was ca 20 min. Apparently, when calculating translocation rates, this time must be deducted to get the true rate. Second, culture conditions could affect translocation. In our translocation experiment, we tried indoor culture after labelling with ^{14}C and could only get an average translocation of 25 cm h^{-1}, in spite of optimal growth conditions (16 400 lx and 15). This is quite similar to previous results with *Laminaria japonica* (Wu et al.,

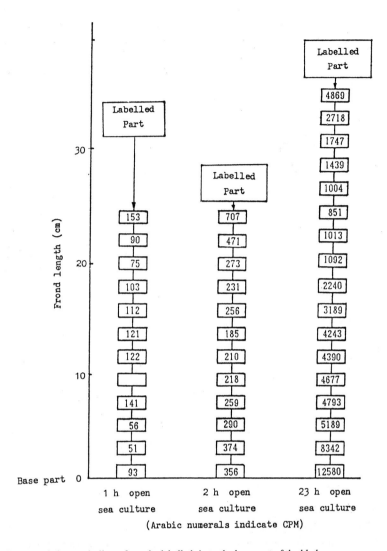

Figure 3. Translocation diagram of photoassimilates from the labelled tip to the base part of the blade.
Standard errors of the measured (N = 10 each) values for 1 h experiment were all less than 6.5%, and that for 2 h and 23 h experiments (N = 10 each) were all less than 7.2%.

1961), where translocation of ^{32}P-labelled metabolites was studied. It is possible that the indoor culture conditions, especially seawater flow, are inferior to those of outdoor conditions. This could be the reason why some phycologists (such as Parker, 1956) could not detect true translocation rates under indoor conditions.

This translocation study on *U. pinnatifida* proved once again the occurrence of the translocation phenomenon in many species in the order Laminariales (Wu, Loc. cit., 1961; Luning et al., 1972; Nicholson & Briggs, 1972). The growing region of *U. pinnatifida* is in blade base. Photoassimilates are translocated from the blade tip to the base part and accumulated there.

This source-sink relation is quite similar to the results of other studies on *Laminaria*. Since most nutrients for growth are assimilated in, and exported from the upper part of the blade in *Laminaria*, it is tempting to assume that the same pattern occurs in *U. pinnatifida*.

Since species in the family of Alariaceae possess a midrib, it is natural to speculate that the midrib is the pathway for translocation. Buggeln (1976) and Schmitz & Srivastava (1975) cut the *Alaria* blade deep to the midrib to prove that translocation occurs through the midrib. *U. pinnatifida* in Qingdao is the subspecies of forma distans Miyabe et Okam. Its blade cuts are close, sometime reaching the midrib. For photoassimi-

292

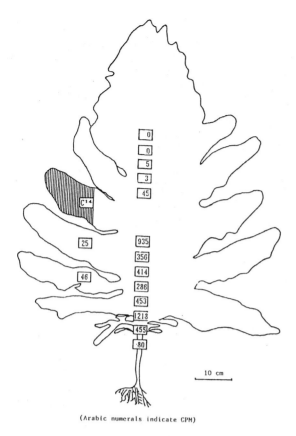

(Arabic numerals indicate CPM)

Figure 4. Translocation of photoassimilates from the labelled branch to other parts of the blade.

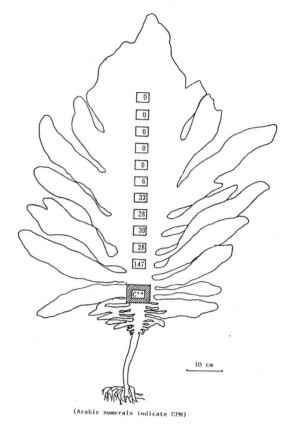

(Arabic numerals indicate CPM)

Figure 5. Translocation of photoassimilates from the labelled base part to other part of the blade.

lates exported from the tip to reach the basal part, where the growing region is, at least most of them have to be translocated through the midrib. Furthermore, Floc'h & Penot (1972) used whole plant autoradiography to show that the midrib is the major site for translocation of labelled products. Similarly, we also observed that when the midrib had high radioactivity, there was much lower level of radioactivity in the neighboring blade part. This is indirect evidence that the midrib is the major site for translocation.

Tip cutting as a means for increasing production

Tip-cutting experiments showed that growth in length was not obviously diminished if 1/4 of the blade was cut from the tip in early February. This is because the photoassimilates transported downward to the intercalary growing region sustain growth. In mid February to early March in North China the *Undaria* under cultivation reach one meter in length and the high blade density led to serious rot of the tip, so about 20 cm of

the old tissue weighing more than 50 g fresh weight per plant was eventually cast off and lost to the sea. Tip-cutting may serve to enhance production. The cut portion is not good for food, but can be used as raw material for the algin industry. It was found that the resistance of the blades to the tidal current in the farm was decreased after cutting, resulting in a change of the floating angle of the blades from 7 to 29 with a current speed at 7 cm s^{-1}. This resulted in better lighting condition for the blades and naturally in a higher photosynthetic rate and higher production. Tip cutting increased production of the main blades including the cuts by about 9%.

Acknowledgment

We wish to sincerely thank Pong Zuaosheng, Zhang Jingpu and Liu Haihang for their help during the course of the experiment. Graphs were made by Fong Minhua.

References

Csavas, I., 1985. Problems of inland fisheries and aquaculture. A. P. O. Symposium on fishing industry. SYP/VII/85. Tokyo: 17–18.

Buggeln, R. G., 1976. The rate of translocation in *Alaria esculenta* (Laminariales, Phaeophyceae). J. Phycol. 12: 439–42.

Floc'h, J. Y. & M. Penot, 1972. Transport du ^{32}P et du ^{86}Rb chez quelques algues brunes: Orientation des migrations et voies de conduction. Physiol. Veg. 10: 677–78.

Liquids Scintillation Group, Institute of Biological Physics, Academia sinica, 1979. Liquid scintillation counting and its applications in biology. Science Publications, Beijing, China, 380–407 pp.

Lobban, C. S., 1974. A simple rapid method of solubilizing algal tissue for scintillation counting. Limnol. Oceanogr. 19: 356–9.

Luning, K., K. Schmitz & J. Willenbrink, 1972. Translocation of ^{14}C-labelled assimilates in two *Laminaria* species. Proc. int. Seaweed Symp. 7: 420–5.

Nicholson, N. L. & W. R. Briggs, 1972. Translocation of photosynthate in the brown alga *Nereocystis*. Am. J. Bot. 59: 97–106.

Parker, J., 1956. Translocation of P^{32} and dye behavior in two species of marine algae. Naturwiss 43: 452.

Schmitz, K. & L. M. Srivastava, 1975. On the fine structure of sieve tubes and the physiology of assimilate transport in *Alaria marginata*. Can. J. Bot. 53: 861–76.

Steinbiss, H. H. & K. Schmitz, 1973. CO$_2$-fixierung und stofftransport in benthischen marinen algen. V. Zur autoradiographischen lokalisation der assimilat-transportbahnen im thallus von *Laminaria hyperborea*. Planta 112: 253–63.

Wu, C. Y., 1962. The relationship between sporophyte growth/development and environment conditions. In Tseng, C. K. & C. Y. Wu (ed.), *Laminaria* Culture. Science Publications, Beijing, China: 34–71.

Wu, C. Y., S. Q. Zheng & C. K. Tseng, 1961. P^{32} translocation in *Laminaria*. Science Communications 44–60 pp.

Hydrobiologia **352**: 295–303, 1997.
Y.-S. Wong & N. F.-Y. Tam (eds), Asia–Pacific Conference on Science and Management of Coastal Environment.
©1997 *Kluwer Academic Publishers. Printed in Belgium.*

The fate of traditional extensive (*gei wai*) shrimp farming at the Mai Po Marshes Nature Reserve, Hong Kong

M. W. Cha[1], L. Young[2] & K. M. Wong[2]
[1]*Swire Institute of Marine Science, University of Hong Kong, Shek O, Hong Kong*
[2]*World Wide Fund for Nature Hong Kong, 1 Tramway Path, Central, Hong Kong*

Key words: Extensive aquaculture, sustainable production, *Metapenaeus ensis*

Abstract

Extensive shrimp farming around Deep Bay, Hong Kong, began in the mid-1940's after the construction of intertidal ponds (*gei wai*) among the coastal mangroves. The ponds are increasingly being seen as an example of how wetlands can be used sustainably since they are naturally stocked with shrimp postlarvae (e.g. *Metapenaeus ensis*) and young fish (e.g. *Mugil cephalus*) flushed into the ponds from Deep Bay. Once inside, these shrimps and fish feed on naturally occurring detritus on the pond floor.

The only *gei wai* remaining in the Territory, are those at the WWF Hong Kong Mai Po Marshes Nature Reserve, adjacent to Deep Bay. Analysis of the shrimp production between 1990–1995 showed that there were two seasonal peaks, from April–June (Recruitment-I) and from July–October (Recruitment-II). The second peak was significantly lower than the first ($p < 0.001$), especially from those *gei wai* in the southern part of the reserve which are much closer to a polluted river. The average harvest from each *gei wai* had also significantly declined from 40.9 ± 6.0 kg ha^{-1} yr^{-1} in 1990 to 15.1 ± 3.6 kg ha^{-1} yr^{-1} in 1995 ($p < 0.01$). This decline can be attributed to the abundance of predatory fish in the *gei wai*, and increasing water pollution in Deep Bay which adversely affects the amount of shrimp larvae for stocking the *gei wai*, as well as the quality of water for flushing the ponds during the rearing and harvesting seasons. Despite this, those *gei wai* which are not-commercially viable can still support many non-commercial, more pollution tolerant fish and shrimp species. As a result, the management of these *gei wai* has been altered such that their objective is to provide feeding habitat for piscivorous waterbirds, which is also in line with the aims of the nature reserve.

Introduction

Onshore traditional shrimp culture using intertidal ponds dug out from coastal mangroves, has been in existence for a long time in many parts of Asia, such as China (Wong, 1986), India (Vanucci, 1987), Indonesia (Naamin, 1987) and Thailand (Menasveta, 1988). This type of traditional aquaculture was usually small-scale, used low inputs and relied on natural tidal action for water-exchange and stocking. The major species were generally *Penaeus merguiensis*, *Penaeus monodon* and *Metapenaeus* spp. This 'natural' system was known as 'extensive aquaculture' and although it produced a relatively smaller harvests, it was sustainable in the long term.

The demand for shrimps in recent years, particularly from Japan, America and Europe, has stimulated an increase in shrimp production. This has lead to the intensification of the industry, with the ponds being stocked with young shrimps from hatcheries, fed using specially formulated feeds, and medication introduced into the ponds to control disease outbreaks (Barraclough & Finger-Stitch, 1996).

Although the intensive systems may lead to greater shrimp production, they may also lead to environmental degradation, such as the destruction of coastal mangroves for intensive shrimp farms (Primavera, 1995). The intensive farms has also caused other environmental problems, such as saltwater intrusion into the watertable because of pumping of the groundwater to

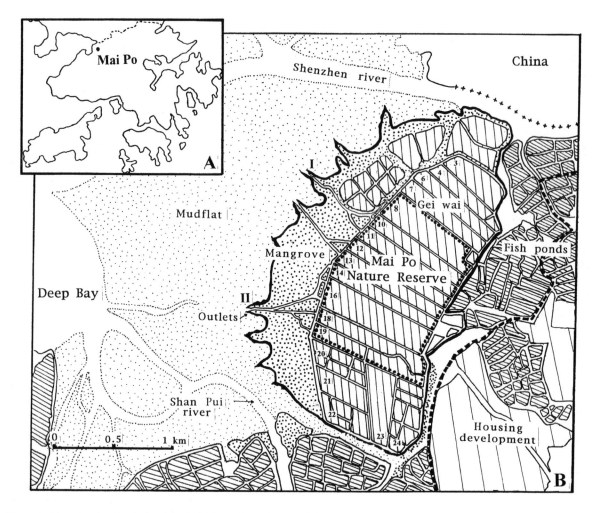

Figure 1. Maps showing (A) the location of Mai Po in Hong Kong and (B), the study site at the Mai Po Marshes Nature Reserve. (Note: I = Channel-I; II = Channel-II and no. 1 to 24 = *gei wai* number)

supply the ponds, introduced pollutants from shrimp farms into neighbouring areas and changing the patterns of water flow in adjacent estuaries (Naamin, 1987). Disease outbreak and/or self-pollution also occurs in intensive ponds in many Asian countries (Liao, 1992; Shariff & Subasinghe, 1992).

This paper describes an example of a traditional, extensive system of (*gei wai*) shrimp farming at the WWF Hong Kong Mai Po Marshes Nature Reserve, and how production from some of the ponds have declined from 1990 to 1995, due principally to pollution. The paper also discusses how the change of management strategy can turn the non-commercial ponds into valuable feeding sites for migratory waterbirds which winter at the site annually, as well as areas for education and recreation.

Study site

The 380 ha Mai Po Marshes Nature Reserve is situated in the north-western corner of Hong Kong (Figure 1). The Reserve, together with the extensive tidal mudflats of Deep Bay and adjacent fishponds, makes up the largest remaining wetland in Hong Kong. A 1500 ha area of this wetland was designated as a Wetland of International Importance under the Ramsar Convention in 1995 because of the large number of waterbirds present during migration (over 68 000 in January 1996). Some of these are endangered or of regional significance (Young & Melville, 1993).

The 24 shrimp ponds (*gei wai*) in the reserve were created in the mid-1940's by impounding areas of intertidal mangroves on the edge of Deep Bay. The 270 ha of *gei wai* (average size ~ 10 ha) at Mai Po are probably one of the only areas of such ponds still being operated traditionally in China (Young & Melville, 1993). Unlike intensive shrimp ponds which have deep water and little vegetation (Pillay, 1993), the Mai Po *gei wai* support extensive stands of mangroves (mainly *Kandelia candel*) and reed (*Phragmites communis*) in the centre of the ponds. These mangroves contribute 6% (17.5 ha) to the total area of mangroves in Deep Bay, and is the sixth largest stand in China (Fan, 1993). The reedbeds (46 ha) are the largest in Hong Kong, and one of the largest in Guangdong Province, China (Gao, pers. comm.). These ponds also support a diverse invertebrate, fish and shrimp community (Lee, 1989), and the vegetation provides feeding and breeding habitats for birds and mammals (Young, 1994).

Each *gei wai* has a single sluice gate which allows water exchange with Deep Bay via a water channel through the coastal mangroves (Figure 1). Apart from a channel (1–1.5 m deep by 10 m wide) around the inner edge of each *gei wai*, there are also channels running across the pond to facilitate water exchange, and to allow a greater area for shrimp production.

Shrimp larvae are flushed into each *gei wai* from August–December on nights when there is a high tide in Deep Bay. The young shrimps feed on naturally occurring detritus on the *gei wai* floor. In late April of the following year, the shrimps are harvested by opening the sluice gate on nights when there is a low tide in Deep Bay, and a funnel net is placed across the sluice gate to catch the outgoing shrimps. In the morning, water from Deep Bay is allowed back into the pond to maintain the water level, and to prevent heat stress which may cause the shrimps to die. Due to the high sediment load of the Pearl River (8.6×10^7 ton yr^{-1} Shen, 1983), the water flushed into the *gei wai* from Deep Bay carries a high silt load. The sedimentation rate in the *gei wai* was estimated to be 1.7 cm yr^{-1}, equivalent to 8.4 ± 8.8 kg dry wt m^{-2} (Lee, 1988). To maintain the channel at a suitable depth for shrimp production, dredging has to be conducted every ten years (Young & Melville, 1993).

Materials and methods

Change in shrimp production from 1990 to 1995

Shrimp production from seven *gei wai* connected to Channel-I (*gei wai* no. 8, 10, and 11) and II (no. 14, 16, 18, 19) were analysed (Figure 1B). The average size of

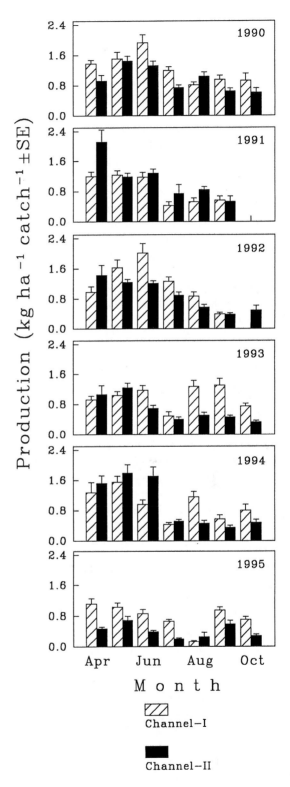

Figure 2. The seasonal change in the average shrimp production (kg ha^{-1} $catch^{-1}$) from Channel I and II between 1990 and 1995.

these *gei wai* was 12 ha and the actual area for shrimps culture generally make up ∼69% of the pond area. The major commercial shrimp species is *Metapenaeus ensis*, their postlarvae being present in the mangroves outside the *gei wai* from June to the end of March in the following year, with peak numbers in August and November (Leung, 1991). They reach marketable size after being trapped in the *gei wai* for 4–5 months. As many species of shrimps remain inactive while the temperature is below 14 °C (Imai, 1982) and the size of most individuals is too small to be marketable during winter and early spring, shrimp harvest at Mai Po takes place mainly between late April and late October.

Shrimps harvesting is done up to 12 times each month during evenings when there is a low tide in Deep Bay. Harvesting takes about 3 hours each night with the catch declining as the evening progresses. Shrimp productions from *gei wai* No. 8, 10, 11, 14, 16, 18 and 19 between 1990 and 1995 were analysed and compared. All data were standardised to the unit of kg ha^{-1} $catch^{-1}$ before analysis. Shrimp production from *gei wai* which connected to different channels and from different recruitment seasons were compared.

Correlation between fish and shrimp production during 1993 to 1995

Fish is another product from the *gei wai*, and the fry are also naturally stocked from Deep Bay through flushing in autumn and winter. Once inside the pond, they feed on natural occurring food in the pond. Lee (1988) has recorded 38 fish species in the *gei wai*, with Grey Mullet (*Mugil cephalus*), Tilapias (*Oreochromis nilotica* and *O. massambicus*) and Lady Fish (*Elops saurus*) being the major commercial and abundant species. These fish are harvested in winter after shrimp production has finished. This is done by completely draining the *gei wai*. As the floor of the pond slopes towards the sluice gate is deeper, all the fish are trapped in the pool of water formed there. The commercial fish are then harvested from this pool and their biomass estimated.

For analysis, the fish were divided into predatory species (mainly Lady Fish) and herbivorous species (mainly Mullet and Tilapias). Correlation between the annual productions of fishes and shrimp from the studied *gei wai* during 1993 to 1995 were calculated.

Results

Change in shrimp production from 1990 to 1995

The production of *Metapenaeus ensis* peaked during April to June (Recruitment-I) and July to October (Recruitment-II; Figure 2). Production was lowest in July and where the two cohorts overlapped. From 1990 to 1995, The shrimp harvest from *gei wai* connected to Channel-I was significantly higher than that to Channel-II (Figure 2; Table 1: $t = 4.05$, $df_{1,1194}$, $p < 0.05$). The lowest harvest rate recorded from Channel-I and II were both in 1995, and represented 86% and 40%, respectively, of that harvested in 1990 (Table 1). The average harvest from Recruitment-I was significantly higher than Recruitment-II, with 1.24 ± 0.03 kg ha^{-1} $catch^{-1}$ in the former and 0.7 ± 0.02 kg ha^{-1} $catch^{-1}$ in the latter (Table 1: $t = 16.66$, $df_{1,1194}$, $p < 0.001$). Productions from Recruitment-I fluctuated, while that from Recruitment-II declined significantly (Table 1; $r = 0.9$, $F = 16.72$, $df_{1,4}$, $p < 0.05$).

From 1990 to 1995, significant declines in shrimp production were recorded from all *gei wai*, except *gei wai* 8 (Figure 3; Table 2: $F = 1.80$, $df_{4,142}$, $p > 0.05$). Table 3 summarised the trends in annual shrimp production from different *gei wai*. Although the annual productions fluctuated within *gei wai*, average production significantly declined from 40.9 ± 6.0 kg ha^{-1} yr^{-1} in 1990 to 15.1 ± 3.6 kg ha^{-1} yr^{-1} in 1995 (Figure 4; $r = 0.96$, $F = 44.08$, $df_{1,4}$, $p < 0.01$). This made up a 63% reduction on the total shrimp production over the last six years.

Correlation between shrimp and fish productions between 1993 and 1995

There were no significant difference in fish production between 1993 and 1995, with the average production rate being 47.38 ± 7.12 kg ha^{-1} yr^{-1} from each *gei wai* ($F = 2.82$, $df_{2,16}$, n.s.). If fish were divided into predatory and herbivorous fish, herbivorous fish were more abundant and contributed an average of 86.3% to the total fish production. Its average production rate was 40.87 ± 7.18 kg ha^{-1} yr^{-1}, with a significantly higher production in 1994 ($F = 5.00$, $df_{2,16}$, $p < 0.05$). The production of predatory fish was significantly lower in 1994, with an average of 1.42 ± 0.63 kg ha^{-1} yr^{-1} compared with 9.68 ± 1.93 kg ha^{-1} yr^{-1} in 1993 and 1995 (Table 4; $F = 10.56$, $df_{2,16}$, $p < 0.01$). Fish production was different between the *gei wai* connected to

Table 1. Results of *t*-tests on the differences between the shrimp productions (kg ha^{-1} catch^{-1} ± SE) (replicates) from different channels and recruitments between 1990 and 1995. The significant level of $p < 0.05$, 0.01, 0.001 and > 0.05 are represented by *, **, *** and n.s., respectively.

Year	Channel			Recruitment		
	I	II	*t*	I	II	*t*
1990	1.19±0.07 (105)	1.02±0.05 (121)	1.99 *	1.41±0.09 (93)	0.88±0.05 (133)	7.15 ***
1991	0.90±0.06 (90)	1.09±0.07 (138)	1.78 n.s.	1.31±0.06 (123)	0.68±0.06 (105)	9.14 ***
1992	1.31±0.10 (62)	0.95±0.05 (135)	3.40 ***	1.40±0.06 (100)	0.72±0.04 (97)	9.81 ***
1993	1.01±0.06 (97)	0.75±0.05 (121)	3.90 ***	1.04±0.06 (111)	0.69±0.05 (107)	5.05 ***
1994	1.03±0.07 (65)	1.21±0.10 (96)	0.66 n.s.	1.56±0.10 (88)	0.63±0.05 (73)	9.40 ***
1995	0.86±0.05 (82)	0.41±0.03 (84)	8.70 ***	0.74±0.05 (88)	0.51±0.04 (78)	3.79 ***
mean	1.05±0.03 (501)	0.93±0.03 (695)	4.05 ***	1.24±0.03 (603)	0.70±0.02 (593)	16.66 ***

Table 2. The trends of shrimp production (kg ha^{-1} catch^{-1}) from different recruitment cycles and *gei wai* at the Mai Po Nature Reserve during 1990 to 1995. The significant level of $p < 0.05$, 0.01, 0.001 and > 0.05 are represented by *, **, *** and n.s., respectively.

	Gei wai	year (x)	production (y)	n	Slope	Intercept	r	F	p
A. Channel-I									
R-I	8	x	log(y)	65	−0.02	−0.04	0.18	2.03	n.s.
	10	log(x)	log(y)	83	−0.26	0.25	0.34	10.41	**
	11	x	log(y)	87	−0.04	0.26	0.37	13.31	***
R-II	8	x	y	82	0.04	0.50	0.17	2.44	n.s.
	10	x	y	84	0.02	0.86	0.07	0.40	n.s.
	11	log(x)	log(y)	100	−0.25	−0.01	0.28	8.43	**
B. Channel-II									
R-I	14	x	log(y)	118	−0.04	0.29	0.44	27.32	***
	16	x	log(y)	92	−0.06	−0.01	0.32	10.23	**
	18	x	log(y)	67	−0.03	0.14	0.14	1.28	n.s.
	19	x	log(y)	91	−0.05	0.24	0.31	9.67	**
R-II	14	log(x)	log(y)	59	−0.59	0.03	0.63	36.80	***
	16	x	log(y)	125	−0.08	−0.17	0.39	22.66	***
	18	x	y	56	−0.09	0.88	0.36	8.14	**
	19	x	log(y)	87	−0.06	−0.09	0.35	12.20	***

Note: R-I = Recruitment-I and R-II = Recruitment-II.

Channel-I and II, with the production of predatory fish ($t = 2.18$, $df_{1,17}$, $p < 0.05$) and total fish ($t = 2.28$, $df_{1,17}$, $p < 0.05$) from Channel-II being significantly higher.

According to all data collected from 1993 to 1995, shrimp production from Recruitment-I and Recruitment-II formed a positive ($r = 0.49$, $n = 19$, $p < 0.05$) and negative ($r = -0.44$, $n = 19$, $p < 0.05$) correlation respectively with fish production. In Channel-II, shrimp production was negatively correlated with the production of predatory fish ($r = -0.56$, $n = 12$, $p < 0.05$), but positively correlated to the production of herbivorous fish ($r = 0.62$, $n = 12$, $p < 0.05$).

Discussion

Shrimp productions from the Mai Po *gei wai* had significantly declined from 40.9 kg ha^{-1} yr^{-1} in 1990 to 15.1 kg ha^{-1} yr^{-1} in 1995. According to a linear regression estimation, these *gei wai* may no longer be productive by 1999 (Figure 4). The phenomenon highlights the threat to this unique system of extensive shrimp farming system in Hong Kong, which is increasingly being seen as an example of the sustainable use of a wetland (Young, in press), and as a way of restoring abandoned intensive shrimp farms (Le Dien Duc, 1996).

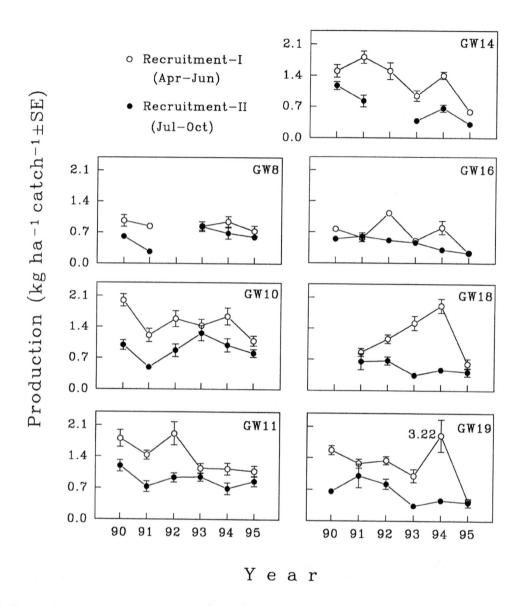

Figure 3. The annual change in shrimp production (kg ha^{-1} catch^{-1}) from different recruitment (Recruitment I and II) within different *gei wai.*

Table 3. The annual shrimp productions (kg ha^{-1} yr^{-1}) from the actively managed *gei wais* at the Mai Po Marshes Nature Reserve during 1990 to 1995.

Year	Gei wai number							Total	Mean±SE
	8	10	11	14	16	18	19		
1990	19.9	44.8	57.9	53.1	26.9	–	42.9	245.5	40.9±6.0
1991	16.6	18.7	46.0	77.0	17.7	15.0	41.9	232.9	33.3±8.7
1992	–	36.7	45.2	19.5	45.6	25.0	37.6	269.6	34.9±4.3
1993	33.4	38.7	26.2	18.1	22.7	35.2	15.1	189.4	27.6±3.4
1994	15.0	32.6	18.6	36.8	13.1	26.1	40.0	182.2	26.0±4.1
1995	19.1	30.2	23.7	9.6	10.1	10.1	9.0	105.8	15.1±3.6

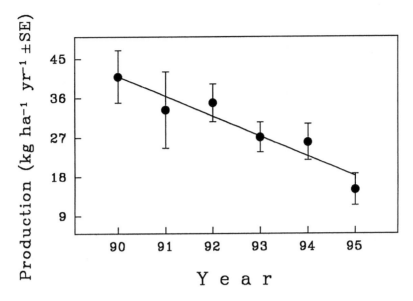

Figure 4. The overall average change in shrimp production (kg ha^{-1} yr^{-1}) from the mai po *gei wai* between 1990 and 1995.

Table 4. The average annual fish productions (kg ha^{-1} yr^{-1} ± SE) from different *gei wai* at the Mai Po Nature Reserve during 1993 and 1995.

		Fish production			
		n	Predatory spp.	Herbivorous spp.	Total
Year	1993	7	9.96±3.10	21.62±6.29	31.59±8.74
	1994	7	1.42±0.63	63.95±13.90	65.37±31.53
	1995	5	8.80±2.64	35.51±9.36	44.30±7.69
Channel	I	7	3.11±0.96	27.10±8.25	30.20±7.38
	II	12	8.80±2.27	48.90±9.94	57.40±9.36
Mean		19	6.51±1.58	40.87±7.18	47.38±7.08

The Mai Po Marshes Nature Reserve is adjacent to Inner Deep Bay, which is one of the most polluted water bodies in Hong Kong (EPD, 1995). With the rapid urbanisation around the north-west New Territories of Hong Kong and the Shenzhen Special Economic Zone, China, large amount of largely untreated sewage mainly from domestic and livestock wastes is entering Deep Bay daily. The annual biochemical oxygen demand (BOD$_5$) loading into the bay is estimated at around 35,000 tons yr^{-1} and *Escherichia coli* counts of $10^4 100$ ml^{-1} have been detected repeatedly in these areas (EPD, 1995). The Mai Po Nature Reserve is bordered by the Shenzhen river at the north and Shan Pui river at the south. These two rivers are seriously polluted and form the major input of waste water into Deep Bay (EPD, 1995). A pollution gradient was detected across the nature reserve, with the northern and southern ends being significantly more polluted than the central region (Chiu, 1992). The annual average BOD$_5$ and dissolved oxygen levels during daytime from one of the southern *gei wai* has been recorded at 22.5 mg O$_2$ l^{-1} and 1.8 mg l^{-1}, respectively (Chiu, 1992). Diurnal DO fluctuation, with the level dropped to around zero at night was recorded in the commercially managed *gei wai*. This phenomenon reflects that the site is overloaded with organic matter which causes the decomposition ability of the pond to decline, and a low DO concentration to form in the bottom layer (Omori et al., 1994).

Penaeid shrimp are a group of estuary-dependent species which breeds at sea but grows and shelters in mangrove areas during its post-larval and juvenile stages. Water quality in the breeding and nursery ground may have a direct effect on its physiology, growth and survival. Aquacop et al. (1988) reported that growth, survival and moulting frequency of

Penaeus vannamei and *P. stylirostris* were depressed when they were maintained at a diurnally fluctuating DO regime. Clark (1986) found that mortality occurred and moulting was inhibited when *P. semisulcatus* was kept at a DO level of 2 mg O_2 l^{-1} for 17 days. Lethal levels in the range 0.2–1.0 mg O_2 l^{-1} have been reported for a number of species of penaeids including *Penaeus schmitti* (Mackay, 1974) and *P. monodon* (Allan & Maguire, 1991). The generally low dissolved oxygen level (<1.8 mg l^{-1}) in the more polluted *gei wai* at Mai Po explains why they are no longer suitable for shrimp farming (Chiu, 1992).

The reduced dissolved oxygen level can also increase shrimp mortality indirectly. Studies have shown that DO level can affect the nocturnal burrowing behaviour of Penaeid shrimp. *Penaeus japonicus* emerge from the burrow regardless of the light intensity while the amount of dissolved oxygen is less than 1 mg O_2 l^{-1} (Imai, 1982). Accordingly, an anoxic environment in a culture ponds will indirectly increase the encounter rate between shrimp and predators and thus the risk of predation. Extensive shrimp farming system generally adopts a polyculture method, with fish and shrimp being trapped into the pond at the same time through natural stocking. Although 38 species of fish have been recorded at Mai Po, only around very few species dominate the whole fish populations (Lee, 1988). At Mai Po, the herbivorous fish, *Mugil cephalus*, is the major economic fish species and contributes 86.3% to the total fish harvest. *Elops saurus* is the second major economic species and preys heavily upon shrimps and other juvenile fishes (Lee, 1988). Total fish production formed a negatively relationship with the shrimp production from Recruitment-II. This probably indicates each pond should have its own carrying capacity, and competition and predation may interact in order to maintain that level. For those *gei wai* connected to Channel-II, production of *Elops saurus* was significantly higher and negatively correlated with the *in situ* shrimp production. Effect of fish predation on shrimp production is thus apparent, especially in the polluted ponds where anoxic condition always occurs. In Asia, some preventive action against pests and predators are applied to eliminate the unwanted pests and predators, e.g. draining and drying the ponds after shrimp harvesting, install screen at the sluice gate, and the use of gill nets to catch predatory fish in the ponds (Apud et al., 1989). These methods can be tried at Mai Po to improve the shrimp production.

At present, only 38% of the *gei wai* are still productive and these are concentrated at the central region where pollution stress is comparatively low. Results of the present study, however, has shown a 63% reduction in average shrimp production from 1990 to 1995. Although water quality in Deep Bay did not show any further deterioration during the last few year, its water body remains at the category of grossly polluted (EPD, 1995). The Mai Po *gei wai* rely on water from Deep Bay to flush and stock the pond. Since flushing constantly with the highly organic polluted water into these ponds will reduce their decomposition ability and encourage anoxic condition (Omori et al., 1994), *gei wai* will continue to deteriorate. Wong (1996) found that sediment nutrient level in the southern *gei wai* which are connected to Channel-II and are close to the polluted Shan Pui river, was significantly higher than those connected to the northern one. This further explains why the overall shrimp production in those *gei wai* were significantly lower than those connected with Channel-I.

Although some of the Mai Po *gei wai* are now no longer productive, they can still support more pollution tolerant non-commercial fish and shrimp, e.g. *Oreochromis mossambicus* and *Exopalaemon styliferus* (Chiu, 1992). In recent years, these ponds are managed for wildlife rather than for shrimp production to take advantage of the availability of these fish and shrimps, which are prey for piscivorous birds. In winter, when large numbers of migratory waterbirds arrive in Deep Bay, the water level in these *gei wai* are lowered gradually so that the fish and shrimps inside become more available, and the pond thus provides a rich feeding site for waterbirds (Young, 1994).

Apart from shrimp production, the productive *gei wai* at Mai Po can be valuable for wildlife. In winter when these *gei wai* are drained for fish harvesting, the pond can attract up to some 700 herons, egrets and other waterbirds to feed in small pools where the non-commercial fish and shrimps are trapped (Kwok, 1993). These *gei wei* are also used as demonstration ponds to raise awareness amongst visiting students and members of the public about the value of this natural, and sustainable method of shrimp farming. Recently, the Mai Po *gei wai* have been adapted as a model system to restore previously intensively managed shrimp ponds in the Red River Delta, Vietnam (Le Dien Duc, 1996). The Mai Po *gei wai* are thus a valuable resource for promoting the sustainability of extensive, as opposed to intensive shrimp farming, and for education and for wildlife conservation. Thus, despite the pollution threats to the Hong Kong's unique traditional shrimp farming, application of the alternative management strategy by WWF Hong Kong can still maintain the conservation and education values of the Mai Po *gei wai*. It is, however, hoped that the pollution of Deep Bay can be controlled and reversed before it causes adverse effects on the remaining *gei wai*.

Acknowledgments

We would like to thank Prof. B. S. Morton, Dr S. Y. Lee and Mr D. S. Melville for their comments on this paper, WWF Hong Kong for providing research facilities at Mai Po and the Croucher Foundation of Hong Kong for providing a Fellowship to Dr M. W. Cha to conduct the research.

References

Apud, F. D., J. H. Primavera & P. L. Torres, 1989. Farming of Prawns and Shrimps. Aquaculture extension manual no. 5. Tigbauan, Philippines.

Allan, G. L. & G. B. Maguire, 1991. Lethal levels of low dissolved oxygen and effects of short term oxygen stress on subsequent growth of juvenile *Penaeus monodon*. Aquaculture 94: 27–37.

Aquacop, E. Bedier & C. Soyez, 1988. Effects of dissolved oxygen concentration on survival and growth of *Penaeus vannamei* and *Penaeus stylirostris*. J. World Aquacult. Soc. 19: 13A.

Barraclough, S. & A. Finger-Stich. 1996. Some ecological and social implications of commercial shrimp farming in Asia. UNDP/WWF.

Chiu, K. T., 1992. An assessment of the water pollution status of the Mai Po Marshes Nature Reserve, Hong Kong. Ph.D. Thesis, University of Hong Kong, Hong Kong.

Clark, J. V., 1986. Inhibition of moulting in *Penaeus semisulcatus* (De Haan) by long-term hypoxia. Aquaculture 52: 253–254.

Environmental Protection Department, 1995. Marine Water Quality of Hong Kong, 1994. Hong Kong Government Press, Hong Kong.

Fan, H. Q., 1993. Necessity of establishing China Mangrove Research Centre and its task. J. Guangxi Acad. Sci. 9: 122–129.

Imai, T., 1982. Aquaculture in Shallow Seas: Progress in Shallow Sea Culture. Pauls Press, New Delhi: 413–462.

Khor, M., 1995. The aquaculture disaster: third world communities fight the 'blue revolution'. Third World Resurgence. 5: 8–10.

Kwok, K., 1993. Effects of water level fluctuation of *gei wai* on distribution of waders in the Mai Po Marshes. Unpub. report. City University of Hong Kong.

Le Dien Duc, 1996. Integrated coastal management in the Tien Hai District, Thai Binh Province. Paper presented at the Ecotone IV meeting, Ho Chi Ming City, Vietnam. January 1996.

Lee, S. Y., 1988. The ecology of a traditionally managed tidal shrimp pond, the production and fate of macro-detritus and implications for management. Ph.D. Thesis, University of Hong Kong, Hong Kong.

Leung, S. F., 1991. The population dynamics of *Metapenaeous ensis* (Penaeidae) and *Exopalaemon styliferus* (Palaemonidae) in a traditional tidal shrimp pond at the Mai Po Marshes Nature Reserve, Hong Kong. Ph.D. Thesis, University of Hong Kong, Hong Kong.

Liao, I. C., 1992. Diseases of Panaeus monodon in Taiwan: a review from 1977 to 1991. In Fulks, W. & K. L. Main (eds), Diseases of Cultured Penaeid Shrimp in Asia and the United States. The Oceanic Institute, Hawaii: 113–138.

Mackay, R. D., 1974. A note on minimal levels of oxygen required to maintain life in *Penaeus schmitti*. Proc. Annu. Meet. World Maricult. Soc. 5: 451–452.

Menasveta, P., 1988. Fishery resources. In Arbhabhirama, A. et al. (eds.), Thailand: Natural Resource Profile. Oxford University Press, Singapore: 327–386.

Naamin, N., 1987. Impact of 'tambak' aquaculture to the mangrove ecosystem and its adjacent areas with special reference to the north coast of West Java. In Umali, R. M. (ed.), Mangroves of Asia and the Pacific: Status and Management. UNDP/UNESCO Technical Report (RAS/79/002), Quezon City: 355–366.

Pillay, T. V. R., 1993. Aquaculture: Principles and Practices. Blackwell Scientific Publications, Cambridge: 425–459.

Omori, K., T. Hirano & H. Takeoka, 1994. The limitations to organic loading on a bottom of a coastal ecosystem. Mar. Pollut. Bull. 28: 73–80.

Primavera, J. H., 1995. Mangroves and brackishwater pond culture in the Philippines. Hydrobiologia. 295: 303–309.

Shariff, M. & R. P. Subasinghe, 1992. Major diseases of cultured shrimp in Asia: an overview. In Fulks, W. & K. L. Main (eds), Diseases of Cultured Penaeid Shrimp in Asia and the United States. The Oceanic Institute, Hawaii: 37–46.

Shen, Y., 1983. Ecological balance in the Pearl River Delta. Occ. Pap. Dep. Geogr. Chin. Univ. Hong Kong, No 1.

Vanucci, M., 1987. Conversion of mangroves to other uses: The Cochin backwaters. In Umali, R. M. (ed.), Mangroves of Asia and the Pacific: Status and Management. UNDP/UNESCO Technical Report (RAS/79/002), Quezon City: 331–336.

Wong, C. M., 1986. Shrimp farming in Deep Bay – the evolution and present situation of a traditional industry. B.A. thesis, University of Hong Kong, Hong Kong.

Wong, W. L., 1996. The physical and chemical characteristics of the sediment and water in Mai Po Gei wai. B.Sc. thesis, City University of Hong Kong.

Young, L., in Press. Mai Po Marshes, Hong Kong. In Wetlands, Biodiversity and the Ramsar Convention. Ramsar Convention Bureau, Switzerland: 52–53.

Young, L., 1994. Conservation activities at the Mai Po Marshes Wildlife Education Centre and Nature Reserve. In Higuchi, H. (ed.), The Future of Cranes and Wetlands. Wild Bird Society of Japan, Japan: 166–175.

Young, L. & D. S. Melville, 1993. Conservation of the Deep Bay environment. In Morton, B. (ed.), The Marine Biology of the South China Seas. University of Hong Kong Press, Hong Kong: 211–232.